CONSISTENT QUANTUM THEORY

Quantum mechanics is one of the most fundamental yet difficult subjects in modern physics. In this book, nonrelativistic quantum theory is presented in a clear and systematic fashion that integrates Born's probabilistic interpretation with Schrödinger dynamics.

Basic quantum principles are illustrated with simple examples requiring no mathematics beyond linear algebra and elementary probability theory, clarifying the main sources of confusion experienced by students when they begin a serious study of the subject. The quantum measurement process is analyzed in a consistent way using fundamental quantum principles that do not refer to measurement. These same principles are used to resolve several of the paradoxes that have long perplexed quantum physicists, including the double slit and Schrödinger's cat. The consistent histories formalism used in this book was first introduced by the author, and extended by M. Gell-Mann, J.B. Hartle, and R. Omnès.

Essential for researchers, yet accessible to advanced undergraduate students in physics, chemistry, mathematics, and computer science, this book may be used as a supplement to standard textbooks. It will also be of interest to physicists and philosophers working on the foundations of quantum mechanics.

ROBERT B. GRIFFITHS is the Otto Stern University Professor of Physics at Carnegie-Mellon University. In 1962 he received his PhD in physics from Stanford University. Currently a Fellow of the American Physical Society and member of the National Academy of Sciences of the USA, he received the Dannie Heineman Prize for Mathematical Physics from the American Physical Society in 1984. He is the author or coauthor of 130 papers on various topics in theoretical physics, mainly statistical and quantum mechanics.

Consistent Quantum Theory

Robert B. Griffiths
Carnegie-Mellon University

CAMBRIDGE
UNIVERSITY PRESS

PUBLISHED BY THE PRESS SYNDICATE OF THE UNIVERSITY OF CAMBRIDGE
The Pitt Building, Trumpington Street, Cambridge, United Kingdom

CAMBRIDGE UNIVERSITY PRESS
The Edinburgh Building, Cambridge CB2 2RU, UK
40 West 20th Street, New York, NY 10011-4211, USA
477 Williamstown Road, Port Melbourne, VIC 3207, Australia
Ruiz de Alarcón 13, 28014, Madrid, Spain
Dock House, The Waterfront, Cape Town 8001, South Africa

http://www.cambridge.org

First published 2002
First published in paperback 2003

Typeface Times 11/14pt.   *System* LATEX 2_ε [DBD]

A catalogue record of this book is available from the British Library

Library of Congress Cataloguing in Publication data

Griffiths, R. B. (Robert Budington)
Consistent quantum theory / Robert B. Griffiths.
p.   cm.
Includes bibliographical references and index.
ISBN 0 521 80349 7 (hc.)
1. Quantum theory. I. Title.
QC174.12.G752   2001
530.12–dc21   2001035048

ISBN 0 521 80349 7   hardback
ISBN 0 521 53929 3   paperback

Transferred to digital printing 2004

To the memory of my parents
examples of integrity and Christian service

Contents

Contents

Preface

Quantum theory is one of the most difficult subjects in the physics curriculum. In part this is because of unfamiliar mathematics: partial differential equations, Fourier transforms, complex vector spaces with inner products. But there is also the problem of relating mathematical objects, such as wave functions, to the physical reality they are supposed to represent. In some sense this second problem is more serious than the first, for even the founding fathers of quantum theory had a great deal of difficulty understanding the subject in physical terms. The usual approach found in textbooks is to relate mathematics and physics through the concept of a *measurement* and an associated wave function collapse. However, this does not seem very satisfactory as the foundation for a fundamental physical theory. Most professional physicists are somewhat uncomfortable with using the concept of measurement in this way, while those who have looked into the matter in greater detail, as part of their research into the foundations of quantum mechanics, are well aware that employing measurement as one of the building blocks of the subject raises at least as many, and perhaps more, conceptual difficulties than it solves.

It is in fact not necessary to interpret quantum mechanics in terms of measurements. The primary mathematical constructs of the theory, that is to say wave functions (or, to be more precise, subspaces of the Hilbert space), can be given a direct physical interpretation whether or not any process of measurement is involved. Doing this in a consistent way yields not only all the insights provided in the traditional approach through the concept of measurement, but much more besides, for it makes it possible to think in a sensible way about quantum systems which are not being measured, such as unstable particles decaying in the center of the earth, or in intergalactic space. Achieving a consistent interpretation is not easy, because one is constantly tempted to import the concepts of classical physics, which fit very well with the mathematics of classical mechanics, into the quantum domain where they sometimes work, but are often in conflict with the very different mathematical structure of Hilbert space that underlies quantum theory. The result of using classical concepts where they do not belong is to generate contradictions and paradoxes of the sort which, especially in more popular expositions of the subject, make quantum physics seem magical. Magic may be good for entertainment, but the resulting confusion is not very helpful to students trying to understand the subject for the first time, or to more mature scientists who want to apply quantum principles to a new domain where there is not yet a well-established set of principles for carrying out and interpreting calculations, or to philosophers interested in the implications of quantum theory for broader questions about human knowledge and the nature of the world.

The basic problem which must be solved in constructing a rational approach to quantum theory that is not based upon measurement as a fundamental principle is to introduce probabilities and stochastic processes as part of the foundations of the subject, and not just an *ad hoc* and somewhat embarrassing addition to Schrödinger's equation. Tools for doing this in a consistent way compatible with the mathematics of Hilbert space first appeared in the scientific research literature about fifteen years ago. Since then they have undergone further developments and refinements although, as with almost all significant scientific advances, there have been some serious mistakes on the part of those involved in the new developments, as well as some serious misunderstandings on the part of their critics. However, the resulting formulation of quantum principles, generally known as *consistent histories* (or as *decoherent histories*), appears to be fundamentally sound. It is conceptually and mathematically "clean": there are a small set of basic principles, not a host of *ad hoc* rules needed to deal with particular cases. And it provides a rational resolution to a number of paradoxes and dilemmas which have troubled some of the foremost quantum physicists of the twentieth century.

The purpose of this book is to present the basic principles of quantum theory with the probabilistic structure properly integrated with Schrödinger dynamics in a coherent way which will be accessible to serious students of the subject (and their teachers). The emphasis is on physical interpretation, and for this reason I have tried to keep the mathematics as simple as possible, emphasizing finite-dimensional vector spaces and making considerable use of what I call "toy models." They are a sort of quantum counterpart to the massless and frictionless pulleys of introductory classical mechanics; they make it possible to focus on essential issues of physics without being distracted by too many details. This approach may seem simplistic, but when properly used it can yield, at least for a certain class of problems, a lot more physical insight for a given expenditure of time than either numerical calculations or perturbation theory, and it is particularly useful for resolving a variety of confusing conceptual issues.

An overview of the contents of the book will be found in the first chapter. In brief, there are two parts: the essentials of quantum theory, in Chs. 2–16, and a variety of applications, including measurements and paradoxes, in Chs. 17–27. References to the literature have (by and large) been omitted from the main text, and will be found, along with a few suggestions for further reading, in the bibliography. In order to make the book self-contained I have included, without giving proofs, those essential concepts of linear algebra and probability theory which are needed in order to obtain a basic understanding of quantum mechanics. The level of mathematical difficulty is comparable to, or at least not greater than, what one finds in advanced undergraduate or beginning graduate courses in quantum theory.

That the book is self-contained does not mean that reading it in isolation from other material constitutes a good way for someone with no prior knowledge to learn the subject. To begin with, there is no reference to the basic phenomenology of blackbody radiation, the photoelectric effect, atomic spectra, etc., which provided the original motivation for quantum theory and still form a very important part of the physical framework of the subject. Also, there is no discussion of a number of standard topics, such as the hydrogen atom, angular momentum, harmonic oscillator wave functions, and perturbation theory, which are part of the usual introductory course. For both of these I can with a clear conscience refer the reader to the many introductory textbooks which provide quite adequate treatments of these topics. Instead, I have concentrated on material which is not yet found in textbooks (hopefully that situation will change), but is very important if one wants to have a clear understanding of basic quantum principles.

It is a pleasure to acknowledge help from a large number of sources. First, I am indebted to my fellow consistent historians, in particular Murray Gell-Mann, James Hartle, and Roland Omnès, from whom I have learned a great deal over the years. My own understanding of the subject, and therefore this book, owes much to their insights. Next, I am indebted to a number of critics, including Angelo Bassi, Bernard d'Espagnat, Fay Dowker, GianCarlo Ghirardi, Basil Hiley, Adrian Kent, and the late Euan Squires, whose challenges, probing questions, and serious efforts to evaluate the claims of the consistent historians have forced me to rethink my own ideas and also the manner in which they have been expressed. Over a number of years I have taught some of the material in the following chapters in both advanced undergraduate and introductory graduate courses, and the questions and reactions by the students and others present at my lectures have done much to clarify my thinking and (I hope) improve the quality of the presentation.

I am grateful to a number of colleagues who read and commented on parts of the manuscript. David Mermin, Roland Omnès, and Abner Shimony looked at particular chapters, while Todd Brun, Oliver Cohen, and David Collins read drafts of the entire manuscript. As well as uncovering many mistakes, they made a large number of suggestions for improving the text, some though not all of which I adopted. For this reason (and in any case) whatever errors of commission or omission are present in the final version are entirely my responsibility.

I am grateful for the financial support of my research provided by the National Science Foundation through its Physics Division, and for a sabbatical year from my duties at Carnegie-Mellon University that allowed me to complete a large part of the manuscript. Finally, I want to acknowledge the encouragement and help I received from Simon Capelin and the staff of Cambridge University Press.

Pittsburgh, Pennsylvania   Robert B Griffiths
March 2001

1

Introduction

1.1 Scope of this book

Quantum mechanics is a difficult subject, and this book is intended to help the reader overcome the main difficulties in the way to understanding it. The first part of the book, Chs. 2–16, contains a systematic presentation of the basic principles of quantum theory, along with a number of examples which illustrate how these principles apply to particular quantum systems. The applications are, for the most part, limited to *toy models* whose simple structure allows one to see what is going on without using complicated mathematics or lengthy formulas. The principles themselves, however, are formulated in such a way that they can be applied to (almost) any nonrelativistic quantum system. In the second part of the book, Chs. 17–25, these principles are applied to quantum measurements and various quantum paradoxes, subjects which give rise to serious conceptual problems when they are not treated in a fully consistent manner.

The final chapters are of a somewhat different character. Chapter 26 on decoherence and the classical limit of quantum theory is a very sketchy introduction to these important topics along with some indication as to how the basic principles presented in the first part of the book can be used for understanding them. Chapter 27 on quantum theory and reality belongs to the interface between physics and philosophy and indicates why quantum theory is compatible with a real world whose existence is not dependent on what scientists think and believe, or the experiments they choose to carry out. The Bibliography contains references for those interested in further reading or in tracing the origin of some of the ideas presented in earlier chapters.

The remaining sections of this chapter provide a brief overview of the material in Chs. 2–25. While it may not be completely intelligible in advance of reading the actual material, the overview should nonetheless be of some assistance to readers who, like me, want to see something of the big picture before plunging into the details. Section 1.2 concerns quantum systems at a single time, and Sec. 1.3 their time development. Sections 1.4 and 1.5 indicate what topics in mathematics are essential for understanding quantum theory, and where the relevant material is located in this book, in case the reader is not already familiar with it. Quantum reasoning as it is developed in the first sixteen chapters is surveyed in Sec. 1.6. Section 1.7 concerns quantum measurements, treated in Chs. 17 and 18. Finally, Sec. 1.8 indicates the motivation behind the chapters, 19–25, devoted to quantum paradoxes.

1.2 Quantum states and variables

Both classical and quantum mechanics describe how physical objects move as a function of time. However, they do this using rather different mathematical structures. In classical mechanics the *state* of a system at a given time is represented by a point in a *phase space*. For example, for a single particle moving in one dimension the phase space is the x, p plane consisting of pairs of numbers (x, p) representing the position and momentum. In quantum mechanics, on the other hand, the state of such a particle is given by a complex-valued *wave function* $\psi(x)$, and, as noted in Ch. 2, the collection of all possible wave functions is a complex linear vector space with an inner product, known as a *Hilbert space*.

The physical significance of wave functions is discussed in Ch. 2. Of particular importance is the fact that two wave functions $\phi(x)$ and $\psi(x)$ represent distinct physical states in a sense corresponding to distinct points in the classical phase space if and only if they are *orthogonal* in the sense that their inner product is zero. Otherwise $\phi(x)$ and $\psi(x)$ represent *incompatible* states of the quantum system (unless they are multiples of each other, in which case they represent the same state). Incompatible states cannot be compared with one another, and this relationship has no direct analog in classical physics. Understanding what incompatibility does and does not mean is essential if one is to have a clear grasp of the principles of quantum theory.

A quantum *property*, Ch. 4, is the analog of a collection of points in a classical phase space, and corresponds to a *subspace* of the quantum Hilbert space, or the *projector* onto this subspace. An example of a (classical or quantum) property is the statement that the energy E of a physical system lies within some specific range, $E_0 \leq E \leq E_1$. Classical properties can be subjected to various logical operations: negation, conjunction (AND), and disjunction (OR). The same is true of quantum properties as long as the projectors for the corresponding subspaces commute with each other. If they do not, the properties are incompatible in much the same way as nonorthogonal wave functions, a situation discussed in Sec. 4.6.

An orthonormal basis of a Hilbert space or, more generally, a decomposition of the identity as a sum of mutually commuting projectors constitutes a *sample space* of *mutually-exclusive possibilities*, one and only one of which can be a correct description of a quantum system at a given time. This is the quantum counterpart of a sample space in ordinary probability theory, as noted in Ch. 5, which discusses how probabilities can be assigned to quantum systems. An important difference between classical and quantum physics is that quantum sample spaces can be mutually incompatible, and probability distributions associated with incompatible spaces cannot be combined or compared in any meaningful way.

In classical mechanics a *physical variable*, such as energy or momentum, corresponds to a real-valued function defined on the phase space, whereas in quantum mechanics, as explained in Sec. 5.5, it is represented by a Hermitian operator. Such an operator can be thought of as a real-valued function defined on a particular sample space, or decomposition of the identity, but not on the entire Hilbert space. In particular, a quantum system can be said to have a value (or at least a precise value) of a physical variable represented by the operator F if and only if the quantum wave function is in an eigenstate of F, and in this case the eigenvalue is the value of the physical variable. Two physical variables whose operators do not commute correspond to incompatible sample spaces, and in general it is not possible to simultaneously assign values of both variables to a single quantum system.

1.3 Quantum dynamics

Both classical and quantum mechanics have *dynamical laws* which enable one to say something about the future (or past) state of a physical system if its state is known at a particular time. In classical mechanics the dynamical laws are *deterministic*: at any given time in the future there is a unique state which corresponds to a given initial state. As discussed in Ch. 7, the quantum analog of the deterministic dynamical law of classical mechanics is the (time-dependent) Schrödinger equation. Given some wave function ψ_0 at a time t_0, integration of this equation leads to a unique wave function ψ_t at any other time t. At two times t and t' these uniquely defined wave functions are related by a unitary map or *time development operator* $T(t', t)$ on the Hilbert space. Consequently we say that integrating the Schrödinger equation leads to *unitary* time development.

However, quantum mechanics also allows for a *stochastic* or *probabilistic* time development, analogous to tossing a coin or rolling a die several times in a row. In order to describe this in a systematic way, one needs the concept of a *quantum history*, introduced in Ch. 8: a sequence of quantum *events* (wave functions or subspaces of the Hilbert space) at successive times. A collection of mutually exclusive histories forms a sample space or *family* of histories, where each history is associated with a projector on a *history Hilbert space*.

The successive events of a history are, in general, not related to one another through the Schrödinger equation. However, the Schrödinger equation, or, equivalently, the time development operators $T(t', t)$, can be used to assign probabilities to the different histories belonging to a particular family. For histories involving only two times, an initial time and a single later time, probabilities can be assigned using the *Born rule*, as explained in Ch. 9. However, if three or more times are involved, the procedure is a bit more complicated, and probabilities can only be assigned in a consistent way when certain *consistency conditions* are satisfied, as explained in Ch. 10. When the consistency conditions hold, the corresponding sample space or event algebra is known as a *consistent family* of histories, or a *framework*. Checking consistency conditions is not a trivial task, but it is made easier by various rules and other considerations discussed in Ch. 11. Chapters 9, 10, 12, and 13 contain a number of simple examples which illustrate how the probability assignments in a consistent family lead to physically reasonable results when one pays attention to the requirement that stochastic time development must be described using a *single* consistent family or framework, and results from incompatible families, as defined in Sec. 10.4, are not combined.

1.4 Mathematics I. Linear algebra

Several branches of mathematics are important for quantum theory, but of these the most essential is *linear algebra*. It is the fundamental mathematical language of quantum mechanics in much the same way that calculus is the fundamental mathematical language of classical mechanics. One cannot even define essential quantum concepts without referring to the quantum Hilbert space, a complex linear vector space equipped with an inner product. Hence a good grasp of what quantum mechanics is all about, not to mention applying it to various physical problems, requires some familiarity with the properties of Hilbert spaces.

Unfortunately, the wave functions for even such a simple system as a quantum particle in one dimension form an *infinite-dimensional* Hilbert space, and the rules for dealing with such spaces with mathematical precision, found in books on functional analysis, are rather complicated and involve concepts, such as Lebesgue integrals, which fall outside the mathematical training of the majority of physicists. Fortunately, one does not have to learn functional analysis in order to understand the basic principles of quantum theory. The majority of the illustrations used in Chs. 2–16 are toy models with a finite-dimensional Hilbert space to which the usual rules of linear algebra apply without any qualification, and for these models there are no mathematical subtleties to add to the conceptual difficulties of quantum theory. To be sure, mathematical simplicity is achieved at a certain cost, as toy models are even less "realistic" than the already artificial one-dimensional models one finds in textbooks. Nevertheless, they provide many useful insights into general quantum principles.

For the benefit of readers not already familiar with them, the concepts of linear algebra in finite-dimensional spaces which are most essential to quantum theory are summarized in Ch. 3, though some additional material is presented later: tensor products in Ch. 6 and unitary operators in Sec. 7.2. Dirac notation, in which elements of the Hilbert space are denoted by $|\psi\rangle$, and their duals by $\langle\psi|$, the inner product $\langle\phi|\psi\rangle$ is linear in the element on the right and antilinear in the one on the left, and matrix elements of an operator A take the form $\langle\phi|A|\psi\rangle$, is used throughout the book. Dirac notation is widely used and universally understood among quantum physicists, so any serious student of the subject will find learning it well-worthwhile. Anyone already familiar with linear algebra will have no trouble picking up the essentials of Dirac notation by glancing through Ch. 3.

It would be much too restrictive and also rather artificial to exclude from this book all references to quantum systems with an infinite-dimensional Hilbert space. As far as possible, quantum principles are stated in a form in which they apply to infinite- as well as to finite-dimensional spaces, or at least can be applied to the former given reasonable qualifications which mathematically sophisticated readers can fill in for themselves. Readers not in this category should simply follow the example of the majority of quantum physicists: go ahead and use the rules you learned for finite-dimensional spaces, and if you get into difficulty with an infinite-dimensional problem, go talk to an expert, or consult one of the books indicated in the bibliography (under the heading of Ch. 3).

1.5 Mathematics II. Calculus, probability theory

It is obvious that *calculus* plays an essential role in quantum mechanics; e.g., the inner product on a Hilbert space of wave functions is defined in terms of an integral, and the time-dependent Schrödinger equation is a partial differential equation. Indeed, the problem of constructing explicit solutions as a function of time to the Schrödinger equation is one of the things which makes quantum mechanics more difficult than classical mechanics. For example, describing the motion of a classical particle in one dimension in the absence of any forces is trivial, while the time development of a quantum wave packet is not at all simple.

Since this book focuses on conceptual rather than mathematical difficulties of quantum theory, considerable use is made of toy models with a simple discretized time dependence, as indicated in Sec. 7.4, and employed later in Chs. 9, 12, and 13. To obtain their unitary time development, one only needs to solve a simple difference equation, and this can be done in closed form on the back of an envelope. Because there is no need for approximation methods or numerical solutions, these toy models can provide a lot of insight into the structure of quantum theory, and once one sees how to use them, they can be a valuable guide in discerning what are the really essential elements in the much more complicated mathematical structures needed in more realistic applications of quantum theory.

Probability theory plays an important role in discussions of the time development of quantum systems. However, the more sophisticated parts of this discipline, those that involve measure theory, are not essential for understanding basic quantum concepts, although they arise in various applications of quantum theory. In particular, when using toy models the simplest version of probability theory, based on a finite discrete sample space, is perfectly adequate. And once the basic strategy for using probabilities in quantum theory has been understood, there is no particular difficulty — or at least no greater difficulty than one encounters in classical physics — in extending it to probabilities of continuous variables, as in the case of $|\psi(x)|^2$ for a wave function $\psi(x)$.

In order to make this book self-contained, the main concepts of probability theory needed for quantum mechanics are summarized in Ch. 5, where it is shown how to apply them to a quantum system at a single time. Assigning probabilities to quantum histories is the subject of Chs. 9 and 10. It is important to note that the basic concepts of probability theory are the same in quantum mechanics as in other branches of physics; one does not need a new "quantum probability". What distinguishes quantum from classical physics is the issue of choosing a suitable sample space with its associated event algebra. There are always many different ways of choosing a quantum sample space, and different sample spaces will often be incompatible, meaning that results cannot be combined or compared. However, in any single quantum sample space the ordinary rules for probabilistic reasoning are valid.

Probabilities in the quantum context are sometimes discussed in terms of a *density matrix*, a type of operator defined in Sec. 3.9. Although density matrices are not really essential for understanding the basic principles of quantum theory, they occur rather often in applications, and Ch. 15 discusses their physical significance and some of the ways in which they are used.

1.6 Quantum reasoning

The Hilbert space used in quantum mechanics is in certain respects quite different from a classical phase space, and this difference requires that one make some changes in classical habits of thought when reasoning about a quantum system. What is at stake becomes particularly clear when one considers the two-dimensional Hilbert space of a spin-half particle, Sec. 4.6, for which it is easy to see that a straightforward use of ideas which work very well for a classical phase space will lead to contradictions. Thinking carefully about this example is well-worthwhile, for if one cannot understand the simplest of all quantum systems, one is not likely to make much progress with more complicated situations. One approach to the problem is to change the rules of ordinary (classical) logic, and this was the route taken by Birkhoff and von Neumann when they proposed a special quantum logic. However, their proposal has not been particularly fruitful for resolving the conceptual difficulties of quantum theory.

The alternative approach adopted in this book, starting in Sec. 4.6 and summarized in Ch. 16, leaves the ordinary rules of propositional logic unchanged, but imposes conditions on what constitutes a *meaningful* quantum description to which these rules can be applied. In particular, it is never meaningful to combine incompatible elements — be they wave functions, sample spaces, or consistent families — into a single description. This prohibition is embodied in the *single-framework* rule stated in Sec. 16.1, but already employed in various examples in earlier chapters.

Because so many mutually incompatible frameworks are available, the strategy used for describing the stochastic time development of a quantum system is quite different from that employed in classical mechanics. In the classical case, if one is given an initial state, it is only necessary to integrate the deterministic equations of motion in order to obtain a unique result at any later time. By contrast, an initial quantum state does not single out a particular framework, or sample space of stochastic histories, much less determine which history in the framework will actually occur. To understand how frameworks are chosen in the quantum case, and why, despite the multiplicity of possible frameworks, the theory still leads to consistent and coherent physical results, it is best to look at specific examples, of which a number will be found in Chs. 9, 10, 12, and 13.

Another aspect of incompatibility comes to light when one considers a tensor product of Hilbert spaces representing the subsystems of a composite system, or events at different times in the history of a single system. This is the notion of a *contextual* or *dependent* property or event. Chapter 14 is devoted to a systematic discussion of this topic, which also comes up in several of the quantum paradoxes considered in Chs. 20–25.

The basic principles of quantum reasoning are summarized in Ch. 16 and shown to be internally consistent. This chapter also contains a discussion of the intuitive significance of multiple incompatible frameworks, one of the most significant ways in which quantum theory differs from classical physics. If the principles stated in Ch. 16 seem rather abstract, readers should work through some of the examples found in earlier or later chapters or, better yet, work out some for themselves.

1.7 Quantum measurements

A quantum theory of measurements is a necessary part of any consistent way of understanding quantum theory for a fairly obvious reason. The phenomena which are specific to quantum theory, which lack any description in classical physics, have to do with the behavior of microscopic objects, the sorts of things which human beings cannot observe directly. Instead we must use carefully constructed instruments to amplify microscopic effects into macroscopic signals of the sort we can see with our eyes, or feed into our computers. Unless we understand how the apparatus works, we cannot interpret its macroscopic output in terms of the microscopic quantum phenomena we are interested in.

The situation is in some ways analogous to the problem faced by astronomers who depend upon powerful telescopes in order to study distant galaxies. If they did not understand how a telescope functions, cosmology would be reduced to pure speculation. There is, however, an important difference between the "telescope problem" of the astronomer and the "measurement problem" of the quantum physicist. No fundamental concepts from astronomy are needed in order to understand the operation of a telescope: the principles of optics are, fortunately, independent of the properties of the object which emits the light. But a piece of laboratory apparatus capable of amplifying quantum effects, such as a spark chamber, is itself composed of an enormous number of atoms, and nowadays we believe (and there is certainly no evidence to the contrary) that the behavior of aggregates of atoms as well as individual atoms is governed by quantum laws. Thus quantum measurements can, at least in principle, be analyzed using quantum theory. If for some reason such an analysis were impossible, it would indicate that quantum theory was wrong, or at least seriously defective.

Measurements as parts of gedanken experiments played a very important role in the early development of quantum theory. In particular, Bohr was able to meet many of Einstein's objections to the new theory by pointing out that quantum principles had to be applied to the measuring apparatus itself, as well as to the particle or other microscopic system of interest. A little later the notion of measurement was incorporated as a fundamental principle in the standard interpretation of quantum mechanics, accepted by the majority of quantum physicists, where it served as a device for introducing stochastic time development into the theory. As von Neumann explained it, a system develops unitarily in time, in accordance with Schrödinger's equation, until it interacts with some sort of measuring apparatus, at which point its wave function undergoes a "collapse" or "reduction" correlated with the outcome of the measurement.

However, employing measurements as a fundamental principle for interpreting quantum theory is not very satisfactory. Nowadays quantum mechanics is applied to processes taking place at the centers of stars, to the decay of unstable particles in intergalactic space, and in many other situations which can scarcely be thought of as involving measurements. In addition, laboratory measurements are often of a sort in which the measured particle is either destroyed or else its properties are significantly altered by the measuring process, and the von Neumann scheme does not provide a satisfactory connection between the measurement outcome (e.g., a pointer position) and the corresponding property of the particle *before* the measurement took place. Numerous attempts have been made to construct a fully consistent measurement-based interpretation of quantum mechanics, thus far without success. Instead, this approach leads to a number of conceptual difficulties which constitute what specialists refer to as the "measurement problem."

In this book all of the *fundamental* principles of quantum theory are developed, in Chs. 2–16, *without* making any reference to measurements, though measurements occur in some of the applications. Measurements are taken up in Chs. 17 and 18, and analyzed using the general principles of quantum mechanics introduced earlier. This includes such topics as how to describe a macroscopic measuring apparatus in quantum terms, the role of thermodynamic irreversibility in the measurement process, and what happens when two measurements are carried out in succession. The result is a consistent theory of quantum measurements based upon fundamental quantum principles, one which is able to reproduce all the results of the von Neumann approach and to go beyond it; e.g., by showing how the outcome of a measurement is correlated with some property of the measured system before the measurement took place.

Wave function collapse or reduction, discussed in Sec. 18.2, is not needed for a consistent quantum theory of measurement, as its role is taken over by a suitable use of conditional probabilities. To put the matter in a different way, wave function collapse is one method for computing conditional probabilities that can be obtained equally well using other methods. Various conceptual difficulties disappear when one realizes that collapse is something which takes place in the theoretical physicist's notebook and not in the experimental physicist's laboratory. In particular, there is no physical process taking place instantaneously over a long distance, in conflict with relativity theory.

1.8 Quantum paradoxes

A large number of quantum paradoxes have come to light since the modern form of quantum mechanics was first developed in the 1920s. A paradox is something which is contradictory, or contrary to common sense, but which seems to follow from accepted principles by ordinary logical rules. That is, it is something which ought to be true, but seemingly is not true. A scientific paradox may indicate that there is something wrong with the underlying scientific theory, which is quantum mechanics in the case of interest to us. But a paradox can also be a prediction of the theory that, while rather surprising when one first hears it, is shown by further study or deeper analysis to reflect some genuine feature of the universe in which we live. For example, in relativity theory we learn that it is impossible for a signal to travel faster than the speed of light. This seems paradoxical in that one can imagine being on a rocket ship traveling at half the speed of light, and then shining a flashlight in the forwards direction. However, this (apparent) paradox can be satisfactorily explained by making consistent use of the principles of relativity theory, in particular those which govern transformations to moving coordinate systems.

A consistent understanding of quantum mechanics should make it possible to resolve quantum paradoxes by locating the points where they involve hidden assumptions or flawed reasoning, or by showing how the paradox embodies some genuine feature of the quantum world which is surprising from the perspective of classical physics. The formulation of quantum theory found in the first sixteen chapters of this book is employed in Chs. 20–25 to resolve a number of quantum paradoxes, including delayed choice, Kochen–Specker, EPR, and Hardy's paradox, among others. (Schrödinger's cat and the double-slit paradox, or at least their toy counterparts, are taken up earlier in the book, in Secs. 9.6 and 13.1, respectively, as part of the discussion of basic quantum principles.) Chapter 19 provides a brief introduction to these paradoxes along with two conceptual tools, quantum coins and quantum counterfactuals, which are needed for analyzing them.

In addition to demonstrating the overall consistency of quantum theory, there are at least three other reasons for devoting a substantial amount of space to these paradoxes. The first is that they provide useful and interesting examples of how to apply the basic principles of quantum mechanics. Second, various quantum paradoxes have been invoked in support of the claim that quantum theory is intrinsically *nonlocal* in the sense that there are mysterious influences which can, to take an example, instantly communicate the choice to carry out one measurement rather than another at point A to a distant point B, in a manner which contradicts the basic requirements of relativity theory. A careful analysis of these paradoxes shows, however, that the apparent contradictions arise from a failure to properly apply some principle of quantum reasoning in a purely local setting. Nonlocal influences are generated by logical mistakes, and when the latter are corrected, the ghosts of nonlocality vanish. Third, these paradoxes have sometimes been used to argue that the quantum world is not real, but is in some way created by human consciousness, or else that reality is a concept which only applies to the macroscopic domain immediately accessible to human experience. Resolving the paradoxes, in the sense of showing them to be in accord with consistent quantum principles, is thus a prelude to the discussion of quantum reality in Ch. 27.

2

Wave functions

2.1 Classical and quantum particles

In classical Hamiltonian mechanics the *state* of a particle at a given instant of time is given by two vectors: $\mathbf{r} = (x, y, z)$ representing its position, and $\mathbf{p} = (p_x, p_y, p_z)$ representing its momentum. One can think of these two vectors together as determining a point in a six-dimensional *phase space*. As time increases the point representing the state of the particle traces out an *orbit* in the phase space. To simplify the discussion, consider a particle which moves in only one dimension, with position x and momentum p. Its phase space is the two-dimensional x, p plane. If, for example, one is considering a harmonic oscillator with angular frequency ω, the orbit of a particle of mass m will be an ellipse of the form

$$x = A \sin(\omega t + \phi), \quad p = m A \omega \cos(\omega t + \phi) \tag{2.1}$$

for some amplitude A and phase ϕ, as shown in Fig. 2.1.

A quantum particle at a single instant of time is described by a *wave function* $\psi(\mathbf{r})$, a complex function of position $\mathbf{r}$. Again in the interests of simplicity we will consider a quantum particle moving in one dimension, so that its wave function $\psi(x)$ depends on only a single variable, the position x. Some examples of real-valued wave functions, which can be sketched as simple graphs, are shown in Figs. 2.2–2.4. It is important to note that *all* of the information required to describe a quantum state is contained in the function $\psi(x)$. Thus this one function is the quantum analog of the pair of real numbers x and p used to describe a classical particle at a particular time.

In order to understand the physical significance of quantum wave functions, one needs to know that they belong to a *linear vector space* $\mathcal{H}$. That is, if $\psi(x)$ and $\phi(x)$ are any two wave functions belonging to $\mathcal{H}$, the *linear combination*

$$\omega(x) = \alpha \psi(x) + \beta \phi(x), \tag{2.2}$$

where α and β are any two complex numbers, also belongs to $\mathcal{H}$. The space $\mathcal{H}$ is

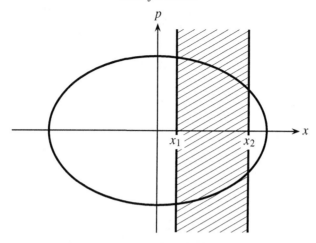

Fig. 2.1. Phase space x, p for a particle in one dimension. The ellipse is a possible orbit for a harmonic oscillator. The cross-hatched region corresponds to $x_1 \leq x \leq x_2$.

equipped with an *inner product* which assigns to any two wave functions $\psi(x)$ and $\phi(x)$ the complex number

$$\langle \phi | \psi \rangle = \int_{-\infty}^{+\infty} \phi^*(x) \psi(x) \, dx. \tag{2.3}$$

Here $\phi^*(x)$ denotes the complex conjugate of the function $\phi(x)$. (The notation used in (2.3) is standard among physicists, and differs in some trivial but annoying details from that generally employed by mathematicians.)

The inner product $\langle \phi | \psi \rangle$ is analogous to the dot product

$$\mathbf{a} \cdot \mathbf{b} = a_x b_x + a_y b_y + a_z b_z \tag{2.4}$$

of two ordinary vectors $\mathbf{a}$ and $\mathbf{b}$. One difference is that a dot product is always a real number, and $\mathbf{a} \cdot \mathbf{b}$ is the same as $\mathbf{b} \cdot \mathbf{a}$. By contrast, the inner product defined in (2.3) is in general a complex number, and interchanging $\psi(x)$ with $\phi(x)$ yields the complex conjugate:

$$\langle \psi | \phi \rangle = \langle \phi | \psi \rangle^*. \tag{2.5}$$

Despite this difference, the analogy between a dot product and an inner product is useful in that it provides an intuitive geometrical picture of the latter.

If $\langle \phi | \psi \rangle = 0$, which in view of (2.5) is equivalent to $\langle \psi | \phi \rangle = 0$, the functions $\psi(x)$ and $\phi(x)$ are said to be *orthogonal* to each other. This is analogous to $\mathbf{a} \cdot \mathbf{b} = 0$, which means that $\mathbf{a}$ and $\mathbf{b}$ are perpendicular to each other. The concept of orthogonal ("perpendicular") wave functions, along with certain generalizations of this notion, plays an extremely important role in the physical interpretation of quantum states. The inner product of $\psi(x)$ with itself,

$$\|\psi\|^2 = \int_{-\infty}^{+\infty} \psi^*(x)\psi(x)\,dx, \qquad (2.6)$$

is a positive number whose (positive) square root $\|\psi\|$ is called the *norm* of $\psi(x)$. The integral must be less than infinity for a wave function to be a member of $\mathcal{H}$. Thus e^{-ax^2} for $a > 0$ is a member of $\mathcal{H}$, whereas e^{-ax^2} is not.

A complex linear space $\mathcal{H}$ with an inner product is known as a *Hilbert space* provided it satisfies some additional conditions which are discussed in texts on functional analysis and mathematical physics, but lie outside the scope of this book (see the remarks in Sec. 1.4). Because of the condition that the norm as defined in (2.6) be finite, the linear space of wave functions is called the *Hilbert space of square-integrable functions*, often denoted by L^2.

2.2 Physical interpretation of the wave function

The intuitive significance of the pair of numbers x, p used to describe a classical particle in one dimension at a particular time is relatively clear: the particle is located at the point x, and its velocity is p/m. The interpretation of a quantum wave function $\psi(x)$, on the other hand, is much more complicated, and an intuition for what it means has to be built up by thinking about various examples. We will begin this process in Sec. 2.3. However, it is convenient at this point to make some very general observations, comparing and contrasting quantum with classical descriptions.

Any point x, p in the classical phase space represents a possible state of the classical particle. In a similar way, almost every wave function in the space $\mathcal{H}$ represents a possible state of a quantum particle. The exception is the state $\psi(x)$ which is equal to 0 for every value of x, and thus has norm $\|\psi\| = 0$. This is an element of the linear space, and from a mathematical point of view it is a very significant element. Nevertheless, it cannot represent a possible state of a physical system. All the other members of $\mathcal{H}$ represent possible quantum states.

A point in the phase space represents the most precise description one can have of the state of a classical particle. If one knows both x and p for a particle in one dimension, that is all there is to know. In the same way, the quantum wave function $\psi(x)$ represents a complete description of a quantum particle, there is nothing more that can be said about it. To be sure, a classical "particle" might possess some sort of internal structure and in such a case the pair x, p, or $\mathbf{r}$, $\mathbf{p}$, would represent the position of the center of mass and the total momentum, respectively, and one would need additional variables in order to describe the internal degrees of freedom. Similarly, a quantum particle can possess an internal structure, in which case $\psi(x)$ or $\psi(\mathbf{r})$ provides a complete description of the center of mass, whereas ψ must also depend upon additional variables if it is to describe the internal structure as well as the center of mass. The quantum description of particles with internal degrees of freedom, and of collections of several particles is taken up in Ch. 6.

An important difference between the classical phase space and the quantum Hilbert space $\mathcal{H}$ has to do with the issue of whether elements which are mathematically distinct describe situations which are physically distinct. Let us begin with the classical case, which is relatively straightforward. Two states (x, p) and (x', p') represent the same physical state if and only if

$$x' = x, \quad p' = p, \tag{2.7}$$

that is, if the two points in phase space coincide with each other. Otherwise they represent *mutually-exclusive possibilities*: a particle cannot be in two different places at the same time, nor can it have two different values of momentum (or velocity) at the same time. To summarize, two states of a classical particle have the same *physical interpretation* if and only if they have the same *mathematical description*.

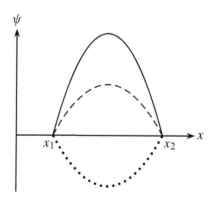

Fig. 2.2. Three wave functions which have the same physical meaning.

The case of a quantum particle is not nearly so simple. There are three different situations one needs to consider.

1. If two functions $\psi(x)$ and $\phi(x)$ are *multiples of each other*, that is, $\phi(x) = \alpha\psi(x)$ for some nonzero complex number α, then these two functions have *precisely the same* physical meaning. For example, all three functions in Fig. 2.2 have the same physical meaning. This is in marked contrast to the waves one is familiar with in classical physics, such as sound waves, or waves on the surface of water. Increasing the amplitude of a sound wave by a factor of 2 means that it carries four

times as much energy, whereas multiplying a quantum wave function by 2 leaves its physical significance unchanged.

Given any $\psi(x)$ with positive norm, it is always possible to introduce another function

$$\bar{\psi}(x) = \psi(x)/\|\psi\| \tag{2.8}$$

which has the same physical meaning as $\psi(x)$, but whose norm is $\|\bar{\psi}\| = 1$. Such *normalized* states are convenient when carrying out calculations, and for this reason quantum physicists often develop a habit of writing wave functions in normalized form, even when it is not really necessary. A normalized wave function remains normalized when it is multiplied by a complex constant $e^{i\phi}$, where the phase ϕ is some real number, and of course its physical meaning is not changed. Thus a normalized wave function representing some physical situation still has an *arbitrary phase*.

Warning! Although multiplying a wave function by a nonzero scalar does not change its physical significance, there are cases in which a careless use of this principle can lead to mistakes. Suppose that one is interested in a wave function which is a linear combination of two other functions,

$$\psi(x) = \phi(x) + \omega(x). \tag{2.9}$$

Multiplying $\phi(x)$ but not $\omega(x)$ by a complex constant α leads to a function

$$\tilde{\psi}(x) = \alpha\phi(x) + \omega(x) \tag{2.10}$$

which does *not*, at least in general, have the same physical meaning as $\psi(x)$, because it is not equal to a constant times $\psi(x)$.

2. Two wave functions $\phi(x)$ and $\psi(x)$ which are orthogonal to each other, $\langle\phi|\psi\rangle = 0$, represent *mutually exclusive* physical states: if one of them is true, in the sense that it is a correct description of the quantum system, the other is false, that is, an incorrect description of the quantum system. For example, the inner product of the two wave functions $\phi(x)$ and $\psi(x)$ sketched in Fig. 2.3 is zero, because at any x where one of them is finite, the other is zero, and thus the integrand in (2.3) is zero. As discussed in Sec. 2.3, if a wave function vanishes outside some finite interval, the quantum particle is located inside that interval. Since the two intervals $[x_1, x_2]$ and $[x_3, x_4]$ in Fig. 2.3 do not overlap, they represent mutually-exclusive possibilities: if the particle is in one interval, it cannot be in the other.

In Fig. 2.4, $\psi(x)$ and $\phi(x)$ are the ground state and first excited state of a quantum particle in a smooth, symmetrical potential well (such as a harmonic oscillator). In this case the vanishing of $\langle\phi|\psi\rangle$ is not quite so obvious, but it follows from the fact that $\psi(x)$ is an even and $\phi(x)$ an odd function of x. Thus their product is an odd function of x, and the integral in (2.3) vanishes. From a physical point

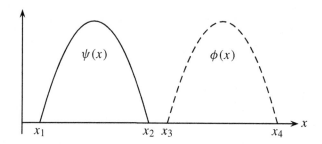

Fig. 2.3. Two orthogonal wave functions.

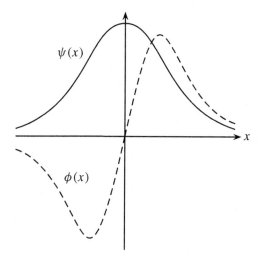

Fig. 2.4. Two orthogonal wave functions.

of view these two states are mutually-exclusive possibilities because if a quantum particle has a definite energy, it cannot have some other energy.

3. If $\phi(x)$ and $\psi(x)$ are *not* multiples of each other, and $\langle\phi|\psi\rangle$ is *not* equal to zero, the two wave functions represent *incompatible* states-of-affairs, a relationship which will be discussed in Sec. 4.6. Figure 2.5 shows a pair of incompatible wave functions. It is obvious that $\phi(x)$ cannot be a multiple of $\psi(x)$, because there are values of x at which ϕ is positive and ψ is zero. On the other hand, it is also obvious that the inner product $\langle\phi|\psi\rangle$ is not zero, for the integrand in (2.3) is positive, and nonzero over a finite interval.

There is nothing in classical physics corresponding to descriptions which are incompatible in the quantum sense of the term. This is one of the main reasons why quantum theory is hard to understand: there is no good classical analogy for

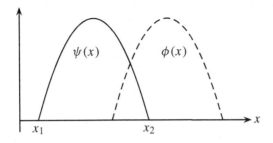

Fig. 2.5. Two incompatible wave functions.

the situation shown in Fig. 2.5. Instead, one has to build up one's physical intuition for this situation using examples that are quantum mechanical. It is important to keep in mind that quantum states which are *incompatible* stand in a very different relationship to each other than states which are *mutually exclusive*; one must not confuse these two concepts!

2.3 Wave functions and position

The quantum wave function $\psi(x)$ is a function of x, and in classical physics x is simply the position of the particle. But what can one say about the position of a quantum particle described by $\psi(x)$? In classical physics wave packets are used to describe water waves, sound waves, radar pulses, and the like. In each of these cases the wave packet does not have a precise position; indeed, one would not recognize something as a wave if it were not spread out to some extent. Thus there is no reason to suppose that a quantum particle possesses a precise position if it is described by a wave function $\psi(x)$, since the wave packet itself, thought of as a mathematical object, is obviously not located at a precise position x.

In addition to waves, there are many objects, such as clouds and cities, which do not have a precise location. These, however, are made up of other objects whose location is more definite: individual water droplets in a cloud, or individual buildings in a city. However, in the case of a quantum wave packet, a more detailed description in terms of smaller (better localized) physical objects or properties is not possible. To be sure, there is a very localized *mathematical* description: at each x the wave packet takes on some precise value $\psi(x)$. But there is no reason to suppose that this represents a corresponding physical "something" located at this precise point. Indeed, the discussion in Sec. 2.2 suggests quite the opposite. To begin with, the value of $\psi(x_0)$ at a particular point x_0 cannot in any direct way represent the value of some physical quantity, since one can always multiply the function $\psi(x)$ by a complex constant to obtain another wave function with the same physical significance, thus altering $\psi(x_0)$ in an arbitrary fashion (unless, of course, $\psi(x_0) = 0$). Furthermore, in order to see that the mathematically distinct wave functions in Fig. 2.2 represent the same physical state of affairs, and that the two functions in Fig. 2.4 represent distinct physical states, one cannot simply carry out a point-by-point comparison; instead it is necessary to consider each wave function "as a whole".

It is probably best to think of a quantum particle as *delocalized*, that is, as not having a position which is more precise than that of the wave function representing its quantum state. The term "delocalized" should be understood as meaning that no precise position can be defined, and not as suggesting that a quantum particle is in two different places at the same time. Indeed, we shall show in Sec. 4.5, there is a well-defined sense in which a quantum particle *cannot* be in two (or more) places at the same time.

Things which do not have precise positions, such as books and tables, can nonetheless often be assigned *approximate* locations, and it is often useful to do so. The situation with quantum particles is similar. There are two different, though related, approaches to assigning an approximate position to a quantum particle in one dimension (with obvious generalizations to higher dimensions). The first is mathematically quite "clean", but can only be applied for a rather limited set of wave functions. The second is mathematically "sloppy", but is often of more use to the physicist. Both of them are worth discussing, since each adds to one's physical understanding of the meaning of a wave function.

It is sometimes the case, as in the examples in Figs. 2.2, 2.3, and 2.5, that the quantum wave function is nonzero only in some finite interval

$$x_1 \le x \le x_2. \tag{2.11}$$

In such a case it is safe to assert that the quantum particle is *not* located *outside* this interval, or, equivalently, that it is inside this interval, provided the latter is not interpreted to mean that there is some precise point inside the interval where the particle is located. In the case of a classical particle, the statement that it is not outside, and therefore inside the interval (2.11) corresponds to asserting that the point x, p representing the state of the particle falls somewhere inside the region of its phase space indicated by the cross-hatching in Fig. 2.1. To be sure, since the actual position of a classical particle must correspond to a single number x, we know that if it is inside the interval (2.11), then it is actually located at a definite point in this interval, even though we may not know what this precise point is. By contrast, in the case of any of the wave functions in Fig. 2.2 it is incorrect to say that the particle has a location which is more precise than is given by the interval (2.11), because the wave packet cannot be located more precisely than this, and the particle cannot be located more precisely than its wave packet.

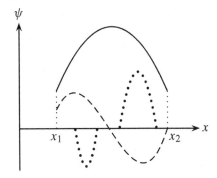

Fig. 2.6. Some of the many wave functions which vanish outside the interval $x_1 \leq x \leq x_2$.

There is a quantum analog of the cross-hatched region of the phase space in Fig. 2.1: it is the collection of all wave functions in $\mathcal{H}$ with the property that they vanish outside the interval $[x_1, x_2]$. There are, of course, a very large number of wave functions of this type, a few of which are indicated in Fig. 2.6. Given a wave function which vanishes outside (2.11), it still has this property if multiplied by an arbitrary complex number. And the sum of two wave functions of this type will also vanish outside the interval. Thus the collection of all functions which vanish outside $[x_1, x_2]$ is itself a linear space. If in addition we impose the condition that the allowable functions have a finite norm, the corresponding collection of functions $\mathcal{X}$ is part of the collection $\mathcal{H}$ of all allowable wave functions, and because $\mathcal{X}$ is a linear space, it is a *subspace* of the quantum Hilbert space $\mathcal{H}$. As we shall see in Ch. 4, a *physical property* of a quantum system can always be associated with a subspace of $\mathcal{H}$, in the same way that a physical property of a classical system corresponds to a subset of points in its phase space. In the case at hand, the physical property of being located inside the interval $[x_1, x_2]$ corresponds in the classical case to the cross-hatched region in Fig. 2.1, and in the quantum case to the subspace $\mathcal{X}$ which has just been defined.

The notion of approximate location discussed above has limited applicability, because one is often interested in wave functions which are never equal to zero, or at least do not vanish outside some finite interval. An example is the Gaussian wave packet

$$\psi(x) = \exp[-(x - x_0)^2/4(\Delta x)^2], \tag{2.12}$$

centered at $x = x_0$, where Δx is a constant, with the dimensions of a length, that provides a measure of the width of the wave packet. The function $\psi(x)$ is never equal to 0. However, when $|x - x_0|$ is large compared to Δx, $\psi(x)$ is very small, and so it seems sensible, at least to a physicist, to suppose that for this quantum state, the particle is located "near" x_0, say within an interval

$$x_0 - \lambda \Delta x \leq x \leq x_0 + \lambda \Delta x, \qquad (2.13)$$

where λ might be set equal to 1 when making a rough back-of-the-envelope calculation, or perhaps 2 or 3 or more if one is trying to be more careful or conservative.

What the physicist is, in effect, doing in such circumstances is approximating the Gaussian wave packet in (2.12) by a function which has been set equal to 0 for x lying outside the interval (2.13). Once the "tails" of the Gaussian packet have been eliminated in this manner, one can employ the ideas discussed above for functions which vanish outside some finite interval. To be sure, "cutting off the tails" of the original wave function involves an approximation, and as with all approximations, this requires the application of some judgment as to whether or not one will be making a serious mistake, and this will in turn depend upon the sort of questions which are being addressed. Since approximations are employed in all branches of theoretical physics (apart from those which are indistinguishable from pure mathematics), it would be quibbling to deny this possibility to the quantum physicist. Thus it makes physical sense to say that the wave packet (2.12) represents a quantum particle with an approximate location given by (2.13), as long as λ is not too small. Of course, similar reasoning can be applied to other wave packets which have long tails.

It is sometimes said that the meaning, or at least one of the meanings, of the wave function $\psi(x)$ is that

$$\rho(x) = |\psi(x)|^2 / \|\psi\|^2 \qquad (2.14)$$

is a probability distribution density for the particle to be located at the position x, or found to be at the position x by a suitable measurement. Wave functions can indeed be used to calculate probability distributions, and in certain circumstances (2.14) is a correct way to do such a calculation. However, in quantum theory it is necessary to differentiate between $\psi(x)$ as representing a *physical property* of a quantum system, and $\psi(x)$ as a *pre-probability*, a mathematical device for calculating probabilities. It is necessary to look at examples to understand this distinction, and we shall do so in Ch. 9, following a general discussion of probabilities in quantum theory in Ch. 5.

2.4 Wave functions and momentum

The state of a classical particle in one dimension is specified by giving both x *and* p, while in the quantum case the wave function $\psi(x)$ depends upon only one of these two variables. From this one might conclude that quantum theory has nothing to say about the momentum of a particle, but this is not correct. The information about the momentum provided by quantum mechanics is contained in $\psi(x)$, but one has to know how to extract it. A convenient way to do so is to define the *momentum wave function*

$$\hat{\psi}(p) = \frac{1}{\sqrt{2\pi\hbar}} \int_{-\infty}^{+\infty} e^{-ipx/\hbar}\,\psi(x)\,dx \qquad (2.15)$$

as the Fourier transform of $\psi(x)$.

Note that $\hat{\psi}(p)$ is completely determined by the position wave function $\psi(x)$. On the other hand, (2.15) can be inverted by writing

$$\psi(x) = \frac{1}{\sqrt{2\pi\hbar}} \int_{-\infty}^{+\infty} e^{+ipx/\hbar}\,\hat{\psi}(p)\,dp, \qquad (2.16)$$

so that, in turn, $\psi(x)$ is completely determined by $\hat{\psi}(p)$. Therefore $\psi(x)$ and $\hat{\psi}(p)$ contain precisely the same information about a quantum state; they simply express this information in two different forms. Whatever may be the physical significance of $\psi(x)$, that of $\hat{\psi}(p)$ is exactly the same. One can say that $\psi(x)$ is the *position representation* and $\hat{\psi}(p)$ the *momentum representation* of the *single* quantum state which describes the quantum particle at a particular instant of time. (As an analogy, think of a novel published simultaneously in two different languages: the two editions represent exactly the same story, assuming the translator has done a good job.) The inner product (2.3) can be expressed equally well using either the position or the momentum representation:

$$\langle\phi|\psi\rangle = \int_{-\infty}^{+\infty} \phi^*(x)\psi(x)\,dx = \int_{-\infty}^{+\infty} \hat{\phi}^*(p)\hat{\psi}(p)\,dp. \qquad (2.17)$$

Information about the momentum of a quantum particle can be obtained from the momentum wave function in the same way that information about its position can be obtained from the position wave function, as discussed in Sec. 2.3. A quantum particle, unlike a classical particle, does not possess a well-defined momentum. However, if $\hat{\psi}(p)$ vanishes outside an interval

$$p_1 \leq p \leq p_2, \qquad (2.18)$$

it possesses an *approximate* momentum in that the momentum does *not* lie *outside* the interval (2.18); equivalently, the momentum lies inside this interval, though it does not have some particular precise value inside this interval.

Even when $\hat{\psi}(p)$ does not vanish outside any interval of the form (2.18), one can still assign an approximate momentum to the quantum particle in the same way that one can assign an approximate position when $\psi(x)$ has nonzero tails, as in (2.12).

In particular, in the case of a Gaussian wave packet

$$\hat{\psi}(p) = \exp[-(p - p_0)^2/4(\Delta p)^2], \qquad (2.19)$$

it is reasonable to say that the momentum is "near" p_0 in the sense of lying in the interval

$$p_0 - \lambda \Delta p \le p \le p_0 + \lambda \Delta p, \qquad (2.20)$$

with λ on the order of 1 or larger. The justification for this is that one is approximating (2.19) with a function which has been set equal to 0 outside the interval (2.20). Whether or not "cutting off the tails" in this manner is an acceptable approximation is a matter of judgment, just as in the case of the position wave packet discussed in Sec. 2.3.

The momentum wave function can be used to calculate a probability distribution density

$$\hat{\rho}(p) = |\hat{\psi}(p)|^2/\|\psi\|^2 \qquad (2.21)$$

for the momentum p in much the same way as the position wave function can be used to calculate a similar density for x, (2.14). See the remarks following (2.14): it is important to distinguish between $\hat{\psi}(p)$ as representing a physical property, which is what we have been discussing, and as a pre-probability, which is its role in (2.21). If one sets $x_0 = 0$ in the Gaussian wave packet (2.12) and carries out the Fourier transform (2.15), the result is (2.19) with $p_0 = 0$ and $\Delta p = \hbar/2\Delta x$. As shown in introductory textbooks, it is quite generally the case that for any given quantum state

$$\Delta p \cdot \Delta x \ge \hbar/2, \qquad (2.22)$$

where $(\Delta x)^2$ is the variance of the probability distribution density (2.14), and $(\Delta p)^2$ the variance of the one in (2.21). Probabilities will be taken up later in the book, but for present purposes it suffices to regard Δx and Δp as convenient, albeit somewhat crude measures of the widths of the wave packets $\psi(x)$ and $\hat{\psi}(p)$, respectively. What the inequality tells us is that the narrower the position wave packet $\psi(x)$, the broader the corresponding momentum wave packet $\hat{\psi}(p)$ has got to be, and vice versa.

The inequality (2.22) expresses the well-known *Heisenberg uncertainty principle*. This principle is often discussed in terms of *measurements* of a particle's position or momentum, and the difficulty of simultaneously measuring both of these quantities. While such discussions are not without merit — and we shall have more to say about measurements later in this book — they tend to put the emphasis in the wrong place, suggesting that the inequality somehow arises out of peculiarities associated with measurements. But in fact (2.22) is a consequence of the decision by quantum physicists to use a Hilbert space of wave packets in order to describe quantum particles, and to make the momentum wave packet for a particular quantum state equal to the Fourier transform of the position wave packet for the same state. In the Hilbert space there are, as a fact of mathematics, no states for which the widths of the position and momentum wave packets violate the inequality (2.22). Hence if this Hilbert space is appropriate for describing the real world, no particles exist for which the position and momentum can even be approximately defined with a precision better than that allowed by (2.22). If measurements can accurately determine the properties of quantum particles — another topic to which we shall later return — then the results cannot, of course, be more precise than the quantities which are being measured. To use an analogy, the fact that the location of the city of Pittsburgh is uncertain by several kilometers has nothing to do with the lack of precision of surveying instruments. Instead a city, as an extended object, does not have a precise location.

2.5 Toy model

The Hilbert space $\mathcal{H}$ for a quantum particle in one dimension is extremely large; viewed as a linear space it is infinite-dimensional. Infinite-dimensional spaces provide headaches for physicists and employment for mathematicians. Most of the conceptual issues in quantum theory have nothing to do with the fact that the Hilbert space is infinite-dimensional, and therefore it is useful, in order to simplify the mathematics, to replace the continuous variable x with a discrete variable m which takes on only a *finite number of integer values*. That is to say, we will assume that the quantum particle is located at one of a finite collection of sites arranged in a straight line, or, if one prefers, it is located in one of a finite number of boxes or cells. It is often convenient to think of this system of sites as having "periodic boundary conditions" or as placed on a circle, so that the last site is adjacent to (just in front of) the first site. If one were representing a wave function numerically on a computer, it would be sensible to employ a discretization of this type. However, our goal is not numerical computation, but physical insight. Temporarily shunting mathematical difficulties out of the way is part of a useful "divide and conquer" strategy for attacking difficult problems. Our aim will not be realistic descriptions, but instead *simple* descriptions which still contain the essential features of quantum theory. For this reason, the term "toy model" seems appropriate.

Let us suppose that the quantum wave function is of the form $\psi(m)$, with m an integer in the range

$$-M_a \leq m \leq M_b, \tag{2.23}$$

where M_a and M_b are fixed integers, so m can take on $M = M_a + M_b + 1$ different values. Such wave functions form an M-dimensional Hilbert space. For example, if $M_a = 1 = M_b$, the particle can be at one of the three sites, $m = -1, 0, 1$, and its wave function is completely specified by the $M = 3$ complex numbers $\psi(-1)$, $\psi(0)$, and $\psi(1)$. The inner product of two wave functions is given by

$$\langle \phi | \psi \rangle = \sum_m \phi^*(m)\psi(m), \qquad (2.24)$$

where the sum is over those values of m allowed by (2.23), and the norm of ψ is the positive square root of

$$\|\psi\|^2 = \sum_m |\psi(m)|^2. \qquad (2.25)$$

The toy wave function χ_n, defined by

$$\chi_n(m) = \delta_{mn} = \begin{cases} 1 & \text{if } m = n, \\ 0 & \text{for } m \neq n, \end{cases} \qquad (2.26)$$

where δ_{mn} is the Kronecker delta function, has the physical significance that the particle is at site n (or in cell n). Now suppose that $M_a = 3 = M_b$, and consider the wave function

$$\psi(m) = \chi_{-1}(m) + 1.5\chi_0(m) + \chi_1(m). \qquad (2.27)$$

It is sketched in Fig. 2.7, and one can think of it as a relatively coarse approximation to a continuous function of the sort shown in Fig. 2.2, with $x_1 = -2$, $x_2 = +2$. What can one say about the location of the particle whose quantum wave function is given by (2.27)?

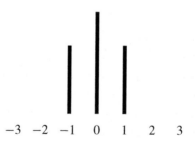

Fig. 2.7. The toy wave packet (2.27).

In light of the discussion in Sec. 2.3 it seems sensible to interpret $\psi(m)$ as signifying that the position of the quantum particle is not outside the interval $[-1, +1]$, where by $[-1, +1]$ we mean the three values -1, 0, and $+1$. The circumlocution "not outside the interval" can be replaced with the more natural "inside the interval" provided the latter is not interpreted to mean "at a particular site inside this interval", since the particle described by (2.27) cannot be said to be at $m = -1$ or at $m = 0$ or at $m = 1$. Instead it is delocalized, and its position cannot be specified any more precisely than by giving the interval $[-1, +1]$. There is no concise way of stating this in English, which is one reason we need a mathematical notation in which quantum properties can be expressed in a precise way — this will be introduced in Ch. 4.

It is important not to look at a wave function written out as a sum of different pieces whose physical significance one understands, and interpret it in physical terms as meaning the quantum system has one or the other of the properties corresponding to the different pieces. In particular, one should not interpret (2.27) to mean that the particle is at $m = -1$ or at $m = 0$ or at $m = 1$. A simple example which illustrates how such an interpretation can lead one astray is obtained by writing χ_0 in the form

$$\chi_0(m) = (1/2)[\chi_0(m) + i\chi_2(m)] + (1/2)[\chi_0(m) + (-i)\chi_2(m)]. \qquad (2.28)$$

If we carelessly interpret "+" to mean "or", then both of the functions in square brackets on the right side of (2.28), and therefore also their sum, have the interpretation that the particle is at 0 or 2, whereas in fact $\chi_0(m)$ means that the particle is at 0 and *not* at 2. The correct quantum mechanical way to use "or" will be discussed in Secs. 4.5, 4.6, and 5.2.

Just as $\psi(m)$ is a discrete version of the position wave function $\psi(x)$, there is also a discrete version $\hat{\psi}(k)$ of the momentum wave function $\hat{\psi}(p)$, given by the formula

$$\hat{\psi}(k) = \frac{1}{\sqrt{M}} \sum_m e^{-2\pi i k m/M} \psi(m), \qquad (2.29)$$

where k is an integer which can take on the same set of values as m, (2.23). The inverse transformation is

$$\psi(m) = \frac{1}{\sqrt{M}} \sum_k e^{2\pi i k m/M} \hat{\psi}(k). \qquad (2.30)$$

The inner product of two states, (2.24), can equally well be written in terms of momentum wave functions:

$$\langle \phi | \psi \rangle = \sum_k \hat{\phi}^*(k)\hat{\psi}(k). \qquad (2.31)$$

These expressions are similar to those in (2.15)–(2.17). The main difference is that integrals have been replaced by sums. The reason $\hbar$ has disappeared from the toy model expressions is that position and momentum are being expressed in dimensionless units.

3

Linear algebra in Dirac notation

3.1 Hilbert space and inner product

In Ch. 2 it was noted that quantum wave functions form a linear space in the sense that multiplying a function by a complex number or adding two wave functions together produces another wave function. It was also pointed out that a particular quantum state can be represented either by a wave function $\psi(x)$ which depends upon the position variable x, or by an alternative function $\hat{\psi}(p)$ of the momentum variable p. It is convenient to employ the Dirac symbol $|\psi\rangle$, known as a "ket", to denote a quantum state without referring to the particular function used to represent it. The kets, which we shall also refer to as *vectors* to distinguish them from *scalars*, which are complex numbers, are the elements of the quantum Hilbert space $\mathcal{H}$. (The real numbers form a subset of the complex numbers, so that when a scalar is referred to as a "complex number", this includes the possibility that it might be a real number.)

If α is any scalar (complex number), the ket corresponding to the wave function $\alpha\psi(x)$ is denoted by $\alpha|\psi\rangle$, or sometimes by $|\psi\rangle\alpha$, and the ket corresponding to $\phi(x)+\psi(x)$ is denoted by $|\phi\rangle+|\psi\rangle$ or $|\psi\rangle+|\phi\rangle$, and so forth. This correspondence could equally well be expressed using momentum wave functions, because the Fourier transform, (2.15) or (2.16), is a linear relationship between $\psi(x)$ and $\hat{\psi}(p)$, so that $\alpha\phi(x)+\beta\psi(x)$ and $\alpha\hat{\phi}(p)+\beta\hat{\psi}(p)$ correspond to the same quantum state $\alpha|\psi\rangle+\beta|\phi\rangle$. The addition of kets and multiplication by scalars obey some fairly obvious rules:

$$\alpha\big(\beta|\psi\rangle\big) = (\alpha\beta)|\psi\rangle, \quad (\alpha+\beta)|\psi\rangle = \alpha|\psi\rangle + \beta|\psi\rangle,$$
$$\alpha\big(|\phi\rangle + |\psi\rangle\big) = \alpha|\phi\rangle + \alpha|\psi\rangle, \quad 1|\psi\rangle = |\psi\rangle. \tag{3.1}$$

Multiplying any ket by the number 0 yields the unique *zero vector* or *zero ket*, which will, because there is no risk of confusion, also be denoted by 0.

The linear space $\mathcal{H}$ is equipped with an *inner product*

$$\mathcal{I}\big(|\omega\rangle, |\psi\rangle\big) = \langle\omega|\psi\rangle \qquad (3.2)$$

which assigns to any pair of kets $|\omega\rangle$ and $|\psi\rangle$ a complex number. While the Dirac notation $\langle\omega|\psi\rangle$, already employed in Ch. 2, is more compact than the one based on $\mathcal{I}(\,,\,)$, it is, for purposes of exposition, useful to have a way of writing the inner product which clearly indicates how it depends on two different ket vectors.

An inner product must satisfy the following conditions:

1. Interchanging the two arguments results in the complex conjugate of the original expression:

$$\mathcal{I}\big(|\psi\rangle, |\omega\rangle\big) = \big[\mathcal{I}\big(|\omega\rangle, |\psi\rangle\big)\big]^{*}. \qquad (3.3)$$

2. The inner product is *linear* as a function of its second argument:

$$\mathcal{I}\big(|\omega\rangle, \alpha|\phi\rangle + \beta|\psi\rangle\big) = \alpha\mathcal{I}\big(|\omega\rangle, |\phi\rangle\big) + \beta\mathcal{I}\big(|\omega\rangle, |\psi\rangle\big). \qquad (3.4)$$

3. The inner product is an *antilinear* function of its first argument:

$$\mathcal{I}\big(\alpha|\phi\rangle + \beta|\psi\rangle, |\omega\rangle\big) = \alpha^{*}\mathcal{I}\big(|\phi\rangle, |\omega\rangle\big) + \beta^{*}\mathcal{I}\big(|\psi\rangle, |\omega\rangle\big). \qquad (3.5)$$

4. The inner product of a ket with itself,

$$\mathcal{I}\big(|\psi\rangle, |\psi\rangle\big) = \langle\psi|\psi\rangle = \|\psi\|^{2} \qquad (3.6)$$

is a positive (greater than 0) real number unless $|\psi\rangle$ is the zero vector, in which case $\langle\psi|\psi\rangle = 0$.

The term "antilinear" in the third condition refers to the fact that the *complex conjugates* of α and β appear on the right side of (3.5), rather than α and β themselves, as would be the case for a linear function. Actually, (3.5) is an immediate consequence of (3.3) and (3.4) — simply take the complex conjugate of both sides of (3.4), and then apply (3.3) — but it is of sufficient importance that it is worth stating separately. The reader can check that the inner products defined in (2.3) and (2.24) satisfy these conditions. (There are some subtleties associated with $\psi(x)$ when x is a continuous real number, but we must leave discussion of these matters to books on functional analysis.)

The positive square root $\|\psi\|$ of $\|\psi\|^{2}$ in (3.6) is called the *norm* of $|\psi\rangle$. As already noted in Ch. 2, $\alpha|\psi\rangle$ and $|\psi\rangle$ have exactly the same physical significance if α is a nonzero complex number. Consequently, as far as the quantum physicist is concerned, the actual norm, as long as it is positive, is a matter of indifference. By multiplying a nonzero ket by a suitable constant, one can always make its norm equal to 1. This process is called *normalizing* the ket, and a ket with norm equal to 1 is said to be *normalized*. Normalizing does not produce a unique result, because

$e^{i\phi}|\psi\rangle$, where ϕ is an arbitrary real number or *phase*, has precisely the same norm as $|\psi\rangle$. Two kets $|\phi\rangle$ and $|\psi\rangle$ are said to be *orthogonal* if $\langle\phi|\psi\rangle = 0$, which by (3.3) implies that $\langle\psi|\phi\rangle = 0$.

3.2 Linear functionals and the dual space

Let $|\omega\rangle$ be some fixed element of $\mathcal{H}$. Then the function

$$\mathcal{J}\big(|\psi\rangle\big) = \mathcal{I}\big(|\omega\rangle, |\psi\rangle\big) \tag{3.7}$$

assigns to every $|\psi\rangle$ in $\mathcal{H}$ a complex number in a linear manner,

$$\mathcal{J}\big(\alpha|\phi\rangle + \beta|\psi\rangle\big) = \alpha\mathcal{J}\big(|\phi\rangle\big) + \beta\mathcal{J}\big(|\psi\rangle\big), \tag{3.8}$$

as a consequence of (3.4). Such a function is called a *linear functional*. There are many different linear functionals of this sort, one for every $|\omega\rangle$ in $\mathcal{H}$. In order to distinguish them we could place a label on $\mathcal{J}$ and, for example, write it as $\mathcal{J}_{|\omega\rangle}\big(|\psi\rangle\big)$. The notation $\mathcal{J}_{|\omega\rangle}$ is a bit clumsy, even if its meaning is clear, and Dirac's $\langle\omega|$, called a "bra", provides a simpler way to denote the same object, so that (3.8) takes the form

$$\langle\omega|\big(\alpha|\phi\rangle + \beta|\psi\rangle\big) = \alpha\langle\omega|\phi\rangle + \beta\langle\omega|\psi\rangle, \tag{3.9}$$

if we also use the compact Dirac notation for inner products.

Among the advantages of (3.9) over (3.8) is that the former looks very much like the distributive law for multiplication if one takes the simple step of replacing $\langle\omega| \cdot |\psi\rangle$ by $\langle\omega|\psi\rangle$. Indeed, a principal virtue of Dirac notation is that many different operations of this general type become "automatic", allowing one to concentrate on issues of physics without getting overly involved in mathematical bookkeeping. However, if one is in doubt about what Dirac notation really means, it may be helpful to check things out by going back to the more awkward but also more familiar notation of functions, such as $\mathcal{I}(,)$ and $\mathcal{J}()$.

Linear functionals can themselves be added together and multiplied by complex numbers, and the rules are fairly obvious. Thus the right side of

$$\big[\alpha\langle\tau| + \beta\langle\omega|\big]\big(|\psi\rangle\big) = \alpha\langle\tau|\psi\rangle + \beta\langle\omega|\psi\rangle \tag{3.10}$$

gives the complex number obtained when the linear functional $\alpha\langle\tau| + \beta\langle\omega|$, formed by addition following multiplication by scalars, and placed inside square brackets for clarity, is applied to the ket $|\psi\rangle$. Thus linear functionals themselves form a linear space, called the *dual* of the space $\mathcal{H}$; we shall denote it by $\mathcal{H}^{\dagger}$.

Although $\mathcal{H}$ and $\mathcal{H}^{\dagger}$ are not identical spaces — the former is inhabited by kets and the latter by bras — the two are closely related. There is a one-to-one map from one to the other denoted by a dagger:

$$\langle\omega| = \big(|\omega\rangle\big)^{\dagger}, \quad |\omega\rangle = \big(\langle\omega|\big)^{\dagger}. \tag{3.11}$$

The parentheses may be omitted when it is obvious what the dagger operation applies to, but including them does no harm. The dagger map is *antilinear*,

$$\begin{aligned}
\big(\alpha|\phi\rangle + \beta|\psi\rangle\big)^{\dagger} &= \alpha^*\langle\phi| + \beta^*\langle\psi|, \\
\big(\gamma\langle\tau| + \delta\langle\omega|\big)^{\dagger} &= \gamma^*|\tau\rangle + \delta^*|\omega\rangle,
\end{aligned} \tag{3.12}$$

reflecting the fact that the inner product $\mathcal{I}$ is antilinear in its left argument, (3.5). When applied twice in a row, the dagger operation is the identity map:

$$\big((|\omega\rangle)^{\dagger}\big)^{\dagger} = |\omega\rangle, \quad \big((\langle\omega|)^{\dagger}\big)^{\dagger} = \langle\omega|. \tag{3.13}$$

There are occasions when the Dirac notation $\langle\omega|\psi\rangle$ is not convenient because it is too compact. In such cases the dagger operation can be useful, because $\big(|\omega\rangle\big)^{\dagger}|\psi\rangle$ means the same thing as $\langle\omega|\psi\rangle$. Thus, for example,

$$\big(\alpha|\tau\rangle + \beta|\omega\rangle\big)^{\dagger}|\psi\rangle = \big(\alpha^*\langle\tau| + \beta^*\langle\omega|\big)|\psi\rangle = \alpha^*\langle\tau|\psi\rangle + \beta^*\langle\omega|\psi\rangle \tag{3.14}$$

is one way to express the fact the inner product is antilinear in its first argument, (3.5), without having to employ $\mathcal{I}(,)$.

3.3 Operators, dyads

A *linear operator*, or simply an *operator* A is a linear function which maps $\mathcal{H}$ into itself. That is, to each $|\psi\rangle$ in $\mathcal{H}$, A assigns another element $A\big(|\psi\rangle\big)$ in $\mathcal{H}$ in such a way that

$$A\big(\alpha|\phi\rangle + \beta|\psi\rangle\big) = \alpha A\big(|\phi\rangle\big) + \beta A\big(|\psi\rangle\big) \tag{3.15}$$

whenever $|\phi\rangle$ and $|\psi\rangle$ are any two elements of $\mathcal{H}$, and α and β are complex numbers. One customarily omits the parentheses and writes $A|\phi\rangle$ instead of $A\big(|\phi\rangle\big)$ where this will not cause confusion, as on the right (but not the left) side of (3.15). In general we shall use capital letters, A, B, and so forth, to denote operators. The letter I is reserved for the *identity operator* which maps every element of $\mathcal{H}$ to itself:

$$I|\psi\rangle = |\psi\rangle. \tag{3.16}$$

The *zero operator* which maps every element of $\mathcal{H}$ to the zero vector will be denoted by 0.

The inner product of some element $|\phi\rangle$ of $\mathcal{H}$ with the ket $A|\psi\rangle$ can be written as

$$\big(|\phi\rangle\big)^{\dagger} A|\psi\rangle = \langle\phi|A|\psi\rangle, \tag{3.17}$$

where the notation on the right side, the "sandwich" with the operator between a bra and a ket, is standard Dirac notation. It is often referred to as a "matrix element", even when no matrix is actually under consideration. (Matrices are discussed in Sec. 3.6.) One can write $\langle\phi|A|\psi\rangle$ as $\big(\langle\phi|A\rangle\big(|\psi\rangle\big)$, and think of it as the linear functional or bra vector

$$\langle\phi|A \tag{3.18}$$

acting on or evaluated at $|\psi\rangle$. In this sense it is natural to think of a linear operator A on $\mathcal{H}$ as inducing a linear map of the dual space $\mathcal{H}^{\dagger}$ onto itself, which carries $\langle\phi|$ to $\langle\phi|A$. This map can also, without risk of confusion, be denoted by A, and while one could write it as $A\big(\langle\phi|\big)$, in Dirac notation $\langle\phi|A$ is more natural. Sometimes one speaks of "the operator A acting to the left".

Dirac notation is particularly convenient in the case of a simple type of operator known as a *dyad*, written as a ket followed by a bra, $|\omega\rangle\langle\tau|$. Applied to some ket $|\psi\rangle$ in $\mathcal{H}$, it yields

$$|\omega\rangle\langle\tau|\big(|\psi\rangle\big) = |\omega\rangle\langle\tau|\psi\rangle = \langle\tau|\psi\rangle|\omega\rangle. \tag{3.19}$$

Just as in (3.9), the first equality is "obvious" if one thinks of the product of $\langle\tau|$ with $|\psi\rangle$ as $\langle\tau|\psi\rangle$, and since the latter is a scalar it can be placed either after or in front of the ket $|\omega\rangle$. Setting A in (3.17) equal to the dyad $|\omega\rangle\langle\tau|$ yields

$$\langle\phi|\big(|\omega\rangle\langle\tau|\big)|\psi\rangle = \langle\phi|\omega\rangle\langle\tau|\psi\rangle, \tag{3.20}$$

where the right side is the product of the two scalars $\langle\phi|\omega\rangle$ and $\langle\tau|\psi\rangle$. Once again the virtues of Dirac notation are evident in that this result is an almost automatic consequence of writing the symbols in the correct order.

The collection of all operators is itself a linear space, since a scalar times an operator is an operator, and the sum of two operators is also an operator. The operator $\alpha A + \beta B$ applied to an element $|\psi\rangle$ of $\mathcal{H}$ yields the result:

$$(\alpha A + \beta B)|\psi\rangle = \alpha\big(A|\psi\rangle\big) + \beta\big(B|\psi\rangle\big), \tag{3.21}$$

where the parentheses on the right side can be omitted, since $\big(\alpha A\big)|\psi\rangle$ is equal to $\alpha\big(A|\psi\rangle\big)$, and both can be written as $\alpha A|\psi\rangle$.

The *product AB* of two operators A and B is the operator obtained by first applying B to some ket, and then A to the ket which results from applying B:

$$AB\big(|\psi\rangle\big) = A\big(B\big(|\psi\rangle\big)\big). \tag{3.22}$$

Normally the parentheses are omitted, and one simply writes $AB|\psi\rangle$. However, it is very important to note that operator multiplication, unlike multiplication of scalars, is *not* commutative: in general, $AB \neq BA$, since there is no particular reason to expect that $A(B(|\psi\rangle))$ will be the same element of $\mathcal{H}$ as $B(A(|\psi\rangle))$.

In the exceptional case in which $AB = BA$, that is, $AB|\psi\rangle = BA|\psi\rangle$ for all $|\psi\rangle$, one says that these two operators *commute with each other*, or (simply) *commute*. The identity operator I commutes with every other operator, $IA = AI = A$, and the same is true of the zero operator, $A0 = 0A = 0$. The operators in a collection $\{A_1, A_2, A_3, \dots\}$ are said to commute with each other provided

$$A_j A_k = A_k A_j \tag{3.23}$$

for every j and k.

Operator products follow the usual distributive laws, and scalars can be placed anywhere in a product, though one usually moves them to the left side:

$$\begin{aligned} A(\gamma C + \delta D) &= \gamma AC + \delta AD, \\ (\alpha A + \beta B)C &= \alpha AC + \beta BC. \end{aligned} \tag{3.24}$$

In working out such products it is important that the order of the operators, from left to right, be preserved: one cannot (in general) replace AC with CA. The operator product of two dyads $|\omega\rangle\langle\tau|$ and $|\psi\rangle\langle\phi|$ is fairly obvious if one uses Dirac notation:

$$|\omega\rangle\langle\tau| \cdot |\psi\rangle\langle\phi| = |\omega\rangle\langle\tau|\psi\rangle\langle\phi| = \langle\tau|\psi\rangle|\omega\rangle\langle\phi|, \tag{3.25}$$

where the final answer is a scalar $\langle\tau|\psi\rangle$ multiplying the dyad $|\omega\rangle\langle\phi|$. Multiplication in the reverse order will yield an operator proportional to $|\psi\rangle\langle\tau|$, so in general two dyads do not commute with each other.

Given an operator A, if one can find an operator B such that

$$AB = I = BA, \tag{3.26}$$

then B is called the *inverse* of the operator A, written as A^{-1}, and A is the inverse of the operator B. On a finite-dimensional Hilbert space one only needs to check one of the equalities in (3.26), as it implies the other, whereas on an infinite-dimensional space both must be checked. Many operators do not possess inverses, but if an inverse exists, it is unique.

The antilinear dagger operation introduced earlier, (3.11) and (3.12), can also be applied to operators. For a dyad one has:

$$\left(|\omega\rangle\langle\tau|\right)^{\dagger} = |\tau\rangle\langle\omega|. \tag{3.27}$$

Note that the right side is obtained by applying † separately to each term in the ket-bra "product" $|\omega\rangle\langle\tau|$ on the left, following the prescription in (3.11), and then writing the results in reverse order. When applying it to linear combinations of dyads, one needs to remember that the dagger operation is antilinear:

$$\left(\alpha |\omega\rangle\langle\tau| + \beta |\phi\rangle\langle\psi| \right)^{\dagger} = \alpha^* |\tau\rangle\langle\omega| + \beta^* |\psi\rangle\langle\phi|. \tag{3.28}$$

By generalizing (3.28) in an obvious way, one can apply the dagger operation to any sum of dyads, and thus to any operator on a finite-dimensional Hilbert space $\mathcal{H}$, since any operator can be written as a sum of dyads. However, the following definition is more useful. Given an operator A, its *adjoint* $(A)^{\dagger}$, usually written as $A^{\dagger}$, is the unique operator such that

$$\langle\psi|A^{\dagger}|\phi\rangle = \langle\phi|A|\psi\rangle^* \tag{3.29}$$

for any $|\phi\rangle$ and $|\psi\rangle$ in $\mathcal{H}$. Note that bra and ket are interchanged on the two sides of the equation. A useful mnemonic for expressions such as (3.29) is to think of complex conjugation as a special case of the dagger operation when that is applied to a scalar. Then the right side can be written and successively transformed,

$$\left(\langle\phi|A|\psi\rangle \right)^{\dagger} = \left(|\psi\rangle \right)^{\dagger} A^{\dagger} \left(\langle\phi| \right)^{\dagger} = \langle\psi|A^{\dagger}|\phi\rangle, \tag{3.30}$$

into the left side of (3.29) using the general rule that a dagger applied to a product is the product of the result of applying it to the individual factors, but written in the reverse order.

The adjoint of a linear combination of operators is what one would expect,

$$(\alpha A + \beta B)^{\dagger} = \alpha^* A^{\dagger} + \beta^* B^{\dagger}, \tag{3.31}$$

in light of (3.28) and the fact that the dagger operation is antilinear. The adjoint of a product of operators is the product of the adjoints *in the reverse order*:

$$(AB)^{\dagger} = B^{\dagger} A^{\dagger}, \quad (ABC)^{\dagger} = C^{\dagger} B^{\dagger} A^{\dagger}, \tag{3.32}$$

and so forth. The dagger operation, see (3.11), applied to a ket of the form $A|\psi\rangle$ yields a linear functional or bra vector

$$\left(A|\psi\rangle \right)^{\dagger} = \langle\psi|A^{\dagger}, \tag{3.33}$$

where the right side should be interpreted in the same way as (3.18): the operator $A^{\dagger}$ on $\mathcal{H}$ induces a map, denoted by the same symbol $A^{\dagger}$, on the space $\mathcal{H}^{\dagger}$ of linear functionals, by "operating to the left". One can check that (3.33) is consistent with (3.29).

An operator which is equal to its adjoint, $A = A^{\dagger}$ is said to be *Hermitian* or *self-adjoint*. (The two terms mean the same thing for operators on finite-dimensional spaces, but have different meanings for infinite-dimensional spaces.) Given that the dagger operation is in some sense a generalization of complex conjugation, one will not be surprised to learn that Hermitian operators behave in many respects like real numbers, a point to which we shall return in Ch. 5.

3.4 Projectors and subspaces

A particular type of Hermitian operator called a *projector* plays a central role in quantum theory. A projector is any operator P which satisfies the two conditions

$$P^2 = P, \quad P^\dagger = P. \tag{3.34}$$

The first of these, $P^2 = P$, defines a *projection operator* which need not be Hermitian. Hermitian projection operators are also called *orthogonal projection operators*, but we shall call them projectors. Associated with a projector P is a linear subspace $\mathcal{P}$ of $\mathcal{H}$ consisting of all kets which are left unchanged by P, that is, those $|\psi\rangle$ for which $P|\psi\rangle = |\psi\rangle$. We shall say that P *projects onto* $\mathcal{P}$, or is the *projector onto* $\mathcal{P}$. The projector P acts like the identity operator on the subspace $\mathcal{P}$. The identity operator I is a projector, and it projects onto the entire Hilbert space $\mathcal{H}$. The zero operator 0 is a projector which projects onto the subspace consisting of nothing but the zero vector.

Any nonzero ket $|\phi\rangle$ generates a one-dimensional subspace $\mathcal{P}$, often called a *ray* or (by quantum physicists) a *pure state*, consisting of all scalar multiples of $|\phi\rangle$, that is to say, the collection of kets of the form $\{\alpha|\phi\rangle\}$, where α is any complex number. The projector onto $\mathcal{P}$ is the dyad

$$P = [\phi] = |\phi\rangle\langle\phi|/\langle\phi|\phi\rangle, \tag{3.35}$$

where the right side is simply $|\phi\rangle\langle\phi|$ if $|\phi\rangle$ is normalized, which we shall assume to be the case in the following discussion. The symbol $[\phi]$ for the projector projecting onto the ray generated by $|\phi\rangle$ is not part of standard Dirac notation, but it is very convenient, and will be used throughout this book. Sometimes, when it will not cause confusion, the square brackets will be omitted: ϕ will be used in place of $[\phi]$. It is straightforward to show that the dyad (3.35) satisfies the conditions in (3.34) and that

$$P(\alpha|\phi\rangle) = |\phi\rangle\langle\phi|(\alpha|\phi\rangle) = \alpha|\phi\rangle\langle\phi|\phi\rangle = \alpha|\phi\rangle, \tag{3.36}$$

so that P leaves the elements of $\mathcal{P}$ unchanged. When it acts on any vector $|\chi\rangle$ orthogonal to $|\phi\rangle$, $\langle\phi|\chi\rangle = 0$, P produces the zero vector:

$$P|\chi\rangle = |\phi\rangle\langle\phi|\chi\rangle = 0|\phi\rangle = 0. \tag{3.37}$$

The properties of P in (3.36) and (3.37) can be given a geometrical interpretation, or at least one can construct a geometrical analogy using real numbers instead of complex numbers. Consider the two-dimensional plane shown in Fig. 3.1, with

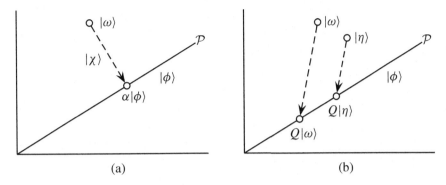

Fig. 3.1. Illustrating: (a) an orthogonal (perpendicular) projection onto $\mathcal{P}$; (b) a nonorthogonal projection represented by Q.

vectors labeled using Dirac kets. The line passing through $|\phi\rangle$ is the subspace $\mathcal{P}$. Let $|\omega\rangle$ be some vector in the plane, and suppose that its projection onto $\mathcal{P}$, along a direction perpendicular to $\mathcal{P}$, Fig. 3.1(a), falls at the point $\alpha|\phi\rangle$. Then

$$|\omega\rangle = \alpha|\phi\rangle + |\chi\rangle, \tag{3.38}$$

where $|\chi\rangle$ is a vector perpendicular (orthogonal) to $|\phi\rangle$, indicated by the dashed line. Applying P to both sides of (3.38), using (3.36) and (3.37), one finds that

$$P|\omega\rangle = \alpha|\phi\rangle. \tag{3.39}$$

That is, P on acting on any point $|\omega\rangle$ in the plane projects it onto $\mathcal{P}$ along a line perpendicular to $\mathcal{P}$, as indicated by the arrow in Fig. 3.1(a). Of course, such a projection applied to a point already on $\mathcal{P}$ leaves it unchanged, corresponding to the fact that P acts as the identity operation for elements of this subspace. For this reason, $P\big(P\big(|\omega\rangle\big)\big)$ is always the same as $P\big(|\omega\rangle\big)$, which is equivalent to the statement that $P^2 = P$. It is also possible to imagine projecting points onto $\mathcal{P}$ along a fixed direction which is *not* perpendicular to $\mathcal{P}$, as in Fig. 3.1(b). This defines a linear operator Q which is again a projection operator, since elements of $\mathcal{P}$ are mapped onto themselves, and thus $Q^2 = Q$. However, this operator is not Hermitian (in the terminology of real vector spaces, it is not symmetrical), so it is not an orthogonal ("perpendicular") projection operator.

Given a projector P, we define its *complement*, written as $\sim P$ or $\tilde{P}$, also called the *negation* of P (see Sec. 4.4), to be the projector defined by

$$\tilde{P} = I - P. \tag{3.40}$$

It is easy to show that $\tilde{P}$ satisfies the conditions for a projector in (3.34) and that

$$P\tilde{P} = 0 = \tilde{P}P. \tag{3.41}$$

From (3.40) it is obvious that P is, in turn, the complement (or negation) of $\tilde{P}$. Let $\mathcal{P}$ and $\mathcal{P}^\perp$ be the subspaces of $\mathcal{H}$ onto which P and $\tilde{P}$ project. Any element $|\omega\rangle$ of $\mathcal{P}^\perp$ is orthogonal to any element $|\phi\rangle$ of $\mathcal{P}$:

$$\langle \omega | \phi \rangle = \left(|\omega\rangle \right)^\dagger |\phi\rangle = \left(\tilde{P} |\omega\rangle \right)^\dagger \left(P |\phi\rangle \right) = \langle \omega | \tilde{P} P |\phi\rangle = 0, \tag{3.42}$$

because $\tilde{P} P = 0$. Here we have used the fact that P and $\tilde{P}$ act as identity operators on their respective subspaces, and the third equality makes use of (3.33). As a consequence, any element $|\psi\rangle$ of $\mathcal{H}$ can be written as the sum of two orthogonal kets, one belonging to $\mathcal{P}$ and one to $\mathcal{P}^\perp$:

$$|\psi\rangle = I |\psi\rangle = P |\psi\rangle + \tilde{P} |\psi\rangle. \tag{3.43}$$

Using (3.43), one can show that $\mathcal{P}^\perp$ is the *orthogonal complement* of $\mathcal{P}$, the collection of all elements of $\mathcal{H}$ which are orthogonal to *every* ket in $\mathcal{P}$. Similarly, $\mathcal{P}$ is the orthogonal complement $(\mathcal{P}^\perp)^\perp$ of $\mathcal{P}^\perp$.

3.5 Orthogonal projectors and orthonormal bases

Two projectors P and Q are said to be (mutually) *orthogonal* if

$$PQ = 0. \tag{3.44}$$

By taking the adjoint of this equation, one can show that $QP = 0$, so that the order of the operators in the product does not matter. An *orthogonal collection of projectors*, or a *collection of (mutually) orthogonal projectors* is a set of *nonzero* projectors $\{P_1, P_2, \dots\}$ with the property that

$$P_j P_k = 0 \text{ for } j \neq k. \tag{3.45}$$

The zero operator never plays a useful role in such collections, and excluding it simplifies various definitions.

Using (3.34) one can show that the sum $P + Q$ of two orthogonal projectors P and Q is a projector, and, more generally, the sum

$$R = \sum_j P_j \tag{3.46}$$

of the members of an orthogonal collection of projectors is a projector. When a projector R is written as a sum of projectors in an orthogonal collection, we shall say that this collection constitutes a *decomposition* or *refinement* of R. In particular, if R is the identity operator I, the collection is a *decomposition (refinement) of the identity*:

$$I = \sum_j P_j. \tag{3.47}$$

We shall often write down a sum in the form (3.47) and refer to it as a "decomposition of the identity." However, it is important to note that the decomposition is not the sum itself, but rather it is the set of summands, the collection of projectors which enter the sum. Whenever a *projector* R can be written as a sum of projectors in the form (3.46), it is necessarily the case that these projectors form an *orthogonal* collection, meaning that (3.45) is satisfied (see the Bibliography). Nonetheless it does no harm to consider (3.45) as part of the definition of a decomposition of the identity, or of some other projector.

If two nonzero kets $|\omega\rangle$ and $|\phi\rangle$ are orthogonal, the same is true of the corresponding projectors $[\omega]$ and $[\phi]$, as is obvious from the definition in (3.35). An *orthogonal collection* of kets is a set $\{|\phi_1\rangle, |\phi_2\rangle, \dots\}$ of *nonzero* elements of $\mathcal{H}$ such that $\langle\phi_j|\phi_k\rangle = 0$ when j is unequal to k. If in addition the kets in such a collection are normalized, so that

$$\langle\phi_j|\phi_k\rangle = \delta_{jk}, \tag{3.48}$$

we shall say that it is an *orthonormal collection*; the word "orthonormal" combines "orthogonal" and "normalized". The corresponding projectors $\{[\phi_1], [\phi_2], \dots\}$ form an orthogonal collection, and

$$[\phi_j]\,|\phi_k\rangle = |\phi_j\rangle\langle\phi_j|\phi_k\rangle = \delta_{jk}|\phi_j\rangle. \tag{3.49}$$

Let $\mathcal{R}$ be the subspace of $\mathcal{H}$ consisting of all linear combinations of kets belonging to an orthonormal collection $\{|\phi_j\rangle\}$, that is, all elements of the form

$$|\psi\rangle = \sum_j \sigma_j|\phi_j\rangle, \tag{3.50}$$

where the σ_j are complex numbers. Then the projector R onto $\mathcal{R}$ is the sum of the corresponding dyad projectors:

$$R = \sum_j |\phi_j\rangle\langle\phi_j| = \sum_j [\phi_j]. \tag{3.51}$$

This follows from the fact that, in light of (3.49), R acts as the identity operator on a sum of the form (3.50), whereas $R|\omega\rangle = 0$ for every $|\omega\rangle$ which is orthogonal to every $|\phi_j\rangle$ in the collection, and thus to every $|\psi\rangle$ of the form (3.50).

If every element of $\mathcal{H}$ can be written in the form (3.50), the orthonormal collection is said to form an *orthonormal basis* of $\mathcal{H}$, and the corresponding decomposition of the identity is

$$I = \sum_j |\phi_j\rangle\langle\phi_j| = \sum_j [\phi_j]. \tag{3.52}$$

A *basis* of $\mathcal{H}$ is a collection of linearly independent kets which span $\mathcal{H}$ in the sense that any element of $\mathcal{H}$ can be written as a linear combination of kets in the collection. Such a collection need not consist of normalized states, nor do they have to be mutually orthogonal. However, in this book we shall for the most part use orthonormal bases, and for this reason the adjective "orthonormal" will sometimes be omitted when doing so will not cause confusion.

3.6 Column vectors, row vectors, and matrices

Consider a Hilbert space $\mathcal{H}$ of dimension n, and a particular orthonormal basis. To make the notation a bit less cumbersome, let us label the basis kets as $|j\rangle$ rather than $|\phi_j\rangle$. Then (3.48) and (3.52) take the forms

$$\langle j|k\rangle = \delta_{jk}, \tag{3.53}$$

$$I = \sum_j |j\rangle\langle j|, \tag{3.54}$$

and any element $|\psi\rangle$ of $\mathcal{H}$ can be written as

$$|\psi\rangle = \sum_j \sigma_j |j\rangle. \tag{3.55}$$

By taking the inner product of both sides of (3.55) with $|k\rangle$, one sees that

$$\sigma_k = \langle k|\psi\rangle, \tag{3.56}$$

and therefore (3.55) can be written as

$$|\psi\rangle = \sum_j \langle j|\psi\rangle |j\rangle = \sum_j |j\rangle\langle j|\psi\rangle. \tag{3.57}$$

The form on the right side with the scalar coefficient $\langle j|\psi\rangle$ following rather than preceding the ket $|j\rangle$ provides a convenient way of deriving or remembering this result since (3.57) is the obvious equality $|\psi\rangle = I|\psi\rangle$ with I replaced with the dyad expansion in (3.54).

Using the basis $\{|j\rangle\}$, the ket $|\psi\rangle$ can be conveniently represented as a *column vector* of the coefficients in (3.57):

$$\begin{pmatrix} \langle 1|\psi\rangle \\ \langle 2|\psi\rangle \\ \cdots \\ \langle n|\psi\rangle \end{pmatrix}. \tag{3.58}$$

Because of (3.57), this column vector uniquely determines the ket $|\psi\rangle$, so as long as the basis is held fixed there is a one-to-one correspondence between kets and column vectors. (Of course, if the basis is changed, the same ket will be represented by a different column vector.) If one applies the dagger operation to both sides of (3.57), the result is

$$\langle \psi | = \sum_j \langle \psi | j \rangle \langle j |, \tag{3.59}$$

which could also be written down immediately using (3.54) and the fact that $\langle \psi | = \langle \psi | I$. The numerical coefficients on the right side of (3.59) form a *row vector*

$$\Big(\langle \psi | 1 \rangle, \langle \psi | 2 \rangle, \dots \langle \psi | n \rangle \Big) \tag{3.60}$$

which uniquely determines $\langle \psi |$, and vice versa. This row vector is obtained by "transposing" the column vector in (3.58) — that is, laying it on its side — and taking the complex conjugate of each element, which is the vector analog of $\langle \psi | = \big(|\psi\rangle \big)^\dagger$. An inner product can be written as a row vector times a column vector, in the sense of matrix multiplication:

$$\langle \phi | \psi \rangle = \sum_j \langle \phi | j \rangle \langle j | \psi \rangle. \tag{3.61}$$

This can be thought of as $\langle \phi | \psi \rangle = \langle \phi | I | \psi \rangle$ interpreted with the help of (3.54).

Given an operator A on $\mathcal{H}$, its *jk matrix element* is

$$A_{jk} = \langle j | A | k \rangle, \tag{3.62}$$

where the usual subscript notation is on the left, and the corresponding Dirac notation, see (3.17), is on the right. The matrix elements can be arranged to form a square matrix

$$\begin{pmatrix} \langle 1|A|1 \rangle & \langle 1|A|2 \rangle & \cdots & \langle 1|A|n \rangle \\ \langle 2|A|1 \rangle & \langle 2|A|2 \rangle & \cdots & \langle 2|A|n \rangle \\ \cdots & \cdots & \cdots & \cdots \\ \cdots & \cdots & \cdots & \cdots \\ \langle n|A|1 \rangle & \langle n|A|2 \rangle & \cdots & \langle n|A|n \rangle \end{pmatrix} \tag{3.63}$$

with the first or left index j of $\langle j|A|k \rangle$ labeling the rows, and the second or right index k labeling the columns. It is sometimes helpful to think of such a matrix as made up of a collection of n row vectors of the form (3.60), or, alternatively, n column vectors of the type (3.58). The matrix elements of the adjoint $A^\dagger$ of the operator A are given by

$$\langle j|A^\dagger|k \rangle = \langle k|A|j \rangle^*, \tag{3.64}$$

which is a particular case of (3.29). Thus the matrix of $A^\dagger$ is the complex conjugate of the transpose of the matrix of A. If the operator $A = A^\dagger$ is Hermitian, (3.64) implies that its diagonal matrix elements $\langle j|A|j \rangle$ are real.

Let us suppose that the result of A operating on $|\psi\rangle$ is a ket

$$|\phi\rangle = A|\psi\rangle. \tag{3.65}$$

By multiplying this on the left with the bra $\langle k|$, and writing A as AI with I in the form (3.54), one obtains

$$\langle k|\phi\rangle = \sum_j \langle k|A|j\rangle\langle j|\psi\rangle. \tag{3.66}$$

That is, the column vector for $|\phi\rangle$ is obtained by multiplying the matrix for A times the column vector for $|\psi\rangle$, following the usual rule for matrix multiplication. This shows, incidentally, that the operator A is uniquely determined by its matrix (given a fixed orthonormal basis), since this matrix determines how A maps any $|\psi\rangle$ of the Hilbert space onto $A|\psi\rangle$. Another way to see that the matrix determines A is to write A as a sum of dyads, starting with $A = IAI$ and using (3.54):

$$A = \sum_j \sum_k |j\rangle\langle j|A|k\rangle\langle k| = \sum_j \sum_k \langle j|A|k\rangle\, |j\rangle\langle k|. \tag{3.67}$$

The matrix of the product AB of two operators is the matrix product of the two matrices, in the same order:

$$\langle j|AB|k\rangle = \sum_i \langle j|A|i\rangle\langle i|B|k\rangle, \tag{3.68}$$

an expression easily derived by writing $AB = AIB$ and invoking the invaluable (3.54).

3.7 Diagonalization of Hermitian operators

Books on linear algebra show that if $A = A^\dagger$ is Hermitian, it is always possible to find a particular orthonormal basis $\{|\alpha_j\rangle\}$ such that in this basis the matrix of A is *diagonal*, that is, $\langle \alpha_j|A|\alpha_k\rangle = 0$ whenever $j \neq k$. The diagonal elements $a_j = \langle \alpha_j|A|\alpha_j\rangle$ must be real numbers in view of (3.64). By using (3.67) one can write A in the form

$$A = \sum_j a_j|\alpha_j\rangle\langle\alpha_j| = \sum_j a_j[\alpha_j], \tag{3.69}$$

a sum of real numbers times projectors drawn from an orthogonal collection. The ket $|\alpha_j\rangle$ is an *eigenvector* or *eigenket* of the operator A with *eigenvalue* a_j:

$$A|\alpha_j\rangle = a_j|\alpha_j\rangle. \tag{3.70}$$

An eigenvalue is said to be *degenerate* if it occurs more than once in (3.69), and its *multiplicity* is the number of times it occurs in the sum. An eigenvalue which only occurs once (multiplicity of 1) is called *nondegenerate*. The identity operator has only one eigenvalue, equal to 1, whose multiplicity is the dimension n of the Hilbert space. A projector has only two distinct eigenvalues: 1 with multiplicity equal to the dimension m of the subspace onto which it projects, and 0 with multiplicity $n - m$.

The basis which diagonalizes A is unique only if all its eigenvalues are non-degenerate. Otherwise this basis is not unique, and it is sometimes more convenient to rewrite (3.69) in an alternative form in which each eigenvalue appears just once. The first step is to suppose that, as is always possible, the kets $|\alpha_j\rangle$ have been indexed in such a fashion that the eigenvalues are a nondecreasing sequence:

$$a_1 \leq a_2 \leq a_3 \leq \cdots . \tag{3.71}$$

The next step is best explained by means of an example. Suppose that $n = 5$, and that $a_1 = a_2 < a_3 < a_4 = a_5$. That is, the multiplicity of a_1 is 2, that of a_3 is 1, and that of a_4 is 2. Then (3.69) can be written in the form

$$A = a_1 P_1 + a_3 P_2 + a_4 P_3, \tag{3.72}$$

where the three projectors

$$P_1 = |\alpha_1\rangle\langle\alpha_1| + |\alpha_2\rangle\langle\alpha_2|, \quad P_2 = |\alpha_3\rangle\langle\alpha_3|, \\ P_3 = |\alpha_4\rangle\langle\alpha_4| + |\alpha_5\rangle\langle\alpha_5| \tag{3.73}$$

form a decomposition of the identity. By relabeling the eigenvalues as

$$a_1' = a_1, \quad a_2' = a_3, \quad a_3' = a_4, \tag{3.74}$$

it is possible to rewrite (3.72) in the form

$$A = \sum_j a_j' P_j, \tag{3.75}$$

where no two eigenvalues are the same:

$$a_j' \neq a_k' \text{ for } j \neq k. \tag{3.76}$$

Generalizing from this example, we see that it is always possible to write a Hermitian operator in the form (3.75) with eigenvalues satisfying (3.76). If all the eigenvalues of A are nondegenerate, each P_j projects onto a ray or pure state, and (3.75) is just another way to write (3.69).

One advantage of using the expression (3.75), in which the eigenvalues are un-equal, in preference to (3.69), where some of them can be the same, is that the decomposition of the identity $\{P_j\}$ which enters (3.75) is uniquely determined by the operator A. On the other hand, if an eigenvalue of A is degenerate, the corresponding eigenvectors are not unique. In the example in (3.72) one could replace

$|\alpha_1\rangle$ and $|\alpha_2\rangle$ by any two normalized and mutually orthogonal kets $|\alpha_1'\rangle$ and $|\alpha_2'\rangle$ belonging to the two-dimensional subspace onto which P_1 projects, and similar considerations apply to $|\alpha_4\rangle$ and $|\alpha_5\rangle$. We shall call the (unique) decomposition of the identity $\{P_j\}$ which allows a Hermitian operator A to be written in the form (3.75) with eigenvalues satisfying (3.76) the *decomposition corresponding to* or *generated by* the operator A.

If $\{A, B, C, \ldots\}$ is a collection of Hermitian operators which commute with each other, (3.23), they can be simultaneously diagonalized in the sense that there is a single orthonormal basis $|\phi_j\rangle$ such that

$$A = \sum_j a_j[\phi_j], \quad B = \sum_j b_j[\phi_j], \quad C = \sum_j c_j[\phi_j], \tag{3.77}$$

and so forth. If instead one writes the operators in terms of the decompositions which they generate, as in (3.75),

$$A = \sum_j a_j' P_j, \quad B = \sum_k b_k' Q_k, \quad C = \sum_l c_l' R_l, \tag{3.78}$$

and so forth, the projectors in each decomposition commute with the projectors in the other decompositions: $P_j Q_k = Q_k P_j$, etc.

3.8 Trace

The trace of an operator A is the sum of its diagonal matrix elements:

$$\mathrm{Tr}(A) = \sum_j \langle j|A|j\rangle. \tag{3.79}$$

While the individual diagonal matrix elements depend upon the orthonormal basis, their sum, and thus the trace, is independent of basis and depends only on the operator A. The trace is a linear operation in that if A and B are operators, and α and β are scalars,

$$\mathrm{Tr}(\alpha A + \beta B) = \alpha\, \mathrm{Tr}(A) + \beta\, \mathrm{Tr}(B). \tag{3.80}$$

The trace of a dyad is the corresponding inner product,

$$\mathrm{Tr}\big(|\phi\rangle\langle\tau|\big) = \sum_j \langle j|\phi\rangle\langle\tau|j\rangle = \langle\tau|\phi\rangle, \tag{3.81}$$

as is clear from (3.61).

The trace of the product of two operators A and B is independent of the order of the product,

$$\text{Tr}(AB) = \text{Tr}(BA), \tag{3.82}$$

and the trace of the product of three or more operators is not changed if one makes a *cyclic permutation* of the factors:

$$\begin{aligned} \text{Tr}(ABC) &= \text{Tr}(BCA) = \text{Tr}(CAB), \\ \text{Tr}(ABCD) &= \text{Tr}(BCDA) = \text{Tr}(CDAB) = \text{Tr}(DABC), \end{aligned} \tag{3.83}$$

and so forth; the cycling is done by moving the operator from the first position in the product to the last, or vice versa. By contrast, $\text{Tr}(ACB)$ is, in general, *not* the same as $\text{Tr}(ABC)$, for ACB is obtained from ABC by interchanging the second and third factor, and this is not a cyclic permutation.

The complex conjugate of the trace of an operator is equal to the trace of its adjoint, as is evident from (3.64), and a similar rule applies to products of operators, where one should remember to reverse the order, see (3.32):

$$\begin{aligned} \left(\text{Tr}(A)\right)^* &= \text{Tr}(A^\dagger), \\ \left(\text{Tr}(ABC)\right)^* &= \text{Tr}(C^\dagger B^\dagger A^\dagger), \end{aligned} \tag{3.84}$$

etc.; additional identities can be obtained using cyclic permutations, as in (3.83).

If $A = A^\dagger$ is Hermitian, one can calculate the trace in the basis in which A is diagonal, with the result

$$\text{Tr}(A) = \sum_{j=1}^{n} a_j. \tag{3.85}$$

That is, the trace is equal to the sum of the eigenvalues appearing in (3.69). In particular, the trace of a projector P is the dimension of the subspace onto which it projects.

3.9 Positive operators and density matrices

A Hermitian operator A is said to be a *positive* operator provided

$$\langle \psi | A | \psi \rangle \geq 0 \tag{3.86}$$

holds for every $|\psi\rangle$ in the Hilbert space or, equivalently, if all its eigenvalues are nonnegative:

$$a_j \geq 0 \text{ for all } j. \tag{3.87}$$

While (3.87) is easily shown to imply (3.86), and vice versa, memorizing both
definitions is worthwhile, as sometimes one is more useful, and sometimes the
other.

If A is a positive operator and a a positive real number, then aA is a positive
operator. Also the sum of any collection of positive operators is a positive operator;
this is an obvious consequence of (3.86). The *support* of a positive operator A is
defined to be the projector A_s, or the subspace onto which it projects, given by the
sum of those $[\alpha_j]$ in (3.69) with $a_j > 0$, or of the P_j in (3.75) with $a'_j > 0$. It
follows from the definition that

$$A_s A = A. \tag{3.88}$$

An alternative definition is that the support of A is the smallest projector A_s, in the
sense of minimizing $\text{Tr}(A_s)$, which satisfies (3.88).

The trace of a positive operator is obviously a nonnegative number, see (3.85)
and (3.87), and is strictly positive unless the operator is the zero operator with all
zero eigenvalues. A positive operator A which is not the zero operator can always
be *normalized* by defining a new operator

$$\bar{A} = A/\text{Tr}(A) \tag{3.89}$$

whose trace is equal to 1. In quantum physics a positive operator with trace equal
to 1 is called a *density matrix*. The terminology is unfortunate, because a density
matrix is an operator, not a matrix, and the matrix for this operator depends on
the choice of orthonormal basis. However, by now the term is firmly embedded
in the literature, and attempts to replace it with something more rational, such as
"statistical operator", have not been successful.

If C is any operator, then $C^\dagger C$ is a positive operator, since for any $|\psi\rangle$,

$$\langle\psi|C^\dagger C|\psi\rangle = \langle\phi|\phi\rangle \geq 0, \tag{3.90}$$

where $|\phi\rangle = C|\psi\rangle$. Consequently, $\text{Tr}(C^\dagger C)$ is nonnegative. If $\text{Tr}(C^\dagger C) = 0$, then
$C^\dagger C = 0$, and $\langle\psi|C^\dagger C|\psi\rangle$ vanishes for every $|\psi\rangle$, which means that $C|\psi\rangle$ is zero
for every $|\psi\rangle$, and therefore $C = 0$. Thus for any operator C it is the case that

$$\text{Tr}(C^\dagger C) \geq 0, \tag{3.91}$$

with equality if and only if $C = 0$.

The product AB of two positive operators A and B is, in general, not Hermitian,
and therefore not a positive operator. However, if A and B commute, AB is positive, as can be seen from the fact that there is an orthonormal basis in which the
matrices of both A and B, and therefore also AB, are diagonal. This result gener-
alizes to the product of any collection of commuting positive operators. Whether

or not A and B commute, the fact that they are both positive means that $\text{Tr}(AB)$ is a real, nonnegative number,

$$\text{Tr}(AB) = \text{Tr}(BA) \geq 0, \tag{3.92}$$

equal to 0 if and only if $AB = BA = 0$. This result does not generalize to a product of three or more operators: if A, B, and C are positive operators that do not commute with each other, there is in general nothing one can say about $\text{Tr}(ABC)$.

To derive (3.92) it is convenient to first define the square root $A^{1/2}$ of a positive operator A by means of the formula

$$A^{1/2} = \sum_j \sqrt{a_j}\,[\alpha_j], \tag{3.93}$$

where $\sqrt{a_j}$ is the positive square root of the eigenvalue a_j in (3.69). Then when A and B are both positive, one can write

$$\begin{aligned} \text{Tr}(AB) &= \text{Tr}(A^{1/2}A^{1/2}B^{1/2}B^{1/2}) \\ &= \text{Tr}(A^{1/2}B^{1/2}B^{1/2}A^{1/2}) = \text{Tr}(C^\dagger C) \geq 0, \end{aligned} \tag{3.94}$$

where $C = B^{1/2}A^{1/2}$. If $\text{Tr}(AB) = 0$, then, see (3.91), $C = 0 = C^\dagger$, and both $BA = B^{1/2}CA^{1/2}$ and $AB = A^{1/2}C^\dagger B^{1/2}$ vanish.

3.10 Functions of operators

Suppose that $f(x)$ is an ordinary numerical function, such as x^2 or e^x. It is sometimes convenient to define a corresponding function $f(A)$ of an operator A, so that the value of the function is also an operator. When $f(x)$ is a polynomial

$$f(x) = a_0 + a_1 x + a_2 x^2 + \cdots a_p x^p, \tag{3.95}$$

one can write

$$f(A) = a_0 I + a_1 A + a_2 A^2 + \cdots a_p A^p, \tag{3.96}$$

since the square, cube, etc. of any operator is itself an operator. When $f(x)$ is represented by a power series, as in $\log(1 + x) = x - x^2/2 + \cdots$, the same procedure will work provided the series in powers of the operator A converges, but this is often not a trivial issue.

An alternative approach is possible in the case of operators which can be diagonalized in some orthonormal basis. Thus if A is written in the form (3.69), one can define $f(A)$ to be the operator

$$f(A) = \sum_j f(a_j)\,[\alpha_j], \tag{3.97}$$

where $f(a_j)$ on the right side is the value of the numerical function. This agrees with the previous definition in the case of polynomials, but allows the use of much more general functions. As an example, the square root $A^{1/2}$ of a positive operator A as defined in (3.93) is obtained by setting $f(x) = \sqrt{x}$ for $x \geq 0$ in (3.97). Note that in order to use (3.97), the numerical function $f(x)$ must be defined for any x which is an eigenvalue of A. For a Hermitian operator these eigenvalues are real, but in other cases, such as the unitary operators discussed in Sec. 7.2, the eigenvalues may be complex, so for such operators $f(x)$ will need to be defined for suitable complex values of x.

4

Physical properties

4.1 Classical and quantum properties

We shall use the term *physical property* to refer to something which can be said to be either *true* or *false* for a particular physical system. Thus "the energy is between 10 and 12 μJ" or "the particle is between x_1 and x_2" are examples of physical properties. One must distinguish between a *physical property* and a *physical variable*, such as the position or energy or momentum of a particle. A physical variable can take on different numerical values, depending upon the state of the system, whereas a physical property is either a true or a false description of a particular physical system at a particular time. A physical variable taking on a particular value, or lying in some range of values, is an example of a physical property.

In the case of a classical mechanical system, a physical property is always associated with some subset of points in its phase space. Consider, for example, a harmonic oscillator whose phase space is the x, p plane shown in Fig. 2.1 on page 12. The property that its energy is equal to some value $E_0 > 0$ is associated with a set of points on an ellipse centered at the origin. The property that the energy is less than E_0 is associated with the set of points inside this ellipse. The property that the position x lies between x_1 and x_2 corresponds to the set of points inside a vertical band shown cross-hatched in this figure, and so forth.

Given a property P associated with a set of points $\mathcal{P}$ in the phase space, it is convenient to introduce an *indicator function*, or *indicator* for short, $P(\gamma)$, where γ is a point in the phase space, defined in the following way:

$$P(\gamma) = \begin{cases} 1 & \text{if } \gamma \in \mathcal{P}, \\ 0 & \text{otherwise.} \end{cases} \tag{4.1}$$

(It is convenient to use the same symbol P for a property and for its indicator, as this will cause no confusion.) Thus if at some instant of time the phase point γ_0 associated with a particular physical system is inside the set $\mathcal{P}$, then $P(\gamma_0) = 1$,

47

meaning that the system possesses this property, or the property is true. Similarly, if $P(\gamma_0) = 0$, the system does not possess this property, so for this system the property is false.

A *physical property of a quantum system* is associated with a *subspace* $\mathcal{P}$ of the quantum Hilbert space $\mathcal{H}$ in much the same way as a physical property of a classical system is associated with a subset of points in its phase space, and the *projector* P onto $\mathcal{P}$, Sec. 3.4, plays a role analogous to the classical indicator function. If the quantum system is described by a ket $|\psi\rangle$ which lies in the subspace $\mathcal{P}$, so that $|\psi\rangle$ is an eigenstate of P with eigenvalue 1,

$$P|\psi\rangle = |\psi\rangle, \tag{4.2}$$

one can say that the quantum system has the property P. On the other hand, if $|\psi\rangle$ is an eigenstate of P with eigenvalue 0,

$$P|\psi\rangle = 0, \tag{4.3}$$

then the quantum system does not have the property P, or, equivalently, it has the property $\tilde{P}$ which is the negation of P, see Sec. 4.4. When $|\psi\rangle$ is *not* an eigenstate of P, a situation with no analog in classical mechanics, we shall say that the property P is *undefined* for the quantum system.

4.2 Toy model and spin half

In this section we will consider various physical properties associated with a toy model and with a spin-half particle, and in Sec. 4.3 properties of a continuous quantum system, such as a particle with a wave function $\psi(x)$. In Sec. 2.5 we introduced a toy model with wave function $\psi(m)$, where the position variable m is an integer restricted to taking on one of the $M = M_a + M_b + 1$ values in the range

$$-M_a \leq m \leq M_b. \tag{4.4}$$

In (2.26) we defined a wave function $\chi_n(m) = \delta_{mn}$ whose physical significance is that the particle is at the site (or in the cell) n. Let $|n\rangle$ be the corresponding Dirac ket. Then (2.24) tells us that

$$\langle k|n\rangle = \delta_{kn}, \tag{4.5}$$

so the kets $\{|n\rangle\}$ form an orthonormal basis of the Hilbert space.

Any scalar multiple $\alpha|n\rangle$ of $|n\rangle$, where α is a nonzero complex number, has precisely the same physical significance as $|n\rangle$. The set of all kets of this form together with the zero ket, that is, the set of all multiples of $|n\rangle$, form a one-dimensional subspace of $\mathcal{H}$, and the projector onto this subspace, see Sec. 3.4, is

$$[n] = |n\rangle\langle n|. \tag{4.6}$$

Thus it is natural to associate the property that the particle is at the site n (something which can be true or false) with this subspace, or, equivalently, with the corresponding projector, since there is a one-to-one correspondence between subspaces and projectors.

Since the projectors [0] and [1] for sites 0 and 1 are orthogonal to each other, their sum is also a projector

$$R = [0] + [1]. \tag{4.7}$$

The subspace $\mathcal{R}$ onto which R projects is two-dimensional and consists of all linear combinations of $|0\rangle$ and $|1\rangle$, that is, all states of the form

$$|\phi\rangle = \alpha|0\rangle + \beta|1\rangle. \tag{4.8}$$

Equivalently, it corresponds to all wave functions $\psi(m)$ which vanish when m is unequal to 0 or 1. The physical significance of $\mathcal{R}$, see the discussion in Sec. 2.3, is that the toy particle is *not outside* the interval [0, 1], where, since we are using a discrete model, the interval [0, 1] consists of the two sites $m = 0$ and $m = 1$. One can interpret "not outside" as meaning "inside", provided that is not understood to mean "at one or the other of the two sites $m = 0$ or $m = 1$."

The reason one needs to be cautious is that a typical state in $\mathcal{R}$ will be of the form (4.8) with both α and β unequal to zero. Such a state does not have the property that it is at $m = 0$, for all states with this property are scalar multiples of $|0\rangle$, and $|\phi\rangle$ is not of this form. Indeed, $|\phi\rangle$ is not an eigenstate of the projector [0] representing the property $m = 0$, and hence according to the definition given at the end of Sec. 4.1, the property $m = 0$ is undefined. The same comments apply to the property $m = 1$. Thus it is certainly incorrect to say that the particle is either at 0 or at 1. Instead, the particle is represented by a delocalized wave, as discussed in Sec. 2.3. There are some states in $\mathcal{R}$ which are localized at 0 or localized at 1, but since $\mathcal{R}$ also contains other, delocalized, states, the property corresponding to $\mathcal{R}$ or its projector R, which holds for *all* states in this subspace, needs to be expressed by some English phrase other than "at 0 or 1". The phrases "not outside the interval [0, 1]" or "no place other than 0 or 1," while they are a bit awkward, come closer to saying what one wants to say. The way to be perfectly precise is to use the projector R itself, since it is a precisely defined mathematical quantity. But of course one needs to build up an intuitive picture of what it is that R means.

The process of building up one's intuition about the meaning of R will be aided by noting that (4.7) is not the only way of writing it as a sum of two orthogonal projectors. Another possibility is

$$R = [\sigma] + [\tau], \tag{4.9}$$

where

$$|\sigma\rangle = (|0\rangle + i|1\rangle)/\sqrt{2}, \quad |\tau\rangle = (|0\rangle - i|1\rangle)/\sqrt{2} \tag{4.10}$$

are two normalized states in $\mathcal{R}$ which are mutually orthogonal. To check that (4.9) is correct, one can work out the dyad

$$
\begin{aligned}
|\sigma\rangle\langle\sigma| &= \tfrac{1}{2}\big(|0\rangle + i|1\rangle\big)\big(\langle 0| - i\langle 1|\big) \\
&= \tfrac{1}{2}\big(|0\rangle\langle 0| + |1\rangle\langle 1| + i|1\rangle\langle 0| - i|0\rangle\langle 1|\big),
\end{aligned} \tag{4.11}
$$

where $\langle\sigma| = (|\sigma\rangle)^\dagger$ has been formed using the rules for the dagger operation (note the complex conjugate) in (3.12), and (4.11) shows how one can conveniently multiply things out to find the resulting projector. The dyad $|\tau\rangle\langle\tau|$ is the same except for the signs of the imaginary terms, so adding this to $|\sigma\rangle\langle\sigma|$ gives R. There are many other ways besides (4.9) to write R as a sum of two orthogonal projectors. In fact, given *any* normalized state in $\mathcal{R}$, one can always find another normalized state orthogonal to it, and the sum of the dyads corresponding to these two states is equal to R. The fact that R can be written as a sum $P + Q$ of two orthogonal projectors P and Q in many different ways is one reason to be cautious in assigning R a physical interpretation of "property P or property Q", although there are occasions, as we shall see later, when such an interpretation is appropriate.

The simplest nontrivial toy model has only $M = 2$ sites, and it is convenient to discuss it using language appropriate to the spin degree of freedom of a particle of spin $1/2$. Its Hilbert space $\mathcal{H}$ consists of all linear combinations of two mutually orthogonal and normalized states which will be denoted by $|z^+\rangle$ and $|z^-\rangle$, and which one can think of as the counterparts of $|0\rangle$ and $|1\rangle$ in the toy model. (In the literature they are often denoted by $|+\rangle$ and $|-\rangle$, or by $|\!\uparrow\rangle$ and $|\!\downarrow\rangle$.) The normalization and orthogonality conditions are

$$\langle z^+|z^+\rangle = 1 = \langle z^-|z^-\rangle, \quad \langle z^+|z^-\rangle = 0. \tag{4.12}$$

The physical significance of $|z^+\rangle$ is that the z-component S_z of the internal or "spin" angular momentum has a value of $+1/2$ in units of $\hbar$, while $|z^-\rangle$ means that $S_z = -1/2$ in the same units. One sometimes refers to $|z^+\rangle$ and $|z^-\rangle$ as "spin up" and "spin down" states.

The two-dimensional Hilbert space $\mathcal{H}$ consists of all linear combinations of the form

$$\alpha|z^+\rangle + \beta|z^-\rangle, \tag{4.13}$$

where α and β are any complex numbers. It is convenient to parameterize these states in the following way. Let w denote a direction in space corresponding to ϑ, φ in spherical polar coordinates. For example, $\vartheta = 0$ is the $+z$ direction, while

$\vartheta = \pi/2$ and $\varphi = \pi$ is the $-x$ direction. Then define the two states

$$|w^+\rangle = +\cos(\vartheta/2)e^{-i\varphi/2}|z^+\rangle + \sin(\vartheta/2)e^{i\varphi/2}|z^-\rangle,$$
$$|w^-\rangle = -\sin(\vartheta/2)e^{-i\varphi/2}|z^+\rangle + \cos(\vartheta/2)e^{i\varphi/2}|z^-\rangle. \tag{4.14}$$

These are normalized and mutually orthogonal,

$$\langle w^+|w^+\rangle = 1 = \langle w^-|w^-\rangle, \quad \langle w^+|w^-\rangle = 0, \tag{4.15}$$

as a consequence of (4.12).

The physical significance of $|w^+\rangle$ is that S_w, the component of spin angular momentum in the w direction, has a value of $1/2$, whereas for $|w^-\rangle$, S_w has the value $-1/2$. For $\vartheta = 0$, $|w^+\rangle$ and $|w^-\rangle$ are the same as $|z^+\rangle$ and $|z^-\rangle$, respectively, apart from a phase factor, $e^{-i\varphi}$, which does not change their physical significance. For $\vartheta = \pi$, $|w^+\rangle$ and $|w^-\rangle$ are the same as $|z^-\rangle$ and $|z^+\rangle$, respectively, apart from phase factors. Suppose that w is a direction which is neither along nor opposite to the z axis, for example, $w = x$. Then both α and β in (4.13) are nonzero, and $|w^+\rangle$ does not have the property $S_z = +1/2$, nor does it have the property $S_z = -1/2$. The same is true if S_z is replaced by S_v, where v is any direction which is not the same as w or opposite to w. The situation is analogous to that discussed earlier for the toy model: think of $|z^+\rangle$ and $|z^-\rangle$ as corresponding to the states $|m = 0\rangle$ and $|m = 1\rangle$.

Any nonzero wave function (4.13) can be written as a complex number times $|w^+\rangle$ for a suitable choice of the direction w. For $\beta = 0$, the choice $w = z$ is obvious, while for $\alpha = 0$ it is $w = -z$. For other cases, write (4.13) in the form

$$\beta\left[(\alpha/\beta)|z^+\rangle + |z^-\rangle\right]. \tag{4.16}$$

A comparison with the expression for $|w^+\rangle$ in (4.14) shows that ϑ and φ are determined by the equation

$$e^{-i\varphi}\cot(\vartheta/2) = \alpha/\beta, \tag{4.17}$$

which, since α/β is finite (neither 0 nor ∞), always has a unique solution in the range

$$0 \le \varphi < 2\pi, \quad 0 < \vartheta < \pi. \tag{4.18}$$

4.3 Continuous quantum systems

This section deals with the quantum properties of a particle in one dimension described by a wave function $\psi(x)$ depending on the continuous variable x. Similar considerations apply to a particle in three dimensions with a wave function $\psi(\mathbf{r})$, and the same general approach can be extended to apply to collections of several particles. Quantum properties are again associated with subspaces of the Hilbert space $\mathcal{H}$, and since $\mathcal{H}$ is infinite-dimensional, these subspaces can be either finite- or infinite-dimensional; we shall consider examples of both. (For infinite-dimensional subspaces one adds the technical requirement that they be closed, as defined in books on functional analysis.)

As a first example, consider the property that a particle lies inside (which is to say not outside) the interval

$$x_1 \leq x \leq x_2, \tag{4.19}$$

with $x_1 < x_2$. As pointed out in Sec. 2.3, the (infinite-dimensional) subspace $\mathcal{X}$ which corresponds to this property consists of all wave functions which vanish for x outside the interval (4.19). The projector X associated with $\mathcal{X}$ is defined as follows. Acting on some wave function $\psi(x)$, X produces a new function $\psi_X(x)$ which is identical to $\psi(x)$ for x inside the interval (4.19), and zero outside this interval:

$$\psi_X(x) = X\psi(x) = \begin{cases} \psi(x) & \text{for } x_1 \leq x \leq x_2, \\ 0 & \text{for } x < x_1 \text{ or } x > x_2. \end{cases} \tag{4.20}$$

Note that X leaves unchanged any function which belongs to the subspace $\mathcal{X}$, so it acts as the identity operator on this subspace. If a wave function $\omega(x)$ vanishes throughout the interval (4.19), it will be orthogonal to all the functions in $\mathcal{X}$, and X applied to $\omega(x)$ will yield a function which is everywhere equal to 0. Thus X has the properties one would expect for a projector as discussed in Sec. 3.4.

One can write (4.20) using Dirac notation as

$$|\psi_X\rangle = X|\psi\rangle, \tag{4.21}$$

where the element $|\psi\rangle$ of the Hilbert space can be represented either as a position wave function $\psi(x)$ or as a momentum wave function $\hat{\psi}(p)$, the Fourier transform of $\psi(x)$, see (2.15). The relationship (4.21) can also be expressed in terms of momentum wave functions as

$$\hat{\psi}_X(p) = \int \hat{\xi}(p - p')\hat{\psi}(p')\,dp', \tag{4.22}$$

where $\hat{\psi}_X(p)$ is the Fourier transform of $\psi_X(x)$, and $\hat{\xi}(p)$ is the Fourier transform of

$$\xi(x) = \begin{cases} 1 & \text{for } x_1 \leq x \leq x_2, \\ 0 & \text{for } x < x_1 \text{ or } x > x_2. \end{cases} \tag{4.23}$$

Although (4.20) is the most straightforward way to define $X|\psi\rangle$, it is important to note that the expression (4.22) is completely equivalent.

As another example, consider the property that the momentum of a particle lies in (that is, not outside) the interval

$$p_1 \le p \le p_2. \tag{4.24}$$

This property corresponds, Sec. 2.4, to the subspace $\mathcal{P}$ of momentum wave functions $\hat{\psi}(p)$ which vanish outside this interval. The projector P corresponding to $\mathcal{P}$ can be defined by

$$\hat{\psi}_P(p) = P\hat{\psi}(p) = \begin{cases} \hat{\psi}(p) & \text{for } p_1 \le p \le p_2, \\ 0 & \text{for } p < p_1 \text{ or } p > p_2, \end{cases} \tag{4.25}$$

and in Dirac notation (4.25) takes the form

$$|\psi_P\rangle = P|\psi\rangle. \tag{4.26}$$

One could also express the position wave function $\psi_P(x)$ in terms of $\psi(x)$ using a convolution integral analogous to (4.22).

As a third example, consider a one-dimensional harmonic oscillator. In textbooks it is shown that the energy E of an oscillator with angular frequency ω takes on a discrete set of values

$$E = n + \tfrac{1}{2}, \ n = 0, 1, 2, \dots \tag{4.27}$$

in units of $\hbar\omega$. Let $\phi_n(x)$ be the normalized wave function for energy $E = n + 1/2$, and $|\phi_n\rangle$ the corresponding ket. The one-dimensional subspace of $\mathcal{H}$ consisting of all scalar multiples of $|\phi_n\rangle$ represents the property that the energy is $n + 1/2$. The corresponding projector is the dyad $[\phi_n]$. When this projector acts on some $|\psi\rangle$ in $\mathcal{H}$, the result is

$$|\bar{\psi}\rangle = [\phi_n]|\psi\rangle = |\phi_n\rangle\langle\phi_n|\psi\rangle = \langle\phi_n|\psi\rangle \, |\phi_n\rangle, \tag{4.28}$$

that is, $|\phi_n\rangle$ multiplied by the scalar $\langle\phi_n|\psi\rangle$. One can write (4.28) using wave functions in the form

$$\bar{\psi}(x) = \int P(x, x')\psi(x') \, dx', \tag{4.29}$$

where

$$P(x, x') = \phi_n(x)\phi_n^*(x') \tag{4.30}$$

corresponds to the dyad $|\phi_n\rangle\langle\phi_n|$.

Since the states $|\phi_n\rangle$ for different n are mutually orthogonal, the same is true of the corresponding projectors $[\phi_n]$. Using this fact makes it easy to write down projectors for the energy to lie in some interval which includes two or more energy eigenvalues. For example, the projector

$$Q = [\phi_1] + [\phi_2] \tag{4.31}$$

onto the two-dimensional subspace of $\mathcal{H}$ consisting of linear combinations of $|\phi_1\rangle$ and $|\phi_2\rangle$ expresses the property that the energy E (in units of $\hbar\omega$) lies inside some interval such as

$$1 < E < 3, \tag{4.32}$$

where the choice of endpoints of the interval is somewhat arbitrary, given that the energy is quantized and takes on only discrete values; any other interval which includes 1.5 and 2.5, but excludes 0.5 and 3.5, would be just as good. The action of Q on a wave function $\psi(x)$ can be written as

$$\bar{\psi}(x) = Q\psi(x) = \int Q(x, x')\psi(x')\,dx', \tag{4.33}$$

with

$$Q(x, x') = \phi_1(x)\phi_1^*(x') + \phi_2(x)\phi_2^*(x'). \tag{4.34}$$

Once again, it is important not to interpret "energy lying inside the interval (4.32)" as meaning that it either has the value 1.5 or that it has the value 2.5. The subspace onto which Q projects also contains states such as $|\phi_1\rangle + |\phi_2\rangle$, for which the energy cannot be defined more precisely than by saying that it does not lie outside the interval, and thus the physical property expressed by Q cannot have a meaning which is more precise than this.

4.4 Negation of properties (NOT)

A physical property can be true or false in the sense that the statement that a particular physical system at a particular time possesses a physical property can be either true or false. Books on logic present simple *logical operations* by which statements which are true or false can be transformed into other statements which are true or false. We shall consider three operations which can be applied to physical properties: negation, taken up in this section, and conjunction and disjunction, taken up in Sec. 4.5. In addition, quantum properties are sometimes incompatible or "noncomparable", a topic discussed in Sec. 4.6.

As noted in Sec. 4.1, a classical property P is associated with a subset $\mathcal{P}$ consisting of those points in the classical phase space for which the property is true. The points of the phase space which do not belong to $\mathcal{P}$ form the *complementary set* $\sim\mathcal{P}$, and this complementary set defines the negation "NOT P" of the property

P. We shall write it as $\sim P$ or as $\tilde{P}$. Alternatively, one can define $\tilde{P}$ as the property which is true if and only if P is false, and false if and only if P is true. From this as well as from the other definition it is obvious that the negation of the negation of a property is the same as the original property: $\sim \tilde{P}$ or $\sim (\sim P)$ is the same property as P. The indicator $\tilde{P}(\gamma)$ of the property $\sim P$, see (4.1), is given by the formula

$$\tilde{P} = I - P, \tag{4.35}$$

or $\tilde{P}(\gamma) = I(\gamma) - P(\gamma)$, where $I(\gamma)$, the indicator of the identity property, is equal to 1 for all values of γ. Thus $\tilde{P}$ is equal to 1 (true) if P is 0 (false), and $\tilde{P} = 0$ when $P = 1$.

Once again consider Fig. 2.1 on page 12, the phase space of a one-dimensional harmonic oscillator, where the ellipse corresponds to an energy E_0. The property P that the energy is less than or equal to E_0 corresponds to the set $\mathcal{P}$ of points inside and on the ellipse. Its negation $\tilde{P}$ is the property that the energy is greater than E_0, and the corresponding region $\sim\mathcal{P}$ is all the points outside the ellipse. The vertical band $\mathcal{Q}$ corresponds to the property Q that the position of the particle is in the interval $x_1 \le x \le x_2$. The negation of Q is the property $\tilde{Q}$ that the particle lies outside this interval, and the corresponding set of points $\sim \mathcal{Q}$ in the phase space consists of the half planes to the left of $x = x_1$ and to the right of $x = x_2$.

A property of a quantum system is associated with a subspace of the Hilbert space, and thus the negation of this property will also be associated with some subspace of the Hilbert space. Consider, for example, a toy model with $M_a = 2 = M_b$. Its Hilbert space consists of all linear combinations of the states $|-2\rangle$, $|-1\rangle$, $|0\rangle$, $|1\rangle$, and $|2\rangle$. Suppose that P is the property associated with the projector

$$P = [0] + [1] \tag{4.36}$$

projecting onto the subspace $\mathcal{P}$ of all linear combinations of $|0\rangle$ and $|1\rangle$. Its physical interpretation is that the quantum particle is confined to these two sites, that is, it is not at some location apart from these two sites. The negation $\tilde{P}$ of P is the property that the particle is not confined to these two sites, but is instead someplace else, so the corresponding projector is

$$\tilde{P} = [-2] + [-1] + [2]. \tag{4.37}$$

This projects onto the orthogonal complement $\mathcal{P}^{\perp}$ of $\mathcal{P}$, see Sec. 3.4, consisting of all linear combinations of $|-2\rangle$, $|-1\rangle$ and $|2\rangle$. Since the identity operator for this Hilbert space is given by

$$I = \sum_{m=-2}^{2} [m], \tag{4.38}$$

see (3.52), it is evident that

$$\tilde{P} = I - P. \tag{4.39}$$

This is precisely the same as (4.35), except that the symbols now refer to quantum projectors rather than to classical indicators.

As a second example, consider a one-dimensional harmonic oscillator, Sec. 4.4. Suppose that P is the property that the energy is less than or equal to 2 in units of $\hbar\omega$. The corresponding projector is

$$P = [\phi_0] + [\phi_1] \tag{4.40}$$

in the notation used in Sec. 4.3. The negation of P is the property that the energy is greater than 2, and its projector is

$$\tilde{P} = [\phi_2] + [\phi_3] + [\phi_4] + \cdots = I - P. \tag{4.41}$$

In this case, P projects onto a finite and $\tilde{P}$ onto an infinite-dimensional subspace of $\mathcal{H}$.

As a third example, consider the property X that a particle in one dimension is located in (that is, not outside) the interval (4.19), $x_1 \leq x \leq x_2$; the corresponding projector X was defined in (4.20). Using the fact that $I\psi(x) = \psi(x)$, it is easy to show that the projector $\tilde{X} = I - X$, corresponding to the property that the particle is located outside (not inside) the interval (4.41) is given by

$$\tilde{X}\psi(x) = \begin{cases} 0 & \text{for } x_1 \leq x \leq x_2, \\ \psi(x) & \text{for all other } x \text{ values.} \end{cases} \tag{4.42}$$

(Note that in this case the action of the projectors X and $\tilde{X}$ is to multiply $\psi(x)$ by the indicator function for the corresponding classical property.)

As a final example, consider a spin-half particle, and let P be the property $S_z = +1/2$ (in units of $\hbar$) corresponding to the projector $[z^+]$. One can think of this as analogous to a toy model with $M = 2$ sites $m = 0, 1$, where $[z^+]$ corresponds to [0]. Then it is evident from the earlier discussion that the negation $\tilde{P}$ of P will be the projector $[z^-]$, the counterpart of [1] in the toy model, corresponding to the property $S_z = -1/2$. Of course, the same reasoning can be applied with z replaced by an arbitrary direction w: The property $S_w = -1/2$ is the negation of $S_w = +1/2$, and vice versa.

The relationship between the projector for a quantum property and the projector for its negation, (4.39), is formally the same as the relationship between the corresponding indicators for a classical property, (4.35). Despite this close analogy, there is actually an important difference. In the classical case, the subset $\sim\mathcal{P}$ corresponding to $\tilde{P}$ is the complement of the subset corresponding to P: any point in the phase space is in one or the other, and the two subsets do not overlap. In the quantum case, the subspaces $\mathcal{P}^\perp$ and $\mathcal{P}$ corresponding to $\tilde{P}$ and P have one element in common, the zero vector. This is different from the classical phase space, but is not important, for the zero vector by itself stands for the property which is always false, corresponding to the empty subset of the classical phase space. Much more significant is the fact that $\mathcal{H}$ contains many nonzero elements which belong neither to $\mathcal{P}^\perp$ nor to $\mathcal{P}$. In particular, the sum of a nonzero vector from $\mathcal{P}^\perp$ and a nonzero vector from $\mathcal{P}$ belongs to $\mathcal{H}$, but does not belong to either of these subspaces. For example, the ket $|x^+\rangle$ for a spin-half particle corresponding to $S_x = +1/2$ belongs neither to the subspace associated with $S_z = +1/2$ nor to that of its negation $S_z = -1/2$. Thus despite the formal parallel, the difference between the mathematics of Hilbert space and that of a classical phase space means that negation is not quite the same thing in quantum physics as it is in classical physics.

4.5 Conjunction and disjunction (AND, OR)

Consider two different properties P and Q of a classical system, corresponding to subsets $\mathcal{P}$ and $\mathcal{Q}$ of its phase space. The system will possess both properties simultaneously if its phase point γ lies in the intersection $\mathcal{P} \cap \mathcal{Q}$ of the sets $\mathcal{P}$ and $\mathcal{Q}$ or, using indicators, if $P(\gamma) = 1 = Q(\gamma)$. See the Venn diagram in Fig. 4.1(a). In this case we can say that the system possesses the property "P AND Q", the *conjunction* of P and Q, which can be written compactly as $P \wedge Q$. The corresponding indicator function is

$$P \wedge Q = PQ, \tag{4.43}$$

that is, $(P \wedge Q)(\gamma)$ is the function $P(\gamma)$ times the function $Q(\gamma)$. In the case of a one-dimensional harmonic oscillator, let P be the property that the energy is less than E_0, and Q the property that x lies between x_1 and x_2. Then the indicator PQ for the combined property $P \wedge Q$, "energy less than E_0 AND x between x_1 and x_2", is 1 at those points in the cross-hatched band in Fig. 2.1 which lie inside the ellipse, and 0 everywhere else.

Given the close correspondence between classical indicators and quantum projectors, one might expect that the projector for the quantum property $P \wedge Q$ (P AND Q) would be the product of the projectors for the separate properties, as in (4.43). This is indeed the case *if P and Q commute with each other*, that is, if

$$PQ = QP. \tag{4.44}$$

In this case it is easy to show that the product PQ is a projector satisfying the two conditions in (3.34). On the other hand, if (4.44) is not satisfied, then PQ will

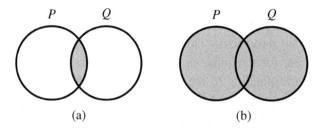

Fig. 4.1. The circles represent the properties P and Q. In (a) the grey region is $P \wedge Q$, and in (b) it is $P \vee Q$.

not be a Hermitian operator, so it cannot be a projector. In this section we will discuss the conjunction and disjunction of properties P and Q assuming that the *two projectors commute*. The case in which they do not commute is taken up in Sec. 4.6.

As a first example, consider the case of a one-dimensional harmonic oscillator in which P is the property that the energy E is less than 3 (in units of $\hbar\omega$), and Q the property that E is greater than 2. The two projectors are

$$P = [\phi_0] + [\phi_1] + [\phi_2], \quad Q = [\phi_2] + [\phi_3] + [\phi_4] + \cdots , \qquad (4.45)$$

and their product is $PQ = QP = [\phi_2]$, the projector onto the state with energy 2.5. As this is the only possible energy of the oscillator which is both greater than 2 and less than 3, the result makes sense.

As a second example, suppose that the property X corresponds to a quantum particle inside (not outside) the interval (4.19), $x_1 \leq x \leq x_2$, and X' to the property that the particle is inside the interval

$$x_1' \leq x \leq x_2'. \qquad (4.46)$$

In addition, assume that the endpoints of these intervals are in the order

$$x_1 < x_1' < x_2 < x_2'. \qquad (4.47)$$

For a classical particle, $X \wedge X'$ clearly corresponds to its being inside the interval

$$x_1' \leq x \leq x_2. \qquad (4.48)$$

In the quantum case, it is easy to show that $XX' = X'X$ is the projector which when applied to a wave function $\psi(x)$ sets it equal to 0 everywhere outside the interval (4.48) while leaving it unchanged inside this interval. This result is sensible, because if a wave packet lies inside the interval (4.48), it will also be inside both of the intervals (4.41) and (4.46).

When two projectors P and Q are mutually orthogonal in the sense defined in Sec. 3.5,

$$PQ = 0 = QP \qquad (4.49)$$

(each equality implies the other), the corresponding properties P and Q are *mutually exclusive* in the sense that if one is true, the other must be false. The reason is that the 0 operator which represents the conjunction $P \wedge Q$ corresponds, as does the 0 indicator for a classical system, to the property which is always false. Hence it is impossible for both P and Q to be true at the same time, for then $P \wedge Q$ would be true. As an example, consider the harmonic oscillator discussed earlier, but change the definitions so that P is the property $E < 2$ and Q the property $E > 3$. Then $PQ = 0$, for there is no energy which is both less than 2 and greater than 3. Similarly, if the intervals corresponding to X and X' for a particle in one dimension do not overlap — e.g., suppose $x_2 < x_1'$ in place of (4.47) — then $XX' = 0$, and if the particle is between x_1 and x_2, it cannot be between x_1' and x_2'. Note that this means that a quantum particle, just like its classical counterpart, can never be in two places at the same time, contrary to some misleading popularizations of quantum theory.

The *disjunction* of two properties P and Q, "P OR Q", where "OR" is understood in the nonexclusive sense of "P or Q or both", can be written in the compact form $P \vee Q$. If P and Q are classical properties corresponding to the subsets $\mathcal{P}$ and $\mathcal{Q}$ of a classical phase space, $P \vee Q$ corresponds to the union $\mathcal{P} \cup \mathcal{Q}$ of these two subsets, see Fig. 4.1(b), and the indicator is given by:

$$P \vee Q = P + Q - PQ, \qquad (4.50)$$

where the final term $-PQ$ on the right makes an appropriate correction at points in $\mathcal{P} \cap \mathcal{Q}$ where the two subsets overlap, and $P + Q = 2$.

The notions of disjunction (OR) and conjunction (AND) are related to each other by formulas familiar from elementary logic:

$$\begin{aligned}
\sim(P \vee Q) &= \tilde{P} \wedge \tilde{Q}, \\
\sim(P \wedge Q) &= \tilde{P} \vee \tilde{Q}.
\end{aligned} \qquad (4.51)$$

The negation of the first of these yields

$$P \vee Q = \sim(\tilde{P} \wedge \tilde{Q}), \qquad (4.52)$$

and one can use this expression along with (4.35) to obtain the right side of (4.50):

$$I - [(I - P)(I - Q)] = P + Q - PQ. \qquad (4.53)$$

Thus if negation and conjunction have already been defined, disjunction does not introduce anything that is really new.

The preceding remarks also apply to the quantum case. In particular, (4.53) is valid if P and Q are projectors. However, $P + Q - PQ$ is a projector if and only if $PQ = QP$. Thus as long as P and Q commute, we can use (4.50) to define the projector corresponding to the property P OR Q. There is, however, something to be concerned about. Suppose, to take a simple example, $P = [0]$ and $Q = [1]$ for a toy model. Then (4.50) gives $[0] + [1]$ for $P \vee Q$. However, as pointed out earlier, the subspace onto which $[0] + [1]$ projects contains kets which do not have either the property P or the property Q. Thus $[0] + [1]$ means something less definite than $[0]$ or $[1]$. A satisfactory resolution of this problem requires the notion of a quantum Boolean event algebra, which will be introduced in Sec. 5.2. In the meantime we will simply adopt (4.50) as a definition of what is meant by the quantum projector $P \vee Q$ when $PQ = QP$, and leave till later a discussion of just how it is to be interpreted.

4.6 Incompatible properties

The situation in which two projectors P and Q do not commute with each other, $PQ \neq QP$, has no classical analog, since the product of two indicator functions on the classical phase space does not depend upon the order of the factors. Consequently, classical physics gives no guidance as to how to think about the conjunction $P \wedge Q$ (P AND Q) of two quantum properties when their projectors do not commute.

Consider the example of a spin-half particle, let P be the property $S_x = +1/2$, and Q the property that $S_z = +1/2$. The projectors are

$$P = [x^+], \quad Q = [z^+], \tag{4.54}$$

and it is easy to show by direct calculation that $[x^+][z^+]$ is unequal to $[z^+][x^+]$, and that neither is a projector. Let us suppose that it is nevertheless possible to define a property $[x^+] \wedge [z^+]$. To what subspace of the two-dimensional spin space might it correspond? Every one-dimensional subspace of the Hilbert space of a spin-half particle corresponds to the property $S_w = +1/2$ for some direction w in space, as discussed in Sec. 4.2. Thus if $[x^+] \wedge [z^+]$ were to correspond to a one-dimensional subspace, it would have to be associated with such a direction. Clearly the direction cannot be x, for $S_x = +1/2$ does not have the property $S_z = +1/2$; see the discussion in Sec. 4.2. By similar reasoning it cannot be z, and all other choices for w are even worse, because then $S_w = +1/2$ possesses neither the property $S_x = +1/2$ nor the property $S_z = +1/2$, much less both of these properties!

If one-dimensional subspaces are out of the question, what is left? There is a two-dimensional "subspace" which is the entire space, with projector I corresponding to the property which is always true. But given that neither $[x^+]$ nor $[z^+]$

is a property which is always true, it seems ridiculous to suppose that $[x^+] \wedge [z^+]$ corresponds to I. There is also the zero-dimensional subspace which contains only the zero vector, corresponding to the property which is always false. Does it make sense to suppose that $[x^+] \wedge [z^+]$, thought of as a particular property possessed by a given spin-half particle at a particular time, is always false in the sense that there are no circumstances in which it could be true? If we adopt this proposal we will, obviously, also want to say that $[x^+] \wedge [z^-]$ is always false. Following the usual rules of logic, the disjunction (OR) of two false propositions is false. Therefore, the left side of

$$([x^+] \wedge [z^+]) \vee ([x^+] \wedge [z^-]) = [x^+] \wedge ([z^+] \vee [z^-]) = [x^+] \wedge I = [x^+]$$
(4.55)

is always false, and thus the right side, the property $S_x = +1/2$, is always false. But this makes no sense, for there are circumstances in which $S_x = +1/2$ is true.

To obtain the first equality in (4.55) requires the use of the distributive identity

$$(P \wedge Q) \vee (P \wedge R) = P \wedge (Q \vee R)$$
(4.56)

of standard logic, with $P = [x^+]$, $Q = [z^+]$, and $R = [z^-]$. One way of avoiding the silly result implied by (4.55) is to modify the laws of logic so that the distributive law does not hold. In fact, Birkhoff and von Neumann proposed a special *quantum logic* in which (4.56) is no longer valid. Despite a great deal of effort, this quantum logic has not turned out to be of much help in understanding quantum theory, and we shall not make use of it.

In conclusion, there seems to be no plausible way to assign a subspace to the conjunction $[x^+] \wedge [z^+]$ of these two properties, or to any other conjunction of two properties of a spin-half particle which are represented by noncommuting projectors. Such conjunctions are therefore meaningless in the sense that the Hilbert space approach to quantum theory, in which properties are associated with subspaces, cannot assign them a meaning. It is sometimes said that it is impossible to *measure* both S_x and S_z simultaneously for a spin-half particle. While this statement is true, it is important to note that the inability to carry out such a measurement reflects the fact that there is no corresponding property which could be measured. How could a measurement tell us, for example, that for a spin-half particle $S_x = +1/2$ and $S_z = +1/2$, if the property $[x^+] \wedge [z^+]$ cannot even be defined?

Guided by the spin-half example, we shall say that two properties P and Q of any quantum system are *incompatible* when their projectors do not commute, $PQ \neq QP$, and that the conjunction $P \wedge Q$ of incompatible properties is meaningless in the sense that quantum theory assigns it no meaning. On the other hand, if $PQ = QP$, the properties are *compatible*, and their conjunction $P \wedge Q$ corresponds to the projector PQ.

To say that $P \wedge Q$ is *meaningless* when $PQ \neq QP$ is very different from saying that it is *false*. The negation of a false statement is a true statement, so if $P \wedge Q$ is false, its negation $\tilde{P} \vee \tilde{Q}$, see (4.51), is true. On the other hand, the negation of a meaningless statement is equally meaningless. Meaningless statements can also occur in ordinary logic. Thus if P and Q are two propositions of an appropriate sort, $P \wedge Q$ is meaningful, but $P \wedge \vee Q$ is meaningless: this last expression cannot be true or false, it just doesn't make any sense. In the quantum case, "$P \wedge Q$" when $PQ \neq QP$ is something like $P \wedge \vee Q$ in ordinary logic. Books on logic always devote some space to the rules for constructing meaningful statements. Physicists when reading books on logic tend to skip over the chapters which give these rules, because the rules seem intuitively obvious. In quantum theory, on the other hand, it is necessary to pay some attention to the rules which separate meaningful and meaningless statements, because they are not the same as in classical physics, and hence they are not intuitively obvious, at least until one has built up some intuition for the quantum world.

When P and Q are incompatible, it makes no sense to ascribe both properties to a single system at the same instant of time. However, this does not exclude the possibility that P might be a meaningful (true or false) property at one instant of time and Q a meaningful property at a *different* time. We will discuss the time dependence of quantum systems starting in Ch. 7. Similarly, P and Q might refer to two distinct physical systems: for example, there is no problem in supposing that $S_x = +1/2$ for one spin-half particle, and $S_z = +1/2$ for a *different* particle.

At the end of Sec. 4.1 we stated that if a quantum system is described by a ket $|\psi\rangle$ which is not an eigenstate of a projector P, then the physical property associated with this projector is undefined. The situation can also be discussed in terms of incompatible properties, for saying that a quantum system is described by $|\psi\rangle$ is equivalent to asserting that it has the property $[\psi]$ corresponding to the ray which contains $|\psi\rangle$. It is easy to show that the projectors $[\psi]$ and P commute if and only if $|\psi\rangle$ is an eigenstate of P, whereas in all other cases $[\psi]P \neq P[\psi]$, so they represent incompatible properties.

It is possible for $|\psi\rangle$ to simultaneously be an eigenstate with eigenvalue 1 of two incompatible projectors P and Q. For example, for the toy model of Sec. 4.2, let

$$|\psi\rangle = |2\rangle, \quad P = [\sigma] + [2], \quad Q = [1] + [2], \tag{4.57}$$

where $|\sigma\rangle$ is defined in (4.10). The definition given in Sec. 4.1 allows us to conclude that the quantum system described by $|\psi\rangle$ has the property P, but we could equally well conclude that it has the property Q. However, it makes no sense to say that it has both properties. Sorting out this issue will require some additional concepts found in later chapters.

If the conjunction of incompatible properties is meaningless, then so is the disjunction of incompatible properties: $P \vee Q$ (P OR Q) makes no sense if $PQ \neq QP$. This follows at once from (4.52), because if P and Q are incompatible, so are their negations $\tilde{P}$ and $\tilde{Q}$, as can be seen by multiplying out $(I - P)(I - Q)$ and comparing it with $(I - Q)(I - P)$. Hence $\tilde{P} \wedge \tilde{Q}$ is meaningless, and so is its negation. Other sorts of logical comparisons, such as the exclusive OR (XOR), are also not possible in the case of incompatible properties.

If $PQ \neq QP$, the question "Does the system have property P or does it have property Q?" makes no sense if understood in a way which requires a comparison of these two incompatible properties. Thus one answer might be, "The system has property P but it does not have property Q". This is equivalent to affirming the truth of P and the falsity of Q, so that P and $\tilde{Q}$ are simultaneously true. But since $P\tilde{Q} \neq \tilde{Q}P$, this makes no sense. Another answer might be that "The system has both properties P and Q", but the assertion that P and Q are simultaneously true also does not make sense. And a question to which one cannot give a meaningful answer is not a meaningful question.

In the case of a spin-half particle it does not make sense to ask whether $S_x = +1/2$ or $S_z = +1/2$, since the corresponding projectors do not commute with each other. This may seem surprising, since it is possible to set up a device which will produce spin-half particles with a definite polarization, $S_w = +1/2$, where w is a direction determined by some property or setting of the device. (This could, for example, be the direction of the magnetic field gradient in a Stern–Gerlach apparatus, Sec. 17.2.) In such a case one can certainly ask whether the setting of the device is such as to produce particles with $S_x = +1/2$ or with $S_z = +1/2$. However, the values of components of spin angular momentum for a particle polarized by this device are then properties *dependent* upon properties of the device in the sense described in Ch. 14, and can only sensibly be discussed with reference to the device.

Along with different components of spin for a spin-half particle, it is easy to find many other examples of incompatible properties of quantum systems. Thus the projectors X and P in Sec. 4.3, for the position of a particle to lie between x_1 and x_2 and its momentum between p_1 and p_2, respectively, do not commute with each other. In the case of a harmonic oscillator, neither X nor P commutes with projectors, such as $[\phi_0] + [\phi_1]$, which define a range for the energy. That quantum operators, including the projectors which represent quantum properties, do not always commute with each other is a consequence of employing the mathematical structure of a quantum Hilbert space rather than that of a classical phase space. Consequently, there is no way to get around the fact that quantum properties cannot always be thought of in the same way as classical properties. Instead, one has to pay attention to the rules for combining them if one wants to avoid inconsistencies and paradoxes.

5

Probabilities and physical variables

5.1 Classical sample space and event algebra

Probability theory is based upon the concept of a *sample space* of mutually exclusive possibilities, one and only one of which actually occurs, or is true, in any given situation. The elements of the sample space are sometimes called *points* or *elements* or *events*. In classical and quantum mechanics the sample space usually consists of various possible states or properties of some physical system. For example, if a coin is tossed, there are two possible outcomes: H (heads) or T (tails), and the sample space $\mathcal{S}$ is $\{H, T\}$. If a die is rolled, the sample space $\mathcal{S}$ consists of six possible outcomes: $s = 1, 2, 3, 4, 5, 6$. If two individuals A and B share an office, the occupancy sample space consists of four possibilities: an empty office, A present, B present, or both A and B present.

Associated with a sample space $\mathcal{S}$ is an *event algebra* $\mathcal{B}$ consisting of subsets of elements of the sample space. In the case of a die, "s is even" is an event in the event algebra. So are "s is odd", "s is less than 4", and "s is equal to 2." It is sometimes useful to distinguish events which are elements of the sample space, such as $s = 2$ in the previous example, and those which correspond to more than one element of the sample space, such as "s is even". We shall refer to the former as *elementary* events and to the latter as *compound* events. If the sample space $\mathcal{S}$ is finite and contains n points, the event algebra contains 2^n possibilities, including the entire sample space $\mathcal{S}$ considered as a single compound event, and the empty set $\emptyset$. For various technical reasons it is convenient to include $\emptyset$, even though it never actually occurs: it is the event which is always false. Similarly, the compound event $\mathcal{S}$, the set of all elements in the sample space, is always true. The subsets of $\mathcal{S}$ form a *Boolean algebra* or *Boolean lattice* $\mathcal{B}$ under the usual set-theoretic relationships: The *complement* $\sim \mathcal{E}$ of a subset $\mathcal{E}$ of $\mathcal{S}$ is the set of elements of $\mathcal{S}$ which are not in $\mathcal{E}$. The *intersection* $\mathcal{E} \cap \mathcal{F}$ of two subsets is the collection of elements they have in common, whereas their *union* $\mathcal{E} \cup \mathcal{F}$ is the collection of elements belonging to one or the other, or possibly both.

The phase space of a classical mechanical system is a sample space, since one and only one point in this space represents the actual state of the system at a particular time. Since this space contains an uncountably infinite number of points, one usually defines the event algebra not as the collection of all subsets of points in the phase space, but as some more manageable collection, such as the Borel sets.

A useful analogy with quantum theory is provided by a *coarse graining* of the classical phase space, a finite or countably infinite collection of nonoverlapping regions or *cells* which together cover the phase space. These cells, which in the notation of Ch. 4 represent properties of the physical system, constitute a sample space $\mathcal{S}$ of mutually exclusive possibilities, since a point γ in the phase space representing the state of the system at a particular time will be in one and only one cell, making this cell a true property, whereas the properties corresponding to all of the other cells in the sample are false. (Note that individual points in the phase space are not, in and of themselves, members of $\mathcal{S}$.) The event algebra $\mathcal{B}$ associated with this coarse graining consists of collections of one or more cells from the sample space, along with the empty set and the collection of all the cells. Each event in $\mathcal{B}$ is associated with a physical property corresponding to the set of all points in the phase space lying in one of the cells making up the (in general compound) event. The negation of an event $\mathcal{E}$ is the collection of cells which are in $\mathcal{S}$ but not in $\mathcal{E}$, the conjunction of two events $\mathcal{E}$ and $\mathcal{F}$ is the collection of cells which they have in common, and their disjunction the collection of cells belonging to $\mathcal{E}$ or to $\mathcal{F}$ or to both.

As an example, consider a one-dimensional harmonic oscillator whose phase space is the x, p plane. One possible coarse graining consists of the four cells

$$x \geq 0,\ p \geq 0; \quad x < 0,\ p \geq 0; \quad x \geq 0,\ p < 0; \quad x < 0,\ p < 0; \tag{5.1}$$

that is, the four quadrants defined so as not to overlap. Another coarse graining is the collection $\{C_n\}$, $n = 1, 2, \ldots$ of cells

$$C_n : (n-1)E_0 \leq E < nE_0 \tag{5.2}$$

defined in terms of the energy E, where $E_0 > 0$ is some constant. Still another coarse graining consists of the rectangles

$$D_{mn} : mx_0 < x \leq (m+1)x_0,\ np_0 < p \leq (n+1)p_0, \tag{5.3}$$

where $x_0 > 0$, $p_0 > 0$ are constants, and m and n are any integers.

As in Sec. 4.1, we define the *indicator* or *indicator function* E for an event $\mathcal{E}$ to be the function on the sample space which takes the value 1 space which is in the set $\mathcal{E}$, and 0 (false) on all other elements:

$$E(s) = \begin{cases} 1 & \text{for } s \in \mathcal{E}, \\ 0 & \text{otherwise.} \end{cases} \tag{5.4}$$

The indicators form an algebra under the operations of negation ($\sim E$), conjunction ($E \wedge F$), and disjunction ($E \vee F$), as discussed in Secs. 4.4 and 4.5:

$$\sim E = \tilde{E} = I - E,$$
$$E \wedge F = EF, \tag{5.5}$$
$$E \vee F = E + F - EF,$$

where the arguments of the indicators have been omitted; one could also write $(E \wedge F)(s) = E(s)F(s)$, etc. Obviously $E \wedge F$ and $E \vee F$ are the counterparts of $\mathcal{E} \cap \mathcal{F}$ and $\mathcal{E} \cup \mathcal{F}$ for the corresponding subsets of $\mathcal{S}$. We shall use the terms "event algebra" and "Boolean algebra" for either the algebra of sets or the corresponding algebra of indicators.

Associated with each element r of a sample space is a special indicator P_r which is zero except at the point r:

$$P_r(s) = \begin{cases} 1 & \text{if } s = r, \\ 0 & \text{if } s \neq r. \end{cases} \tag{5.6}$$

Indicators of this type will be called *elementary* or *minimal*, and it is easy to see that

$$P_r P_s = \delta_{rs} P_s. \tag{5.7}$$

The vanishing of the product of two elementary indicators associated with distinct elements of the sample space reflects the fact that these events are mutually-exclusive possibilities: if one of them occurs (is true), the other cannot occur (is false), since the zero indicator denotes the "event" which never occurs (is always false). An indicator R on the sample space corresponding to the (in general compound) event $\mathcal{R}$ can be written as a sum of elementary indicators,

$$R = \sum_{s \in \mathcal{S}} \pi_s P_s, \tag{5.8}$$

where π_s is equal to 1 if s is in $\mathcal{R}$, and 0 otherwise. The indicator I, which takes the value 1 everywhere, can be written as

$$I = \sum_{s \in \mathcal{S}} P_s, \tag{5.9}$$

which is (5.8) with $\pi_s = 1$ for every s.

5.2 Quantum sample space and event algebra

In Sec. 3.5 a decomposition of the identity was defined to be an orthogonal collection of projectors $\{P_j\}$,

$$P_j P_k = \delta_{jk} P_j, \tag{5.10}$$

which sum to the identity

$$I = \sum_j P_j. \tag{5.11}$$

Any decomposition of the identity of a quantum Hilbert space $\mathcal{H}$ can be thought of as a *quantum sample space* of mutually-exclusive properties associated with the projectors or with the corresponding subspaces. That the properties are mutually exclusive follows from (5.10), see the discussion in Sec. 4.5, which is the quantum counterpart of (5.7). The fact that the projectors sum to I is the counterpart of (5.9), and expresses the fact that one of these properties must be true. Thus the usual requirement that a sample space consist of a collection of mutually-exclusive possibilities, one and only one of which is correct, is satisfied by a quantum decomposition of the identity.

The *quantum event algebra* $\mathcal{B}$ corresponding to the sample space (5.11) consists of all projectors of the form

$$R = \sum_j \pi_j P_j, \tag{5.12}$$

where each π_j is either 0 or 1; note the analogy with (5.8). Setting all the π_j equal to 0 yields the zero operator 0 corresponding to the property that is always false; setting them all equal to 1 yields the identity I, which is always true. If there are n elements in the sample space, there are 2^n elements in the event algebra, just as in ordinary probability theory. The *elementary* or *minimal* elements of $\mathcal{B}$ are the projectors $\{P_j\}$ which belong to the sample space, whereas the *compound* elements are those for which two or more of the π_j in (5.12) are equal to 1.

Since the different projectors which make up the sample space commute with each other, (5.10), so do all projectors of the form (5.12). And because of (5.10), the projectors which make up the event algebra $\mathcal{B}$ form a *Boolean algebra* or *Boolean lattice* under the operations of $\cap$ and $\cup$ interpreted as $\wedge$ and $\vee$; see (5.5), which applies equally to classical indicators and quantum projectors. Any collection of commuting projectors forms a Boolean algebra provided the negation $\tilde{P}$ of any projector P in the collection is also in the collection, and the product PQ $(= QP)$ of two elements in the collection is also in the collection. (Because of (4.50), these rules ensure that $P \vee Q$ is also a member of the collection, so this does not have to be stated as a separate rule.) Note that a Boolean algebra of projectors is a much simpler object (in algebraic terms) than the noncommutative algebra of all operators on the Hilbert space.

A *trivial* decomposition of the identity contains just one projector, I; *nontrivial* decompositions contain two or more projectors. For a spin-half particle, the only nontrivial decompositions of the identity are of the form

$$I = [w^+] + [w^-], \tag{5.13}$$

where w is some direction in space, such as the x axis or the z axis. Thus the sample space consists of two elements, one corresponding to $S_w = +1/2$ and one to $S_w = -1/2$. These are mutually-exclusive possibilities: one and only one can be a correct description of this component of spin angular momentum. The event algebra $\mathcal{B}$ consists of the four elements: 0, I, $[w^+]$, and $[w^-]$.

Next consider a toy model, Sec. 2.5, in which a particle can be located at one of $M = 3$ sites, $m = -1, 0, 1$. The three kets $|-1\rangle$, $|0\rangle$, $|1\rangle$ form an orthonormal basis of the Hilbert space. A decomposition of the identity appropriate for discussing the particle's position contains the three projectors

$$[-1], \quad [0], \quad [1] \tag{5.14}$$

corresponding to the property that the particle is at $m = -1$, $m = 0$, and $m = 1$, respectively. The Boolean event algebra has $2^3 = 8$ elements: 0, I, the three projectors in (5.14), and three projectors

$$[-1] + [0], \quad [0] + [1], \quad [-1] + [1] \tag{5.15}$$

corresponding to compound events. An alternative decomposition of the identity for the same Hilbert space consists of the two projectors

$$[-1], \quad [0] + [1], \tag{5.16}$$

which generate an event algebra with only $2^2 = 4$ elements: the projectors in (5.16) along with 0 and I.

Although the same projector $[0] + [1]$ occurs both in (5.15) and in (5.16), its physical interpretation or meaning in the two cases is actually somewhat different, and discussing the difference will throw light upon the issue raised at the end of Sec. 4.5 about the meaning of a quantum disjunction $P \vee Q$. In (5.15), $[0] + [1]$ represents a *compound* event whose physical interpretation is that the particle is at $m = 0$ or at $m = 1$, in much the same way that the compound event $\{3, 4\}$ in the case of a die would be interpreted to mean that either $s = 3$ or $s = 4$ spots turned up. On the other hand, in (5.16) the projector $[0] + [1]$ represents an *elementary* event which cannot be thought of as the disjunction of two different possibilities. In quantum mechanics, each Boolean event algebra constitutes what is in effect a "language" out of which one can construct a quantum description of some physical system, and a fundamental rule of quantum theory is that a description (which may but need not be couched in terms of probabilities) referring to a single system at a single time must be constructed using a single Boolean algebra, a single "language". (This is a particular case of a more general "single-framework rule" which will be introduced later on, and discussed in some detail in Ch. 16.) The language based on (5.14) contains among its elementary constituents the projector [0] and the projector [1], and its grammatical rules allow one to combine such elements with "and" and "or" in a meaningful way. Hence in this language "[0] or [1]" makes sense, and it is convenient to represent it using the projector $[0] + [1]$ in (5.15). On the other hand, the language based on (5.16) contains neither [0] nor [1] — they are not in the sample space, nor are they among the four elements which constitute its Boolean algebra. Consequently, in this somewhat impoverished language it is impossible to express the idea "[0] or [1]", because both [0] and [1] are meaningless constructs.

The reader may be tempted to dismiss all of this as needless nitpicking better suited to mathematicians and philosophers than to physical scientists. Is it not obvious that one can always replace the impoverished language based upon (5.16) with the richer language based upon (5.14), and avoid all this quibbling? The answer is that one can, indeed, replace (5.16) with (5.14) in appropriate circumstances; the process of doing so is known as "refinement", and will be discussed in Sec. 5.3. However, in quantum theory there can be many different refinements. In particular, a second and rather different refinement of (5.16) will be found in (5.19). Because of the multiple possibilities for refinement, one must pay attention to what one is doing, and it is especially important to be explicit about the sample space ("language") that one is using. Shortcuts in reasoning which never cause difficulty in classical physics can lead to enormous headaches in quantum theory, and avoiding these requires that one take into account the rules which govern meaningful quantum descriptions.

As an example of a sample space associated with a continuous quantum system, consider the decomposition of the identity

$$I = \sum_n [\phi_n] \tag{5.17}$$

corresponding to the energy eigenstates of a quantum harmonic oscillator, in the notation of Sec. 4.3. The elementary event $[\phi_n]$ can be interpreted as the energy having the value $n + 1/2$ in units of $\hbar\omega$. These events are mutually-exclusive possibilities: if the energy is 3.5, it cannot be 0.5 or 2.5, etc. The projector $[\phi_2] + [\phi_3]$ in the Boolean algebra generated by (5.17) means that the energy is equal to 2.5 or 3.5. If, on the other hand, one were to replace (5.17) with an alternative decomposition of the identity consisting of the projectors $\{([\phi_{2m}] + [\phi_{2m+1}]), m = 0, 1, 2, \ldots\}$, each projecting onto a two-dimensional subspace of $\mathcal{H}$, $[\phi_2] + [\phi_3]$ could not be interpreted as an energy equal to 2.5 or 3.5, since states without a well-defined energy are also present in the corresponding subspace. See the preceding discussion of the toy model.

5.3 Refinement, coarsening, and compatibility

Suppose there are two decompositions of the identity, $\mathcal{E} = \{E_j\}$ and $\mathcal{F} = \{F_k\}$, with the property that each F_k can be written as a sum of one of more of the E_j. In such a case we will say that the decomposition $\mathcal{E}$ is a *refinement* of $\mathcal{F}$, or $\mathcal{E}$ is *finer* than $\mathcal{F}$, or $\mathcal{E}$ is obtained by *refining* $\mathcal{F}$. Equivalently, $\mathcal{F}$ is a *coarsening* of $\mathcal{E}$, is *coarser* than $\mathcal{E}$, and is obtained by *coarsening* $\mathcal{E}$. For example, the decomposition (5.14) is a refinement of (5.16) obtained by replacing the single projector $[0] + [1]$ in the latter with the two projectors $[0]$ and $[1]$.

According to this definition, any decomposition of the identity is its own refinement (or coarsening), and it is convenient to allow the possibility of such a *trivial* refinement (or coarsening). If the two decompositions are actually different, one is a *nontrivial* or *proper* refinement/coarsening of the other. An *ultimate* decomposition of the identity is one in which each projector projects onto a one-dimensional subspace, so no further refinement (of a nontrivial sort) is possible. Thus (5.13), (5.14), and (5.17) are ultimate decompositions, whereas (5.16) is not.

Two or more decompositions of the identity are said to be (mutually) *compatible* provided they have a *common refinement*, that is, provided there is a single decomposition $\mathcal{R}$ which is finer than each of the decompositions under consideration. When no common refinement exists the decompositions are said to be (mutually) *incompatible*. If $\mathcal{E}$ is a refinement of $\mathcal{F}$, the two are obviously compatible, because $\mathcal{E}$ is itself the common refinement.

The toy model with $M = 3$ considered in Sec. 5.2 provides various examples of compatible and incompatible decompositions of the identity. The decomposition

$$([-1] + [0]), \quad [1] \tag{5.18}$$

is compatible with (5.16) because (5.14) is a common refinement. The decomposition

$$[-1], \quad [p], \quad [q], \tag{5.19}$$

where the projectors $[p]$ and $[q]$ correspond to the kets

$$|p\rangle = (|0\rangle + |1\rangle)/\sqrt{2}, \quad |q\rangle = (|0\rangle - |1\rangle)/\sqrt{2}, \tag{5.20}$$

is a refinement of (5.16), as is (5.14), so both (5.14) and (5.19) are compatible with (5.16). However, (5.14) and (5.19) are incompatible with each other: since each is an ultimate decomposition, and they are not identical, there is no common refinement. In addition, (5.19) is incompatible with (5.18), though this is not quite so obvious. As another example, the two decompositions

$$I = [x^+] + [x^-], \quad I = [z^+] + [z^-] \tag{5.21}$$

for a spin-half particle are incompatible, because each is an ultimate decomposition, and they are not identical.

If $\mathcal{E}$ and $\mathcal{F}$ are compatible, then each projector E_j can be written as a combination of projectors from the common refinement $\mathcal{R}$, and the same is true of each F_k. That is to say, the projectors $\{E_j\}$ and $\{F_k\}$ belong to the Boolean event algebra generated by $\mathcal{R}$. As all the operators in this algebra commute with each other, it follows that every projector E_j commutes with every projector F_k. Conversely, if every E_j in $\mathcal{E}$ commutes with every F_k in $\mathcal{F}$, there is a common refinement: all nonzero projectors of the form $\{E_j F_k\}$ constitute the decomposition *generated* by $\mathcal{E}$ and $\mathcal{F}$, and it is the coarsest common refinement of $\mathcal{E}$ and $\mathcal{F}$. The same argument can be extended to a larger collection of decompositions, and leads to the general rule that *decompositions of the identity are mutually compatible if and only if all the projectors belonging to all of the decompositions commute with each other*. If any pair of projectors fail to commute, the decompositions are incompatible. Using this rule it is immediately evident that the decompositions in (5.16) and (5.18) are compatible, whereas those in (5.18) and (5.19) are incompatible. The two decompositions in (5.21) are incompatible, as are any two decompositions of the identity of the form (5.13) if they correspond to two directions in space that are neither the same nor opposite to each other. Since it arises from projectors failing to commute with each other, incompatibility is a feature of the quantum world with no close analog in classical physics. Different sample spaces associated with a single classical system are always compatible, they always possess a common refinement. For example, a common refinement of two coarse grainings of a classical phase space is easily constructed using the nonempty intersections of cells taken from the two sample spaces.

As noted in Sec. 5.2, a fundamental rule of quantum theory is that a description of a particular quantum system must be based upon a *single sample space* or decomposition of the identity. If one wants to use two or more *compatible* sample spaces, this rule can be satisfied by employing a common refinement, since its Boolean algebra will include the projectors associated with the individual spaces. On the other hand, trying to combine descriptions based upon two (or more) *incompatible* sample spaces can lead to serious mistakes. Consider, for example, the two incompatible decompositions in (5.21). Using the first, one can conclude that for a spin-half particle, either $S_x = +1/2$ or $S_x = -1/2$. Similarly, by using the second one can conclude that either $S_z = +1/2$ or else $S_z = -1/2$. However, combining these in a manner which would be perfectly correct for a classical spinning object leads to the conclusion that one of the four possibilities

$$S_x = +1/2 \wedge S_z = +1/2, \quad S_x = +1/2 \wedge S_z = -1/2,$$
$$S_x = -1/2 \wedge S_z = +1/2, \quad S_x = -1/2 \wedge S_z = -1/2 \tag{5.22}$$

must be a correct description of the particle. But in fact all four possibilities are meaningless, as discussed previously in Sec. 4.6, because none of them corresponds to a subspace of the quantum Hilbert space.

5.4 Probabilities and ensembles

Given a sample space, a probability distribution assigns a nonnegative number or *probability* p_s, also written $\Pr(s)$, to each point s of the sample space in such a way that these numbers sum to 1. For example, in the case of a six-sided die, one often assigns equal probabilities to each of the six possibilities for the number of spots s; thus $p_s = 1/6$. However, this assignment is not a fundamental law of probability theory, and there exist dice for which a different set of probabilities would be more appropriate. Each compound event E in the event algebra is assigned a probability $\Pr(E)$ equal to the sum of the probabilities of the elements of the sample space which it contains. Thus "s is even" in the case of a die is assigned a probability $p_2 + p_4 + p_6$, which is $1/2$ if each p_s is $1/6$. The assignment of probabilities in the case of continuous variables, e.g., a classical phase space, can be quite a bit more complicated. However, the simpler discrete case will be quite adequate for this book; we will not need sophisticated concepts from measure theory.

Along with a formal definition, one needs an intuitive idea of the meaning of probabilities. One approach is to imagine an *ensemble*: a collection of N nominally identical systems, where N is a very large number, with each system in one of the states which make up the sample space S, and with the fraction of members of the ensemble in state s given by the corresponding probability p_s. For example, the ensemble could be a large number of dice, each displaying a certain number of spots, with $1/6$ of the members of the ensemble displaying one spot, $1/6$ displaying two spots, etc. One should always think of N as such a large number that $p_s N$ is also very large for any p_s that is greater than 0, to get around any concerns about whether the fraction of systems in state s is precisely equal to p_s. One says that the probability that a *single system* chosen at random from such an ensemble is in state s is given by p_s. Of course, any particular system will be in some definite state, but this state is not known before the system is selected from the ensemble. Thus the probability represents "partial information" about a system when its actual state is not known. For example, if the probability of some state is close to 1, one can be fairly confident, but not absolutely certain, that a system chosen at random will be in this state and not in some other state.

Rather than imagining the ensemble to be a large collection of systems, it is sometimes useful to think of it as made up of the outcomes of a large number of experiments carried out at successive times, with care being taken to ensure that these are independent in the sense that the outcome of any one experiment is not allowed to influence the outcome of later experiments. Thus instead of a large collection of dice, one can think of a single die which is rolled a large number of times. The fraction of experiments leading to the result s is then the probability p_s. The outcome of any particular experiment in the sequence is not known in advance, but a knowledge of the probabilities provides partial information.

Probability theory as a mathematical discipline does not tell one how to choose a probability distribution. Probabilities are sometimes obtained directly from experimental data. In other cases, such as the Boltzmann distribution for systems in thermal equilibrium, the probabilities express well-established physical laws. In some cases they simply represent a guess. Later we shall see how to use the dynamical laws of quantum theory to calculate various quantum probabilities. The true meaning of probabilities is a subject about which there continue to be disputes, especially among philosophers. These arguments need not concern us, for probabilities in quantum theory, when properly employed with a well-defined sample space, obey the same rules as in classical physics. Thus the situation in quantum physics is no worse (or better) than in the everyday classical world.

Conditional probabilities play a fundamental role in probabilistic reasoning and in the application of probability theory to quantum mechanics. Let A and B be two events, and suppose that $\Pr(B) > 0$. The *conditional probability of A given B* is defined to be

$$\Pr(A \mid B) = \Pr(A \wedge B)/\Pr(B), \tag{5.23}$$

where $A \wedge B$ is the event "*A* AND *B*" represented by the product AB of the classical indicators, or of the quantum projectors. Hence one can also write $\Pr(AB)$ in place of $\Pr(A \wedge B)$. The intuitive idea of a conditional probability can be expressed in the following way. Given an ensemble, consider only those members in which B occurs (is true). These comprise a *subensemble* of the original ensemble, and in this subensemble the fraction of systems with property A is given by $\Pr(A \mid B)$ rather than by $\Pr(A)$, as in the original ensemble. For example, in the case of a die, let B be the property that s is even, and A the property $s \le 3$. Assuming equal probabilities for all outcomes, $\Pr(A) = 1/2$. However, $\Pr(A \mid B) = 1/3$, corresponding to the fact that of the three possibilities $s = 2, 4, 6$ which constitute the compound event B, only one is less than or equal to 3.

If B is held fixed, $\Pr(A \mid B)$ as a function of its first argument A behaves like an "ordinary" probability distribution. For example, if we use s to indicate points in the sample space, the numbers $\Pr(s \mid B)$ are nonnegative, and $\sum_s \Pr(s \mid B) = 1$. One can think of $\Pr(A \mid B)$ with B fixed as obtained by setting to zero the probabilities of all elements of the sample space for which B is false (does not occur), and multiplying the probabilities of those elements for which B is true by a common factor, $1/\Pr(B)$, to renormalize them, so that the probabilities of mutually-exclusive sets of events sum to one. That this is a reasonable procedure is evident if one imagines an ensemble and thinks about the subensemble of cases in which B occurs. It makes no sense to define a probability conditioned on B if $\Pr(B) = 0$, as there is no way to renormalize zero probability by multiplying it by a constant in order to get something finite.

In the case of quantum systems, once an appropriate sample space has been defined the rules for manipulating probabilities are *precisely the same* as for any other ("classical") probabilities. The probabilities must be nonnegative, they must sum to 1, and conditional probabilities are defined in precisely the manner discussed above. Sometimes it seems as if quantum probabilities obey different rules from what one is accustomed to in classical physics. The reason is that quantum theory allows a multiplicity of sample spaces, that is, decompositions of the identity, which are often incompatible with one another. In classical physics a single sample space is usually sufficient, and in cases in which one may want to use more than one, for example alternative coarse grainings of the phase space, the different possibilities are always compatible with each other. However, in quantum theory different sample spaces are generally incompatible with one another, so one has to learn how to choose the correct sample space needed for discussing a particular physical problem, and how to avoid carelessly combining results from incompatible sample spaces. Thus the difficulties one encounters in quantum mechanics have to do with choosing a sample space. Once the sample space has been specified, the quantum rules are the same as the classical rules.

There have been, and no doubt will continue to be, a number of proposals for introducing special "quantum probabilities" with properties which violate the usual rules of probability theory: probabilities which are negative numbers, or complex numbers, or which are not tied to a Boolean algebra of projectors, etc. Thus far, none of these proposals has proven helpful in untangling the conceptual difficulties of quantum theory. Perhaps someday the situation will change, but until then there seems to be no reason to abandon standard probability theory, a mode of reasoning which is quite well understood, both formally and intuitively, and replace it with some scheme which is deficient in one or both of these respects.

5.5 Random variables and physical variables

In ordinary probability theory a *random variable* is a real-valued function V defined everywhere on the sample space. For example, if s is the number of spots when a die is rolled, $V(s) = s$ is an example of a random variable, as is $V(s) = s^2/6$. For coin tossing, $V(\text{H}) = +1/2$, $V(\text{T}) = -1/2$ is an example of a random variable.

If one regards the x, p phase plane for a particle in one dimension as a sample space, then any real-valued function $V(x, p)$ is a random variable. Examples of physical interest include the position, the momentum, the kinetic energy, the potential energy, and the total energy. For a particle in three dimensions the various components of angular momentum relative to some origin are also examples of random variables.

In classical mechanics the term *physical variable* is probably more descriptive than "random variable" when referring to a function defined on the phase space, and we shall use it for both classical and quantum systems. However, thinking of physical variables as random variables, that is, as functions defined on a sample space, is particularly helpful in understanding what they mean in quantum theory.

The quantum counterpart of the function V representing a physical variable in classical mechanics is a Hermitian or self-adjoint operator $V = V^{\dagger}$ on the Hilbert space. Thus position, energy, angular momentum, and the like all correspond to specific quantum operators. Generalizing from this, we shall think of any self-adjoint operator on the Hilbert space as representing some (not necessarily very interesting) physical variable. A quantum physical variable is often called an *observable*. While this term is not ideal, given its association with somewhat confused and contradictory ideas about quantum measurements, it is widely used in the literature, and in this book we shall employ it to refer to any quantum physical variable, that is, to any self-adjoint operator on the quantum Hilbert space, without reference to whether it could, in practice or in principle, be measured.

To see how self-adjoint operators can be thought of as random variables in the sense of probability theory, one can make use of a fact discussed in Sec. 3.7: if $V = V^{\dagger}$, then there is a unique decomposition of the identity $\{P_j\}$, determined by the operator V, such that, see (3.75),

$$V = \sum_j v'_j P_j, \tag{5.24}$$

where the v'_j are eigenvalues of V, and $v'_j \neq v'_k$ for $j \neq k$. Since any decomposition of the identity can be regarded as a quantum sample space, one can think of the collection $\{P_j\}$ as the "natural" sample space for the physical variable or operator V. On this sample space the operator V behaves very much like a real-valued function: to P_1 it assigns the value v'_1, to P_2 the value v'_2, and so forth. That

(5.24) can be interpreted in this way is suggested by the fact that for a discrete sample space $\mathcal{S}$, an ordinary random variable V can always be written as a sum of numbers times the elementary indicators defined in (5.6),

$$V(s) = \sum_r v_r P_r(s), \tag{5.25}$$

where $v_r = V(r)$. Since quantum projectors are analogous to classical indicators, and the indicators on the right side of (5.25) are associated with the different elements of the sample space, there is an obvious and close analogy between (5.24) and (5.25).

The only possible values for a quantum observable V are the eigenvalues v'_j in (5.24) or, equivalently, the v_j in (5.32), just as the only possible values of a classical random variable are the v_r in (5.25). In order for a quantum system to possess the value v for the observable V, the property "$V = v$" must be true, and this means that the system is in an eigenstate of V. That is to say, the quantum system is described by a nonzero ket $|\psi\rangle$ such that

$$V|\psi\rangle = v|\psi\rangle, \tag{5.26}$$

or, more generally, by a nonzero projector Q such that

$$VQ = vQ. \tag{5.27}$$

In order for (5.27) to hold for a projector Q onto a space of dimension 2 or more, the eigenvalue v must be degenerate, and if $v = v'_j$, then

$$P_j Q = Q, \tag{5.28}$$

where P_j is the projector in (5.24) corresponding to v'_j.

Let us consider some examples, beginning with a one-dimensional harmonic oscillator. Its (total) energy corresponds to the Hamiltonian operator H, which can be written in the form

$$H = \sum_n (n + 1/2)\hbar\omega [\phi_n], \tag{5.29}$$

where the corresponding decomposition of the energy was introduced in (5.17). The Hamiltonian can thus be thought of as a function which assigns to the projector $[\phi_n]$, or to the subspace of multiples of $|\phi_n\rangle$, the energy $(n + 1/2)\hbar\omega$. In the case of a spin-half particle the operator for the z component of spin angular momentum divided by $\hbar$ is

$$S_z = +\tfrac{1}{2}[z^+] - \tfrac{1}{2}[z^-]. \tag{5.30}$$

It assigns to $[z^+]$ the value $+1/2$, and to $[z^-]$ the value $-1/2$. Next think of a toy model in which the sites are labeled by an integer m, and suppose that the distance between adjacent sites is the length b. Then the position operator will be given by

$$B = \sum_m mb\,[m].$$ (5.31)

The position operator x for a "real" quantum particle in one dimension is a complicated object, and writing it in a form equivalent to (5.24) requires replacing the sum with an integral, using mathematics which is outside the scope of this book.

In all the examples considered thus far, the P_j are projectors onto one-dimensional subspaces, so they can be written as dyads, and (5.24) is equivalent to writing

$$V = \sum_j v_j |v_j\rangle\langle v_j| = \sum_j v_j [v_j],$$ (5.32)

where the eigenvalues in (5.32) are identical to those in (5.24), except that the subscript labels may be different. As discussed in Sec. 3.7, (5.24) and (5.32) will be different if one or more of the eigenvalues of V are degenerate, that is, if a particular eigenvalue occurs more than once on the right side of (5.32). For instance, the energy eigenvalues of atoms are often degenerate due to spherical symmetry, and in this case the projector P_j for the jth energy level projects onto a space whose dimension is equal to the multiplicity (or degeneracy) of the level. When such degeneracies occur, it is possible to construct nontrivial *refinements* of the decomposition $\{P_j\}$ in the sense discussed in Sec. 5.3, by writing one or more of the P_j as a sum of two or more nonzero projectors. If $\{Q_k\}$ is such a refinement, it is obviously possible to write

$$V = \sum_k v_k'' Q_k,$$ (5.33)

where the extra prime allows the eigenvalues in (5.33) to carry different subscripts from those in (5.24). One can again think of V as a random variable, that is, a function, on the finer sample space $\{Q_k\}$. Note that when it is possible to refine a quantum sample space in this manner, it is always possible to refine it in many different ways which are mutually incompatible. Whereas any one of these sample spaces is perfectly acceptable so far as the physical variable V is concerned, one will make mistakes if one tries to combine two or more incompatible sample spaces in order to describe a single physical system; see the comments in Sec. 5.3.

On the other hand, V cannot be defined as a physical ("random") variable on a decomposition which is *coarser* than $\{P_j\}$, since one cannot assign two different eigenvalues to the same projector or subspace. (To be sure, one might define a "coarse" version of V, but that would be a different physical variable.) Nor can V be defined as a physical or random variable on a decomposition which is incompatible with $\{P_j\}$, in the sense discussed in Sec. 5.3. It may, of course, be possible to approximate V with an operator which is a function on an alternative decomposition, but such approximations are outside the scope of the present discussion.

5.6 Averages

The average $\langle V \rangle$ of a random variable $V(s)$ on a sample space S is defined by the formula

$$\langle V \rangle = \sum_{s \in S} p_s V(s). \tag{5.34}$$

That is, the probabilities are used to weight the values of V at the different sample points before adding them together. One way to justify the weights in (5.34) is to imagine an ensemble consisting of a very large number N of systems. If V is evaluated for each system, and the results are then added together and divided by N, the outcome will be (5.34), because the fraction of systems in the ensemble in state s is equal to p_s.

Random variables form a real linear space in the sense that if $U(s)$ and $V(s)$ are two random variables, so is the linear combination

$$u U(s) + v V(s), \tag{5.35}$$

where u and v are real numbers. The average operation $\langle \rangle$ defined in (5.34) is a linear functional on this space, since

$$\langle u U(s) + v V(s) \rangle = u \langle U \rangle + v \langle V \rangle. \tag{5.36}$$

Another property of $\langle \rangle$ is that when it is applied to a *positive* random variable $W(s) \geq 0$, the result cannot be negative:

$$\langle W \rangle \geq 0. \tag{5.37}$$

In addition, the average of the identity is 1,

$$\langle I \rangle = 1, \tag{5.38}$$

because the probabilities $\{p_s\}$ sum to 1.

The linear functional $\langle \rangle$ is obviously determined once the probabilities $\{p_s\}$ are given. Conversely, a functional $\langle \rangle$ defined on the linear space of random variables determines a unique probability distribution, since one can use averages of the elementary indicators in (5.6),

$$p_s = \langle P_s \rangle, \tag{5.39}$$

in order to define positive probabilities which sum to 1 in view of (5.9) and (5.38).

In a similar way, the probability of a compound event A is equal to the average of its indicator:

$$\Pr(A) = \langle A \rangle. \tag{5.40}$$

Averages for quantum mechanical physical (random) variables follow precisely the same rules; the only differences are in notation. One starts with a sample space $\{P_j\}$ of projectors which sum to I, and a set of nonnegative probabilities $\{p_j\}$ which sum to 1. A random variable on this space is a Hermitian operator which can be written in the form

$$V = \sum_j v_j P_j, \tag{5.41}$$

where the different eigenvalues appearing in the sum need not be distinct. That is, the sample space could be either the "natural" space associated with the operator V as discussed in Sec. 5.5, or some refinement. The average

$$\langle V \rangle = \sum_j p_j v_j \tag{5.42}$$

is formally equivalent to (5.34).

A probability distribution on a given sample space can only be used to calculate averages of random variables defined on this sample space; it cannot be used, at least directly, to calculate averages of random variables which are defined on some *other* sample space. While this is rather obvious in ordinary probability theory, its quantum counterpart is sometimes overlooked. In particular, the probability distribution associated with $\{P_j\}$ cannot be used to calculate the average of a self-adjoint operator S whose natural sample space is a decomposition $\{Q_k\}$ incompatible with $\{P_j\}$. Instead one must use a probability distribution for the decomposition $\{Q_k\}$.

An alternative way of writing (5.41) is the following. The positive operator

$$\rho = \sum_j p_j P_j / \mathrm{Tr}(P_j) \tag{5.43}$$

has a trace equal to 1, so it is a density matrix, as defined in Sec. 3.9. It is easy to show that

$$\langle V \rangle = \mathrm{Tr}(\rho V) \tag{5.44}$$

by applying the orthogonality conditions (5.10) to the product ρV. Note that ρ and V commute with each other. The formula (5.44) is sometimes used in situations in which ρ and V do *not* commute with each other. In such a case ρ is functioning as a pre-probability, as will be explained in Ch. 15.

6

Composite systems and tensor products

6.1 Introduction

A *composite* system is one involving more than one particle, or a particle with internal degrees of freedom in addition to its center of mass. In classical mechanics the phase space of a composite system is a *Cartesian product* of the phase spaces of its constituents. The Cartesian product of two sets A and B is the set of (ordered) pairs $\{(a, b)\}$, where a is any element of A and b is any element of B. For three sets A, B, and C the Cartesian product consists of triples $\{(a, b, c)\}$, and so forth. Consider two classical particles in one dimension, with phase spaces x_1, p_1 and x_2, p_2. The phase space for the composite system consists of pairs of points from the two phase spaces, that is, it is a collection of quadruples of the form x_1, p_1, x_2, p_2, which can equally well be written in the order x_1, x_2, p_1, p_2. This is formally the same as the phase space of a single particle in two dimensions, a collection of quadruples x, y, p_x, p_y. Similarly, the six-dimensional phase space of a particle in three dimensions is formally the same as that of three one-dimensional particles.

In quantum theory the analog of a Cartesian product of classical phase spaces is a *tensor product* of Hilbert spaces. A particle in three dimensions has a Hilbert space which is the tensor product of three spaces, each corresponding to motion in one dimension. The Hilbert space for two particles, as long as they are not identical, is the tensor product of the two Hilbert spaces for the separate particles. The Hilbert space for a particle with spin is the tensor product of the Hilbert space of wave functions for the center of mass, appropriate for a particle without spin, with the spin space, which is two-dimensional for a spin-half particle.

Not only are tensor products used in quantum theory for describing a composite system at a single time, they are also very useful for describing the time development of a quantum system, as we shall see in Ch. 8. Hence any serious student of quantum mechanics needs to become familiar with the basic facts about tensor products, and the corresponding notation, which is summarized in Sec. 6.2.

Special rules apply to the tensor product spaces used for identical quantum particles. For identical bosons one uses the symmetrical subspace of the Hilbert space formed by taking a tensor product of the spaces for the individual particles, while for identical fermions one uses the antisymmetrical subspace. The basic procedure for constructing these subspaces is discussed in various introductory and more advanced textbooks (see references in the Bibliography), but the idea behind it is probably easiest to understand in the context of quantum field theory, which lies outside the scope of this book. While we shall not discuss the subject further, it is worth pointing out that there are a number of circumstances in which the fact that the particles are identical can be ignored — that is, one makes no significant error by treating them as distinguishable — because they are found in different locations or in different environments. For example, identical nuclei in a solid can be regarded as distinguishable as long as it is a reasonable physical approximation to assume that they are approximately localized, e.g., found in a particular unit cell, or in a particular part of a unit cell. In such cases one can construct the tensor product spaces in a straightforward manner using the principles described below.

6.2 Definition of tensor products

Given two Hilbert spaces $\mathcal{A}$ and $\mathcal{B}$, their tensor product $\mathcal{A} \otimes \mathcal{B}$ can be defined in the following way, where we assume, for simplicity, that the spaces are finite-dimensional. Let $\{|a_j\rangle : j = 1, 2, \ldots m\}$ be an orthonormal basis for the m-dimensional space $\mathcal{A}$, and $\{|b_p\rangle : p = 1, 2, \ldots n\}$ an orthonormal basis for the n-dimensional space $\mathcal{B}$, so that

$$\langle a_j | a_k \rangle = \delta_{jk}, \quad \langle b_p | b_q \rangle = \delta_{pq}. \tag{6.1}$$

Then the collection of mn elements

$$|a_j\rangle \otimes |b_p\rangle \tag{6.2}$$

forms an orthonormal basis of the tensor product $\mathcal{A} \otimes \mathcal{B}$, which is the set of all linear combinations of the form

$$|\psi\rangle = \sum_j \sum_p \gamma_{jp} \left(|a_j\rangle \otimes |b_p\rangle \right), \tag{6.3}$$

where the γ_{jp} are complex numbers.

Given kets

$$|a\rangle = \sum_j \alpha_j |a_j\rangle, \quad |b\rangle = \sum_p \beta_p |b_p\rangle \tag{6.4}$$

in $\mathcal{A}$ and $\mathcal{B}$, respectively, their tensor product is defined as

$$|a\rangle \otimes |b\rangle = \sum_j \sum_p \alpha_j \beta_p (|a_j\rangle \otimes |b_p\rangle), \tag{6.5}$$

which is of the form (6.3) with

$$\gamma_{jp} = \alpha_j \beta_p. \tag{6.6}$$

The parentheses in (6.3) and (6.5) are not really essential, since $(\alpha|a\rangle) \otimes |b\rangle$ is equal to $\alpha(|a\rangle \otimes |b\rangle)$, and we shall henceforth omit them when this gives rise to no ambiguities. The definition (6.5) implies that the tensor product operation $\otimes$ is distributive:

$$\begin{aligned}|a\rangle \otimes (\beta'|b'\rangle + \beta''|b''\rangle) &= \beta'|a\rangle \otimes |b'\rangle + \beta''|a\rangle \otimes |b''\rangle, \\ (\alpha'|a'\rangle + \alpha''|a''\rangle) \otimes |b\rangle &= \alpha'|a'\rangle \otimes |b\rangle + \alpha''|a''\rangle \otimes |b\rangle.\end{aligned} \tag{6.7}$$

An element of $\mathcal{A} \otimes \mathcal{B}$ which can be written in the form $|a\rangle \otimes |b\rangle$ is called a *product* state, and states which are not product states are said to be *entangled*. When several coefficients in (6.3) are nonzero, it may not be readily apparent whether the corresponding state is a product state or entangled, that is, whether or not γ_{jp} can be written in the form (6.6). For example,

$$1.0|a_1\rangle \otimes |b_1\rangle + 0.5|a_1\rangle \otimes |b_2\rangle - 1.0|a_2\rangle \otimes |b_1\rangle - 0.5|a_2\rangle \otimes |b_2\rangle \tag{6.8}$$

is a product state $(|a_1\rangle - |a_2\rangle) \otimes (|b_1\rangle + 0.5|b_2\rangle)$, whereas changing the sign of the last coefficient yields an entangled state:

$$1.0|a_1\rangle \otimes |b_1\rangle + 0.5|a_1\rangle \otimes |b_2\rangle - 1.0|a_2\rangle \otimes |b_1\rangle + 0.5|a_2\rangle \otimes |b_2\rangle. \tag{6.9}$$

The linear functional or bra vector corresponding to the product state $|a\rangle \otimes |b\rangle$ is written as

$$(|a\rangle \otimes |b\rangle)^\dagger = \langle a| \otimes \langle b|, \tag{6.10}$$

where the $\mathcal{A} \otimes \mathcal{B}$ order of the factors on either side of $\otimes$ does not change when the dagger operation is applied. The result for a general linear combination (6.3) follows from (6.10) and the antilinearity of the dagger operation:

$$\langle \psi| = (|\psi\rangle)^\dagger = \sum_{jp} \gamma_{jp}^* \langle a_j| \otimes \langle b_p|. \tag{6.11}$$

Consistent with these formulas, the inner product of two product states is given by

$$(|a\rangle \otimes |b\rangle)^\dagger (|a'\rangle \otimes |b'\rangle) = \langle a|a'\rangle \cdot \langle b|b'\rangle, \tag{6.12}$$

and of a general state $|\psi\rangle$, (6.3), with another state

$$|\psi'\rangle = \sum_{jp} \gamma'_{jp} |a_j\rangle \otimes |b_p\rangle, \tag{6.13}$$

by the expression

$$\langle \psi | \psi' \rangle = \sum_{jp} \gamma_{jp}^* \gamma_{jp}'. \tag{6.14}$$

Because the definition of a tensor product given above employs specific orthonormal bases for $\mathcal{A}$ and $\mathcal{B}$, one might suppose that the space $\mathcal{A} \otimes \mathcal{B}$ somehow depends on the choice of these bases. But in fact it does not, as can be seen by considering alternative bases $\{|a_k'\rangle\}$ and $\{|b_q'\rangle\}$. The kets in the new bases can be written as linear combinations of the original kets,

$$|a_k'\rangle = \sum_j \langle a_j | a_k' \rangle \cdot |a_j\rangle, \quad |b_q'\rangle = \sum_p \langle b_p | b_q' \rangle \cdot |b_p\rangle, \tag{6.15}$$

and (6.5) then allows $|a_k'\rangle \otimes |b_q'\rangle$ to be written as a linear combination of the kets $|a_j\rangle \otimes |b_p\rangle$. Hence the use of different bases for $\mathcal{A}$ or $\mathcal{B}$ leads to the same tensor product space $\mathcal{A} \otimes \mathcal{B}$, and it is easily checked that the property of being a product state or an entangled state does not depend upon the choice of bases.

Just as for any other Hilbert space, it is possible to choose an orthonormal basis of $\mathcal{A} \otimes \mathcal{B}$ in a large number of different ways. We shall refer to a basis of the type used in the original definition, $\{|a_j\rangle \otimes |b_p\rangle\}$, as a *product of bases*. An orthonormal basis of $\mathcal{A} \otimes \mathcal{B}$ may consist entirely of product states without being a product of bases; see the example in (6.22). Or it might consist entirely of entangled states, or of some entangled states and some product states.

Physicists often omit the $\otimes$ and write $|a\rangle \otimes |b\rangle$ in the form $|a\rangle |b\rangle$, or more compactly as $|a, b\rangle$, or even as $|ab\rangle$. Any of these notations is perfectly adequate when it is clear from the context that a tensor product is involved. We shall often use one of the more compact notations, and occasionally insert the $\otimes$ symbol for the sake of clarity, or for emphasis. Note that while a double label inside a ket, as in $|a, b\rangle$, often indicates a tensor product, this is not always the case; for example, the double label $|l, m\rangle$ for orbital angular momentum kets does not signify a tensor product.

The tensor product of three or more Hilbert spaces can be obtained by an obvious generalization of the ideas given above. In particular, the tensor product $\mathcal{A} \otimes \mathcal{B} \otimes \mathcal{C}$ of three Hilbert spaces $\mathcal{A}, \mathcal{B}, \mathcal{C}$, consists of all linear combinations of states of the form

$$|a_j\rangle \otimes |b_p\rangle \otimes |c_s\rangle, \tag{6.16}$$

using the bases introduced earlier, together with $\{|c_s\rangle: s = 1, 2, \dots\}$, an orthonormal basis for $\mathcal{C}$. One can think of $\mathcal{A} \otimes \mathcal{B} \otimes \mathcal{C}$ as obtained by first forming the tensor product of two of the spaces, and then taking the tensor product of this space with the third. The final result does not depend upon which spaces form the initial pairing:

$$\mathcal{A} \otimes \mathcal{B} \otimes \mathcal{C} = (\mathcal{A} \otimes \mathcal{B}) \otimes \mathcal{C} = \mathcal{A} \otimes (\mathcal{B} \otimes \mathcal{C}) = (\mathcal{A} \otimes \mathcal{C}) \otimes \mathcal{B}. \tag{6.17}$$

In what follows we shall usually focus on tensor products of two spaces, but for the most part the discussion can be generalized in an obvious way to tensor products of three or more spaces. Where this is not the case it will be pointed out explicitly.

Given *any* state $|\psi\rangle$ in $\mathcal{A} \otimes \mathcal{B}$, it is always possible to find particular orthonormal bases $\{|\hat{a}_j\rangle\}$ for $\mathcal{A}$ and $\{|\hat{b}_p\rangle\}$ for $\mathcal{B}$ such that $|\psi\rangle$ takes the form

$$|\psi\rangle = \sum_j \lambda_j |\hat{a}_j\rangle \otimes |\hat{b}_j\rangle. \tag{6.18}$$

Here the λ_j are complex numbers, but by choosing appropriate phases for the basis states, one can make them real and nonnegative. The summation index j takes values between 1 and the minimum of the dimensions of $\mathcal{A}$ and $\mathcal{B}$. The result (6.18) is known as the *Schmidt decomposition* of $|\psi\rangle$; it is also referred to as the *biorthogonal* or *polar expansion* of $|\psi\rangle$. It does *not* generalize, at least in any simple way, to a tensor product of three or more Hilbert spaces.

Given an arbitrary Hilbert space $\mathcal{H}$ of dimension mn, with m and n integers greater than 1, it is possible to "decompose" it into a tensor product $\mathcal{A} \otimes \mathcal{B}$, with m the dimension of $\mathcal{A}$ and n the dimension of $\mathcal{B}$; indeed, this can be done in many different ways. Let $\{|h_l\rangle\}$ be any orthonormal basis of $\mathcal{H}$, with $l = 1, 2, \dots mn$. Rather than use a single label for the kets, we can associate each l with a pair j, p, where j takes values between 1 and m, and p values between 1 and n. Any association will do, as long as it is unambiguous (one-to-one). Let $\{|h_{jp}\rangle\}$ denote precisely the same basis using this new labeling. Now write

$$|h_{jp}\rangle = |a_j\rangle \otimes |b_p\rangle, \tag{6.19}$$

where the $\{|a_j\rangle\}$ for j between 1 and m are defined to be an orthonormal basis of a Hilbert space $\mathcal{A}$, and the $\{|b_p\rangle\}$ for p between 1 and n the orthonormal basis of a Hilbert space $\mathcal{B}$. By this process we have turned $\mathcal{H}$ into a tensor product $\mathcal{A} \otimes \mathcal{B}$, or it might be better to say that we have imposed a tensor product structure $\mathcal{A} \otimes \mathcal{B}$ upon the Hilbert space $\mathcal{H}$. In the same way, if the dimension of $\mathcal{H}$ is the product of three or more integers greater than 1, it can always be thought of as a tensor product of three or more spaces, and the decomposition can be carried out in many different ways.

6.3 Examples of composite quantum systems

Figure 6.1(a) shows a toy model involving two particles. The first particle can be at any one of the $M = 6$ sites indicated by circles, and the second particle can be at one of the two sites indicated by squares. The states $|m\rangle$ for m between 0 and 5 span the Hilbert space $\mathcal{M}$ for the first particle, and $|n\rangle$ for $n = 0, 1$ the Hilbert space $\mathcal{N}$ for the second particle. The tensor product space $\mathcal{M} \otimes \mathcal{N}$ is $6 \times 2 = 12$ dimensional, with basis states $|m\rangle \otimes |n\rangle = |m, n\rangle$. (In Sec. 7.4 we shall put this arrangement to good use: the second particle will be employed as a detector to detect the passage of the first particle.) One must carefully distinguish the case of *two* particles, one located on the circles and one on the squares in Fig. 6.1(a), from that of a *single* particle which can be located on either the circles or the squares. The former has a Hilbert space of dimension 12, and the latter a Hilbert space of dimension $6 + 2 = 8$.

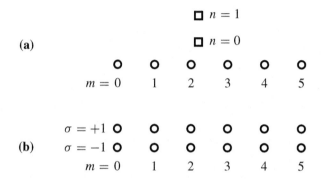

Fig. 6.1. Toy model for: (a) two particles, one located on the circles and one on the squares; (b) a particle with an internal degree of freedom.

A second toy model, Fig. 6.1(b), consists of a *single* particle with an internal degree of freedom represented by a "spin" variable which can take on two possible values. The center of mass of the particle can be at any one of six sites corresponding to a six-dimensional Hilbert space $\mathcal{M}$, whereas the spin degree of freedom is represented by a two-dimensional Hilbert space $\mathcal{S}$. The basis kets of $\mathcal{M} \otimes \mathcal{S}$ have the form $|m, \sigma\rangle$, with $\sigma = \pm 1$. The figure shows two circles at each site, one corresponding to $\sigma = +1$ ("spin up"), and the other to $\sigma = -1$ ("spin down"), so one can think of each basis state as the particle being "at" one of the circles. A general element $|\psi\rangle$ of the Hilbert space $\mathcal{M} \otimes \mathcal{S}$ is a linear combination of the basis kets, so it can be written in the form

$$|\psi\rangle = \sum_m \sum_\sigma \psi(m, \sigma)|m, \sigma\rangle, \tag{6.20}$$

where the complex coefficients $\psi(m, \sigma)$ form a toy wave function; this is simply an alternative way of writing the complex coefficients γ_{jp} in (6.3). The toy wave function $\psi(m, \sigma)$ can be thought of as a discrete analog of the wave function $\psi(\mathbf{r}, \sigma)$ used to describe a spin-half particle in three dimensions. Just as in the toy model, the Hilbert space to which $\psi(\mathbf{r}, \sigma)$ belongs is a tensor product of the space of wave functions $\psi(\mathbf{r})$, appropriate for a spinless quantum particle, with the two-dimensional spin space.

Consider two spin-half particles a and b, such as an electron and a proton, and ignore their center of mass degrees of freedom. The tensor product $\mathcal{H}$ of the two 2-dimensional spin spaces is a four-dimensional space spanned by the orthonormal basis

$$|z_a^+\rangle \otimes |z_b^+\rangle, \quad |z_a^+\rangle \otimes |z_b^-\rangle, \quad |z_a^-\rangle \otimes |z_b^+\rangle, \quad |z_a^-\rangle \otimes |z_b^-\rangle \tag{6.21}$$

in the notation of Sec. 4.2. This is a product of bases in the terminology of Sec. 6.2. By contrast, the basis

$$|z_a^+\rangle \otimes |z_b^+\rangle, \quad |z_a^+\rangle \otimes |z_b^-\rangle, \quad |z_a^-\rangle \otimes |x_b^+\rangle, \quad |z_a^-\rangle \otimes |x_b^-\rangle, \tag{6.22}$$

even though it consists of product states, is *not* a product of bases, because one basis for $\mathcal{B}$ is employed along with $|z_a^+\rangle$, and a different basis along with $|z_a^-\rangle$. Still other bases are possible, including cases in which some or all of the basis vectors are entangled states.

The spin space for three spin-half particles a, b, and c is an eight-dimensional tensor product space, and the state $|z_a^+\rangle \otimes |z_b^+\rangle \otimes |z_c^+\rangle$ along with the seven other states in which some of the pluses are replaced by minuses forms a product basis. For N spins, the tensor product space is of dimension 2^N.

6.4 Product operators

Since $\mathcal{A} \otimes \mathcal{B}$ is a Hilbert space, operators on it obey all the usual rules, Sec. 3.3. What we are interested in is how these operators are related to the tensor product structure, and, in particular, to operators on the separate factor spaces $\mathcal{A}$ and $\mathcal{B}$. In this section we discuss the special case of product operators, while general operators are considered in the next section. The considerations which follow can be generalized in an obvious way to a tensor product of three or more spaces.

If A is an operator on $\mathcal{A}$ and B an operator on $\mathcal{B}$, the *(tensor) product operator* $A \otimes B$ acting on a product state $|a\rangle \otimes |b\rangle$ yields another product state:

$$(A \otimes B)(|a\rangle \otimes |b\rangle) = (A|a\rangle) \otimes (B|b\rangle). \tag{6.23}$$

Since $A \otimes B$ is by definition a linear operator on $\mathcal{A} \otimes \mathcal{B}$, one can use (6.23) to define its action on a general element $|\psi\rangle$, (6.3), of $A \otimes B$:

$$(A \otimes B)\left[\sum_{jp} \gamma_{jp}(|a_j\rangle \otimes |b_p\rangle)\right] = \sum_{jp} \gamma_{jp}(A|a_j\rangle \otimes B|b_p\rangle). \tag{6.24}$$

The tensor product of two operators which are themselves sums of other operators can be written as a sum of product operators using the usual distributive rules. Thus:

$$(\alpha A + \alpha' A') \otimes B = \alpha(A \otimes B) + \alpha'(A' \otimes B),$$
$$A \otimes (\beta B + \beta' B') = \beta(A \otimes B) + \beta'(A \otimes B'). \tag{6.25}$$

The parentheses on the right side are not essential, as there is no ambiguity when $\alpha(A \otimes B) = (\alpha A) \otimes B$ is written as $\alpha A \otimes B$.

If $|\psi\rangle = |a\rangle \otimes |b\rangle$ and $|\phi\rangle = |a'\rangle \otimes |b'\rangle$ are both product states, the dyad $|\psi\rangle\langle\phi|$ is a product operator:

$$\bigl(|a\rangle \otimes |b\rangle\bigr)\bigl(\langle a'| \otimes \langle b'|\bigr) = \bigl(|a\rangle\langle a'|\bigr) \otimes \bigl(|b\rangle\langle b'|\bigr). \tag{6.26}$$

Notice how the terms on the left are rearranged in order to arrive at the expression on the right. One can omit the parentheses on the right side, since $|a\rangle\langle a'| \otimes |b\rangle\langle b'|$ is unambiguous.

The adjoint of a product operator is the tensor product of the adjoints *in the same order* relative to the symbol $\otimes$:

$$(A \otimes B)^\dagger = A^\dagger \otimes B^\dagger. \tag{6.27}$$

Of course, if the operators on $\mathcal{A}$ and $\mathcal{B}$ are themselves products, one must reverse their order when taking the adjoint:

$$(A_1 A_2 \otimes B_1 B_2 B_3)^\dagger = A_2^\dagger A_1^\dagger \otimes B_3^\dagger B_2^\dagger B_1^\dagger. \tag{6.28}$$

The ordinary operator product of two tensor product operators is given by

$$(A \otimes B) \cdot (A' \otimes B') = AA' \otimes BB', \tag{6.29}$$

where it is important that the order of the operators be preserved: A is to the left of A' on both sides of the equation, and likewise B is to the left of B'. An operator product of sums of tensor products can be worked out using the usual distributive law, e.g.,

$$(A \otimes B) \cdot (A' \otimes B' + A'' \otimes B'') = AA' \otimes BB' + AA'' \otimes BB''. \tag{6.30}$$

An operator A on $\mathcal{A}$ can be *extended* to an operator $A \otimes I_B$ on $\mathcal{A} \otimes \mathcal{B}$, where I_B is the identity on $\mathcal{B}$. It is customary to use the same symbol, A, for both the original operator and its extension; indeed, in practice it would often be quite awkward to do anything else. Similarly, B is used to denote either an operator on $\mathcal{B}$, or its extension $I_A \otimes B$. Consider, for example, two spin-half particles, an electron and a proton. It is convenient to use the symbol S_{ez} for the operator corresponding to the z-component of the spin of the electron, whether one is thinking of the two-dimensional Hilbert space associated with the electron spin by itself, the four-dimensional spin space for both particles, the infinite-dimensional space of electron space-and-spin wave functions, or the space needed to describe the spin and position of both the electron and the proton.

Using the same symbol for an operator and its extension normally causes no confusion, since the space to which the operator is applied will be evident from the context. However, it is sometimes useful to employ the longer notation for clarity or emphasis, in which case one can (usually) omit the subscript from the identity operator: in the operator $A \otimes I$ it is clear that I is the identity on $\mathcal{B}$. Note that (6.29) allows one to write

$$A \otimes B = (A \otimes I) \cdot (I \otimes B) = (I \otimes B) \cdot (A \otimes I), \qquad (6.31)$$

and hence if we use A for $A \otimes I$ and B for $I \otimes B$, $A \otimes B$ can be written as the operator product AB or BA. This is perfectly correct and unambiguous as long as it is clear from the context that A is an operator on $\mathcal{A}$ and B an operator on $\mathcal{B}$. However, if $\mathcal{A}$ and $\mathcal{B}$ are identical (isomorphic) spaces, and B denotes an operator which also makes sense on $\mathcal{A}$, then AB could be interpreted as the ordinary product of two operators on $\mathcal{A}$ (or on $\mathcal{B}$), and to avoid confusion it is best to use the unabbreviated $A \otimes B$.

6.5 General operators, matrix elements, partial traces

Any operator on a Hilbert space is uniquely specified by its matrix elements in some orthonormal basis, Sec. 3.6, and thus a general operator D on $\mathcal{A} \otimes \mathcal{B}$ is determined by its matrix elements in the orthonormal basis (6.2). These can be written in a variety of different ways:

$$\langle jp|D|kq \rangle = \langle j, p|D|k, q \rangle = \langle a_j b_p|D|a_k b_q \rangle$$
$$= \big(\langle a_j| \otimes \langle b_p| \big) D \big(|a_k\rangle \otimes |b_q\rangle \big). \qquad (6.32)$$

The most compact notation is on the left, but it is not always the clearest. Note that it corresponds to writing bras and kets with a "double label", and this needs to be taken into account in standard formulas, such as

$$I = I \otimes I = \sum_j \sum_p |jp\rangle\langle jp| \qquad (6.33)$$

and

$$\mathrm{Tr}(D) = \sum_j \sum_p \langle jp|D|jp \rangle, \qquad (6.34)$$

which correspond to (3.54) and (3.79), respectively.

Any operator can be written as a sum of dyads multiplied by appropriate matrix elements, (3.67), which allows us to write

$$D = \sum_{jk} \sum_{pq} \langle a_j b_p | D | a_k b_q \rangle \big(|a_j\rangle \langle a_k| \otimes |b_p\rangle \langle b_q| \big), \qquad (6.35)$$

where we have used (6.26) to rewrite the dyads as product operators. This shows, incidentally, that while not all operators on $\mathcal{A} \otimes \mathcal{B}$ are product operators, any operator can be written as a sum of product operators. The adjoint of D is then given by the formula

$$D^\dagger = \sum_{jk} \sum_{pq} \langle a_j b_p | D | a_k b_q \rangle^* \big(|a_k\rangle \langle a_j| \otimes |b_q\rangle \langle b_p| \big), \qquad (6.36)$$

using (6.27) and the fact the dagger operation is antilinear. If one replaces $\langle a_j b_p | D | a_k b_q \rangle^*$ by $\langle a_k b_q | D^\dagger | a_j b_p \rangle$, see (3.64), (6.36) is simply (6.35) with D replaced by $D^\dagger$ on both sides, aside from dummy summation indices.

The matrix elements of a product operator using the basis (6.2) are the products of the matrix elements of the factors:

$$\langle a_j b_p | A \otimes B | a_k b_q \rangle = \langle a_j | A | a_k \rangle \cdot \langle b_p | B | b_q \rangle. \qquad (6.37)$$

From this it follows that the trace of a product operator is the product of the traces of its factors:

$$\mathrm{Tr}[A \otimes B] = \sum_j \langle a_j | A | a_j \rangle \cdot \sum_p \langle b_p | B | b_p \rangle = \mathrm{Tr}_{\mathcal{A}}[A] \cdot \mathrm{Tr}_{\mathcal{B}}[B]. \qquad (6.38)$$

Here the subscripts on $\mathrm{Tr}_{\mathcal{A}}$ and $\mathrm{Tr}_{\mathcal{B}}$ indicate traces over the spaces $\mathcal{A}$ and $\mathcal{B}$, respectively, while the trace over $\mathcal{A} \otimes \mathcal{B}$ is written without a subscript, though one could denote it by $\mathrm{Tr}_{\mathcal{A}\mathcal{B}}$ or $\mathrm{Tr}_{\mathcal{A}\otimes\mathcal{B}}$. Thus if $\mathcal{A}$ and $\mathcal{B}$ are spaces of dimension m and n, $\mathrm{Tr}_{\mathcal{A}}[I] = m$, $\mathrm{Tr}_{\mathcal{B}}[I] = n$, and $\mathrm{Tr}[I] = mn$.

Given an operator D on $\mathcal{A} \otimes \mathcal{B}$, and two basis states $|b_p\rangle$ and $|b_q\rangle$ of $\mathcal{B}$, one can define $\langle b_p | D | b_q \rangle$ to be the (unique) operator on $\mathcal{A}$ which has matrix elements

$$\langle a_j | \big(\langle b_p | D | b_q \rangle \big) | a_k \rangle = \langle a_j b_p | D | a_k b_q \rangle. \qquad (6.39)$$

The *partial trace* over $\mathcal{B}$ of the operator D is defined to be a sum of operators of this type:

$$D_{\mathcal{A}} = \mathrm{Tr}_{\mathcal{B}}[D] = \sum_p \langle b_p | D | b_p \rangle. \qquad (6.40)$$

Alternatively, one can define $D_{\mathcal{A}}$ to be the operator on $\mathcal{A}$ with matrix elements

$$\langle a_j | D_{\mathcal{A}} | a_k \rangle = \sum_p \langle a_j b_p | D | a_k b_p \rangle. \qquad (6.41)$$

Note that the $\mathcal{B}$ state labels are the same on both sides of the matrix elements on the right sides of (6.40) and (6.41), while those for the $\mathcal{A}$ states are (in general) different. Even though we have employed a specific orthonormal basis of $\mathcal{B}$ in (6.40) and (6.41), it is not hard to show that the partial trace D_A is independent of this basis; that is, one obtains precisely the same operator if a different orthonormal basis $\{|b_p'\rangle\}$ is used in place of $\{|b_p\rangle\}$.

If D is written in the form (6.35), its partial trace is

$$D_A = \mathrm{Tr}_{\mathcal{B}}[D] = \sum_{jk} d_{jk} |a_j\rangle\langle a_k|, \tag{6.42}$$

where

$$d_{jk} = \sum_{p} \langle a_j b_p | D | a_k b_p \rangle, \tag{6.43}$$

since the trace over $\mathcal{B}$ of $|b_p\rangle\langle b_q|$ is $\langle b_p | b_q \rangle = \delta_{pq}$. In the special case of a product operator $A \otimes B$, the partial trace over $\mathcal{B}$ yields an operator

$$\mathrm{Tr}_{\mathcal{B}}[A \otimes B] = (\mathrm{Tr}_{\mathcal{B}}[B])\, A \tag{6.44}$$

proportional to A.

In a similar way, the partial trace of an operator D on $\mathcal{A} \otimes \mathcal{B}$ over $\mathcal{A}$ yields an operator

$$D_{\mathcal{B}} = \mathrm{Tr}_A[D] \tag{6.45}$$

acting on the space $\mathcal{B}$, with matrix elements

$$\langle b_p | D_{\mathcal{B}} | b_q \rangle = \sum_{j} \langle a_j b_p | D | a_j b_q \rangle. \tag{6.46}$$

Note that the full trace of D over $\mathcal{A} \otimes \mathcal{B}$ can be written as a trace of either of its partial traces:

$$\mathrm{Tr}[D] = \mathrm{Tr}_A[D_A] = \mathrm{Tr}_{\mathcal{B}}[D_{\mathcal{B}}]. \tag{6.47}$$

All of the above can be generalized to a tensor product of three or more spaces in an obvious way. For example, if E is an operator on $\mathcal{A} \otimes \mathcal{B} \otimes \mathcal{C}$, its matrix elements using the orthonormal product of bases in (6.16) are of the form

$$\langle jpr | E | kqs \rangle = \langle a_j b_p c_r | E | a_k b_q c_s \rangle. \tag{6.48}$$

The partial trace of E over $\mathcal{C}$ is an operator on $\mathcal{A} \otimes \mathcal{B}$, while its partial trace over $\mathcal{B} \otimes \mathcal{C}$ is an operator on $\mathcal{A}$, etc.

6.6 Product properties and product of sample spaces

Let A and B be projectors representing properties of two physical systems $\mathcal{A}$ and $\mathcal{B}$, respectively. It is easy to show that

$$P = A \otimes B = (A \otimes I) \cdot (I \otimes B) \tag{6.49}$$

is a projector, which therefore represents some property on the tensor product space $\mathcal{A} \otimes \mathcal{B}$ of the combined system. (Note that if A projects onto a pure state $|a\rangle$ and B onto a pure state $|b\rangle$, then P projects onto the pure state $|a\rangle \otimes |b\rangle$.) The physical significance of P is that $\mathcal{A}$ has the property A and $\mathcal{B}$ has the property B. In particular, the projector $A \otimes I$ has the significance that $\mathcal{A}$ has the property A without reference to the system $\mathcal{B}$, since the identity $I = I_B$ operator in $A \otimes I$ is the property which is always true for $\mathcal{B}$, and thus tells us nothing whatsoever about $\mathcal{B}$. Similarly, $I \otimes B$ means that $\mathcal{B}$ has the property B without reference to the system $\mathcal{A}$. The product of $A \otimes I$ with $I \otimes B$ — note that the two operators commute with each other — represents the conjunction of the properties of the separate subsystems, in agreement with the discussion in Sec. 4.5, and consistent with the interpretation of P given previously. As an example, consider two spin-half particles a and b. The projector $[z_a^+] \otimes [x_b^-]$ means that $S_{az} = +1/2$ for particle a and $S_{bx} = -1/2$ for particle b.

The interpretation of projectors on $\mathcal{A} \otimes \mathcal{B}$ which are *not* products of projectors is more subtle. Consider, for example, the entangled state

$$|\psi\rangle = \big(|z_a^+\rangle|z_b^-\rangle - |z_a^-\rangle|z_b^+\rangle\big)/\sqrt{2} \tag{6.50}$$

of two spin-half particles, and let $[\psi]$ be the corresponding dyad projector. Since $[\psi]$ projects onto a subspace of $\mathcal{A} \otimes \mathcal{B}$, it represents some property of the combined system. However, if we ask what this property means in terms of the a spin by itself, we run into the difficulty that the only projectors on the two-dimensional spin space $\mathcal{A}$ which commute with $[\psi]$ are 0 and the identity I. Consequently, any "interesting" property of $\mathcal{A}$, something of the form $S_{aw} = +1/2$ for some direction w, is incompatible with $[\psi]$. Thus $[\psi]$ cannot be interpreted as meaning that the a spin has some property, and likewise it cannot mean that the b spin has some property.

The same conclusion applies to *any* entangled state of two spin-half particles. The situation is not quite as bad if one goes to higher-dimensional spaces. For example, the projector $[\phi]$ corresponding to the entangled state

$$|\phi\rangle = \big(|1\rangle \otimes |0\rangle + |2\rangle \otimes |1\rangle\big)/\sqrt{2} \tag{6.51}$$

of the toy model with two particles shown in Fig. 6.1(a) commutes with the projector

$$([1] + [2]) \otimes I \tag{6.52}$$

for the first particle, and thus if the combined system is described by $[\phi]$, one can say that the first particle is not outside the interval containing the sites $m = 1$ and $m = 2$, although it cannot be assigned a location at one or the other of these sites. However, one can say nothing interesting about the second particle.

A *product of sample spaces* or *product of decompositions* is a collection of projectors $\{A_j \otimes B_p\}$ which sum to the identity

$$I = \sum_{jp} A_j \otimes B_p \tag{6.53}$$

of $\mathcal{A} \otimes \mathcal{B}$, where $\{A_j\}$ is decomposition of the identity for $\mathcal{A}$, and $\{B_p\}$ a decomposition of the identity for $\mathcal{B}$. Note that the event algebra corresponding to (6.53) contains all projectors of the form $\{A_j \otimes I\}$ or $\{I \otimes B_p\}$, so these properties of the individual systems make sense in a description of the composite system based upon this decomposition. A particular example of a product of sample spaces is the collection of dyads corresponding to the product of bases in (6.2):

$$I = \sum_{jp} |a_j\rangle\langle a_j| \otimes |b_p\rangle\langle b_p|. \tag{6.54}$$

A decomposition of the identity can consist of products of projectors without being a product of sample spaces. An example is provided by the four projectors

$$[z_a^+] \otimes [z_b^+], \quad [z_a^+] \otimes [z_b^-], \quad [z_a^-] \otimes [x_b^+], \quad [z_a^-] \otimes [x_b^-] \tag{6.55}$$

corresponding to the states in the basis (6.22) for two spin-half particles. (As noted earlier, (6.22) is not a product of bases.) The event algebra generated by (6.55) contains the projectors $[z_a^+] \otimes I$ and $[z_a^-] \otimes I$, but it does not contain the projectors $I \otimes [z_b^+]$, $I \otimes [z_b^-]$, $I \otimes [x_b^+]$ or $I \otimes [x_b^-]$. Consequently one has the odd situation that if the state $[z_a^-] \otimes [x_b^+]$, which would normally be interpreted to mean $S_{az} = -1/2$ AND $S_{bx} = +1/2$, is a correct description of the system, then using the event algebra based upon (6.55), one can infer that $S_{az} = -1/2$ for spin a, but one cannot infer that $S_{bx} = +1/2$ is a property of spin b by itself, independent of any reference to spin a. Further discussion of this peculiar state of affairs, which arises when one is dealing with *dependent* or *contextual* properties, will be found in Ch. 14.

7

Unitary dynamics

7.1 The Schrödinger equation

The equations of motion of classical Hamiltonian dynamics are of the form

$$\frac{dx_i}{dt} = \frac{\partial H}{\partial p_i}, \quad \frac{dp_i}{dt} = -\frac{\partial H}{\partial x_i}, \tag{7.1}$$

where x_1, x_2, etc. are the (generalized) coordinates, p_1, p_2, etc. their conjugate momenta, and $H(x_1, p_1, x_2, p_2, \dots)$ is the Hamiltonian function on the classical phase space.

In the case of a particle moving in one dimension, there is only a single coordinate x and a single momentum p, and the Hamiltonian is the total energy

$$H = p^2/2m + V(x), \tag{7.2}$$

with $V(x)$ the potential energy. The two equations of motion are then:

$$dx/dt = p/m, \quad dp/dt = -dV/dx. \tag{7.3}$$

For a harmonic oscillator $V(x)$ is $\frac{1}{2}Kx^2$, and the general solution of (7.3) is given in (2.1), where $\omega = \sqrt{K/m}$.

The set of equations (7.1) is deterministic in that there is a unique trajectory or orbit $\gamma(t)$ in the phase space as a function of time which passes through γ_0 at $t = 0$. Of course, the orbit is also determined by giving the point in phase space through which it passes at some time other than $t = 0$. The orbit for a harmonic oscillator is an ellipse in the phase plane; see Fig. 2.1 on page 12.

The quantum analog of (7.1) is *Schrödinger's equation*, which in Dirac notation can be written as

$$i\hbar \frac{d}{dt} |\psi_t\rangle = H |\psi_t\rangle, \tag{7.4}$$

where H is the quantum Hamiltonian for the system, a Hermitian operator which may itself depend upon the time. This is a linear equation, so that if $|\phi_t\rangle$ and $|\omega_t\rangle$ are any two solutions, the linear combination

$$|\chi_t\rangle = \alpha|\phi_t\rangle + \beta|\omega_t\rangle \qquad (7.5)$$

is also a solution, where α and β are arbitrary (time-independent) constants. Equation (7.4) is deterministic in the same sense as (7.1): a given $|\psi_0\rangle$ at $t = 0$ gives rise to a unique solution $|\psi_t\rangle$ for all values of t. The result is a *unitary dynamics* for the quantum system in a sense made precise in Sec. 7.3.

The Hamiltonian H in (7.4) must be an operator defined on the Hilbert space $\mathcal{H}$ of the system one is interested in. This will be true for an *isolated system*, one which does not interact with anything else — imagine something inside a completely impermeable box. It will also be true if the interaction of the system with the outside world can be approximated by an operator acting only on $\mathcal{H}$. For example, the system may be located in an external magnetic field which is effectively "classical", that is, does not have to be assigned its own quantum mechanical degrees of freedom, and thus enters the Hamiltonian H simply as a parameter.

One is sometimes interested in the dynamics of an *open* (in contrast to isolated) subsystem $\mathcal{A}$ of a composite system $\mathcal{A} \otimes \mathcal{B}$ when there is a significant interaction between $\mathcal{A}$ and $\mathcal{B}$. Of course (7.4) can be applied to the total composite system, assuming that it is isolated. However, there is no comparable equation for the subsystem $\mathcal{A}$, as it cannot, at least in general, be described by its own wave function, and its dynamical evolution is influenced by that of the other subsystem $\mathcal{B}$, often referred to as the *environment* of $\mathcal{A}$. Constructing dynamical equations for open subsystems is a topic which lies outside the scope of this book, although Ch. 15 on density matrices provides some preliminary hints on how to think about open subsystems.

For a particle in one dimension moving in a potential $V(x)$, (7.4) is equivalent to the partial differential equation

$$i\hbar\frac{\partial\psi}{\partial t} = -\frac{\hbar^2}{2m}\frac{\partial^2\psi}{\partial x^2} + V(x)\,\psi \qquad (7.6)$$

for a wave packet $\psi(x, t)$ which depends upon the time as well as the position variable x. The Hamiltonian in this case is the linear differential operator

$$H = -\frac{\hbar^2}{2m}\frac{\partial^2}{\partial x^2} + V(x). \qquad (7.7)$$

In general it is much more difficult to find solutions to (7.6) than it is to integrate (7.3). A formal solution to (7.6) for a harmonic oscillator is given in (7.23).

One way to think about (7.4) is to choose an orthonormal basis $\{|j\rangle, \ j = 1, 2, \ldots\}$ of the Hilbert space $\mathcal{H}$ which is *independent of the time t*. Then (7.4) is equivalent to a set of ordinary differential equations, one for each j:

$$i\hbar \frac{d}{dt} \langle j|\psi_t \rangle = \langle j|H|\psi_t \rangle. \tag{7.8}$$

This is somewhat less abstract than (7.4), because both $\langle j|\psi_t \rangle$ and $\langle j|H|\psi_t \rangle$ are simply complex numbers which depend upon t, and thus are complex-valued functions of the time. By writing $|\psi_t \rangle$ as a linear combination of the basis vectors,

$$|\psi_t \rangle = \sum_j |j\rangle \langle j|\psi_t \rangle = \sum_j c_j(t)|j\rangle, \tag{7.9}$$

with time-dependent coefficients $c_j(t)$, and expressing the right side of (7.8) in the form

$$\langle j|H|\psi_t \rangle = \sum_k \langle j|H|k\rangle \langle k|\psi_t \rangle, \tag{7.10}$$

one finds that the Schrödinger equation is equivalent to a collection of coupled linear differential equations

$$i\hbar \, dc_j/dt = \sum_k \langle j|H|k\rangle c_k \tag{7.11}$$

for the $c_j(t)$. The operator H, and therefore also its matrix elements, can be a function of the time, but it must be a Hermitian operator at every time, that is, for any j and k,

$$\langle j|H(t)|k\rangle = \langle k|H(t)|j\rangle^*. \tag{7.12}$$

When $\mathcal{H}$ is two-dimensional, (7.11) has the form

$$\begin{aligned} i\hbar \, dc_1/dt &= \langle 1|H|1\rangle c_1 + \langle 1|H|2\rangle c_2, \\ i\hbar \, dc_2/dt &= \langle 2|H|1\rangle c_1 + \langle 2|H|2\rangle c_2. \end{aligned} \tag{7.13}$$

These are linear equations, and if H, and thus its matrix elements, is independent of time, one can find the general solution by diagonalizing the matrix of coefficients on the right side. Let us assume that this has already been done, since we earlier made no assumptions about the basis $\{|j\rangle\}$, apart from the fact that it is independent of time. That is, assume that

$$H = E_1|1\rangle \langle 1| + E_2|2\rangle \langle 2|, \tag{7.14}$$

so that the off-diagonal terms $\langle 1|H|2\rangle$ and $\langle 2|H|1\rangle$ in (7.13) vanish, while the diagonal terms are E_1 and E_2. Then the general solution of (7.13) is of the form

$$c_1(t) = b_1 e^{-iE_1 t/\hbar}, \quad c_2(t) = b_2 e^{-iE_2 t/\hbar}, \tag{7.15}$$

where b_1 and b_2 are arbitrary (complex) constants.

The precession of the spin angular momentum of a spin-half particle placed in a constant magnetic field is an example of a two-level system with a time-independent Hamiltonian. If the magnetic field is $\mathbf{B} = (B_x, B_y, B_z)$, the Hamiltonian is

$$H = -\gamma(B_x S_x + B_y S_y + B_z S_z), \tag{7.16}$$

where the spin operators S_x, etc., are defined in the manner indicated in (5.30), and γ is the gyromagnetic ratio of the particle in suitable units. This Hamiltonian will be diagonal in the basis $|w^+\rangle$, $|w^-\rangle$, see (4.14), where w is in the direction of the magnetic field $\mathbf{B}$.

The preceding example has an obvious generalization to the case in which a time-independent Hamiltonian is diagonal in an orthonormal basis $\{|e_n\rangle\}$:

$$H = \sum_n E_n |e_n\rangle\langle e_n|. \tag{7.17}$$

Then a general solution of Schrödinger's equation has the form

$$|\psi_t\rangle = \sum_n b_n e^{-iE_n t/\hbar} |e_n\rangle, \tag{7.18}$$

where the b_n are complex constants. One can check this by evaluating the time derivative

$$i\hbar \frac{d}{dt}|\psi_t\rangle = \sum_n b_n E_n e^{-iE_n t/\hbar} |e_n\rangle, \tag{7.19}$$

and verifying that it is equal to $H|\psi_t\rangle$. An alternative way of writing (7.18) is

$$|\psi_t\rangle = e^{-itH/\hbar}|\psi_0\rangle, \tag{7.20}$$

where $|\psi_0\rangle$ is $|\psi_t\rangle$ when $t = 0$. The operator $e^{-itH/\hbar}$ is defined in the manner indicated in Sec. 3.10, see (3.97). It can be written down explicitly as

$$e^{-itH/\hbar} = \sum e^{-iE_n t/\hbar} |e_n\rangle\langle e_n|. \tag{7.21}$$

In the case of a harmonic oscillator, with

$$E_n = (n + 1/2)\hbar\omega \tag{7.22}$$

the energy and $|\phi_n\rangle$ the eigenstate of the nth level, (7.18) is equivalent to

$$\psi(x, t) = (e^{-i\omega t/2}) \sum_n b_n e^{-in\omega t} \phi_n(x), \tag{7.23}$$

where $\phi_n(x)$ is the position wave function corresponding to the ket $|\phi_n\rangle$.

A particular case of (7.18) is that in which $b_n = \delta_{np}$, that is, all except one of the b_n vanish, so that

$$|\psi_t\rangle = e^{-iE_pt/\hbar}|e_p\rangle. \tag{7.24}$$

The only time dependence comes in the phase factor, but since two states which differ by a phase factor have exactly the same physical significance, a quantum state with a precisely defined energy, known as a *stationary state*, represents a physical situation which is completely independent of time. By contrast, a classical system with a precisely defined energy will typically have a nontrivial time dependence; e.g., a harmonic oscillator tracing out an ellipse in the classical phase plane.

The inner product $\langle \omega_t|\psi_t\rangle$ of any two solutions of Schrödinger's equation is independent of time. Thus $|\psi_t\rangle$ satisfies (7.8), while the complex conjugate of this equation with ψ_t replaced by ω_t is

$$-i\hbar\frac{d}{dt}\langle\omega_t|j\rangle = \langle\omega_t|H|j\rangle, \tag{7.25}$$

where $H^\dagger$ on the right side has been replaced with H, since the Hamiltonian (which could depend upon the time) is Hermitian. Using (7.8) along with (7.25), one arrives at the result

$$i\hbar\frac{d}{dt}\langle\omega_t|\psi_t\rangle = i\hbar\sum_j\frac{d}{dt}\langle\omega_t|j\rangle\langle j|\psi_t\rangle$$
$$= \sum_j\langle\omega_t|j\rangle\langle j|H|\psi_t\rangle - \sum_j\langle\omega_t|H|j\rangle\langle j|\psi_t\rangle = 0, \tag{7.26}$$

since both of the last two sums are equal to $\langle\omega_t|H|\psi_t\rangle$. This means, in particular, that the norm $\|\psi_t\|$ of a solution $|\psi_t\rangle$ of Schrödinger's equation is independent of time, since it is the square root of the inner product of the ket with itself.

The fact that the Schrödinger equation preserves inner products and norms means that its action on the ket vectors in the Hilbert space is analogous to rigidly rotating a collection of vectors in ordinary three-dimensional space about the origin. If one thinks of these vectors as arrows directed outwards from the origin, the rotation will leave the lengths of the vectors and the angles between them, and hence the dot product of any two of them, unchanged, in the same way that inner products of vectors in the Hilbert space are left unchanged by the Schrödinger equation. An operator on the Hilbert space which leaves inner products unchanged is called an *isometry*. If, in addition, it maps the space onto itself, it is a *unitary operator*. Some important properties of unitary operators are stated in the next section, and we shall return to the topic of time development in Sec. 7.3.

7.2 Unitary operators

An operator U on a Hilbert space $\mathcal{H}$ is said to be *unitary* provided (i) it is an isometry, and (ii) it maps $\mathcal{H}$ onto itself. An isometry preserves inner products, so condition (i) is equivalent to

$$\left(U|\omega\rangle\right)^{\dagger} U|\phi\rangle = \langle\omega|U^{\dagger}U|\phi\rangle \tag{7.27}$$

for any pair of kets $|\omega\rangle$ and $|\phi\rangle$ in $\mathcal{H}$, and this in turn will be true if and only if

$$U^{\dagger}U = I. \tag{7.28}$$

Condition (ii) means that given any $|\eta\rangle$ in $\mathcal{H}$, there is some $|\psi\rangle$ in $\mathcal{H}$ such that $|\eta\rangle = U|\psi\rangle$. This will be the case if, in addition to (7.28),

$$UU^{\dagger} = I. \tag{7.29}$$

The two equalities (7.28) and (7.29) tell us that $U^{\dagger}$ is the same as the *inverse* U^{-1} of the operator U. For a finite-dimensional Hilbert space, condition (ii) for a unitary operator is automatically satisfied in the case of an isometry, so (7.28) implies (7.29), and vice versa, and it suffices to check one or the other in order to show that U is unitary.

If U is unitary, then so is $U^{\dagger}$. In addition, the operator product of two or more unitary operators is a unitary operator. This follows at once from (7.28) and (7.29) and the rule giving the adjoint of a product of operators, (3.32). Thus if both U and V satisfy (7.28), so does their product,

$$(UV)^{\dagger}UV = V^{\dagger}U^{\dagger}UV = V^{\dagger}IV = I, \tag{7.30}$$

and the same is true for (7.29).

A second, equivalent definition of a unitary operator is the following: Let $\{|j\rangle\}$ be some orthonormal basis of $\mathcal{H}$. Then U is unitary if and only if $\{U|j\rangle\}$ is also an orthonormal basis. If $\mathcal{H}$ is of finite dimension, one need only check that $\{U|j\rangle\}$ is an orthonormal collection, for then it will also be an orthonormal basis, given that $\{|j\rangle\}$ is such a basis.

The matrix $\{\langle j|U|k\rangle\}$ of a unitary operator in an orthonormal basis can be thought of as a collection of column vectors which are normalized and mutually orthogonal, a result which follows at once from (7.28) and the usual rule for matrix multiplication. Similarly, (7.29) tells one that the row vectors which make up this matrix are normalized and mutually orthogonal. Any 2×2 unitary matrix can be written in the form

$$e^{i\phi} \begin{pmatrix} \alpha & \beta \\ -\beta^{*} & \alpha^{*} \end{pmatrix}, \tag{7.31}$$

where α and β are complex numbers satisfying

$$|\alpha|^2 + |\beta|^2 = 1 \tag{7.32}$$

and ϕ is an arbitrary phase. It is obvious from (7.32) that the two column vectors making up this matrix are normalized, and their orthogonality is easily checked. The same is true of the two row vectors.

Given a unitary operator, one can find an orthonormal basis $\{|u_j\rangle\}$ in which it can be written in diagonal form

$$U = \sum_j \lambda_j |u_j\rangle\langle u_j|, \tag{7.33}$$

where the eigenvalues λ_j of U are complex numbers with $|\lambda_j| = 1$. Just as Hermitian operators can be thought of as somewhat analogous to real numbers, since their eigenvalues are real, unitary operators are analogous to complex numbers of unit modulus. (In an infinite-dimensional space the sum in (7.33) may have to be replaced by an appropriate integral.) As in the case of Hermitian operators, Sec. 3.7, if some of the eigenvalues in (7.33) are degenerate, the sum can be rewritten in the form

$$U = \sum_k \lambda_k' S_k, \tag{7.34}$$

where the S_k are projectors which form a decomposition of the identity, and $\lambda_k' \neq \lambda_l'$ for $k \neq l$.

All operators in a collection $\{U, V, W, \ldots\}$ of unitary operators which *commute* with each other can be simultaneously diagonalized using a single orthonormal basis. That is, there is some basis $\{|u_j\rangle\}$ in which U, V, W, and so forth can be expressed using the same collection of dyads $|u_j\rangle\langle u_j|$, as in (7.33), but with different eigenvalues for the different operators. If one writes down expressions of the form (7.34) for V, W, etc., the decompositions of the identity need not be identical with the $\{S_j\}$ appropriate for U, but the different decompositions will all be compatible in the sense that the projectors will all commute with one another.

7.3 Time development operators

Consider integrating Schrödinger's equation from time $t = 0$ to $t = \tau$ starting from an arbitrary initial state $|\psi_0\rangle$. Because the equation is linear, the dependence of the state $|\psi_\tau\rangle$ at time τ upon the initial state $|\psi_0\rangle$ can be written in the form

$$|\psi_\tau\rangle = T(\tau, 0)|\psi_0\rangle, \tag{7.35}$$

where $T(\tau, 0)$ is a linear operator. And because Schrödinger's equation preserves inner products, (7.26), $T(\tau, 0)$ is an isometry. In addition, it maps $\mathcal{H}$ onto itself, because if $|\eta\rangle$ is any ket in $\mathcal{H}$, we can treat it as a "final" condition at time τ and integrate Schrödinger's equation backwards to time 0 in order to obtain a ket $|\zeta\rangle$ such that $|\eta\rangle = T(\tau, 0)|\zeta\rangle$. Therefore $T(\tau, 0)$ is a unitary operator, since it satisfies the conditions given Sec. 7.2.

Of course there is nothing special about the times 0 and τ, and the same argument could be applied equally well to the integration of Schrödinger's equation between two arbitrary times t' and t, where t can be earlier or later than t'. That is to say, there is a collection of unitary *time development operators* $T(t, t')$, labeled by the two times t and t', such that if $|\psi_t\rangle$ is any solution of Schrödinger's equation, then

$$|\psi_t\rangle = T(t, t')|\psi_{t'}\rangle. \tag{7.36}$$

These time development operators satisfy a set of fairly obvious conditions. First, if t' is equal to t,

$$T(t, t) = I. \tag{7.37}$$

Next, since

$$|\psi_t\rangle = T(t, t'')|\psi_{t''}\rangle = T(t, t')|\psi_{t'}\rangle = T(t, t')T(t', t'')|\psi_{t''}\rangle, \tag{7.38}$$

it follows that

$$T(t, t')T(t', t'') = T(t, t'') \tag{7.39}$$

for any three times t, t', t''. In particular, if we set $t'' = t$ in this expression and use (7.37), the result is

$$T(t, t')T(t', t) = I. \tag{7.40}$$

Since $T(t, t')$ is a unitary operator, this tells us that

$$T(t', t) = T(t, t')^{\dagger} = T(t, t')^{-1}. \tag{7.41}$$

Thus the adjoint of a time development operator, which is the same as its inverse, is obtained by interchanging its two arguments.

If one applies the dagger operation to (7.36), see (3.33), the result is

$$\langle\psi_t| = \langle\psi_{t'}|T(t, t')^{\dagger} = \langle\psi_{t'}|T(t', t). \tag{7.42}$$

Consequently, the projectors $[\psi_t]$ and $[\psi_{t'}]$ onto the rays containing $|\psi_t\rangle$ and $|\psi_{t'}\rangle$ are related by

$$[\psi_t] = |\psi_t\rangle\langle\psi_t| = T(t, t')|\psi_{t'}\rangle\langle\psi_{t'}|T(t', t) = T(t, t')[\psi_{t'}]T(t', t). \tag{7.43}$$

This formula can be generalized to the case in which $P_{t'}$ is any projector onto some subspace $\mathcal{P}_{t'}$ of the Hilbert space. Then P_t defined by

$$P_t = T(t, t')P_{t'}T(t', t) \tag{7.44}$$

is a projector onto a subspace $\mathcal{P}_t$ with the property that if $|\psi_{t'}\rangle$ is any ket in $\mathcal{P}_{t'}$, its image $T(t, t')|\psi_{t'}\rangle$ under the time translation operator lies in $\mathcal{P}_t$, and $\mathcal{P}_t$ is composed of kets of this form. That is to say, the same unitary dynamics which "moves" one ket onto another through (7.36) "moves" subspaces in the manner indicated in (7.44). The difference is that only a single T operator is needed to move kets, while two are necessary in order to move a projector.

Since $|\psi_t\rangle$ in (7.36) satisfies Schrödinger's equation (7.4), it follows that

$$i\hbar \frac{\partial T(t, t')}{\partial t} = H(t)T(t, t'), \tag{7.45}$$

where one can write H in place of $H(t)$ if the Hamiltonian is independent of time. There is a similar equation in which the first argument of $T(t, t')$ is held fixed,

$$-i\hbar \frac{\partial T(t, t')}{\partial t'} = T(t, t')H(t'), \tag{7.46}$$

obtained by taking the adjoint of (7.45) with the help of (7.41), and then interchanging t and t'. Given a time-independent orthonormal basis $\{|j\rangle\}$, (7.45) is equivalent to a set of coupled ordinary differential equations for the matrix elements of $T(t, t')$,

$$i\hbar \frac{\partial}{\partial t}\langle j|T(t, t')|k\rangle = \sum_m \langle j|H(t)|m\rangle\langle m|T(t, t')|k\rangle, \tag{7.47}$$

and one can write down an analogous expression corresponding to (7.46).

Obtaining explicit forms for the time development operators is in general a very difficult task, since it is equivalent to integrating the Schrödinger equation for all possible initial conditions. However, if the Hamiltonian is independent of time, one can write

$$T(t, t') = e^{-i(t-t')H/\hbar} = \sum_n e^{-iE_n(t-t')/\hbar}|e_n\rangle\langle e_n|, \tag{7.48}$$

where the E_n and $|e_n\rangle$ are the eigenvalues and eigenfunctions of H, (7.17). Thus when the Hamiltonian is independent of time, $T(t, t')$ depends only on the difference $t - t'$ of its two arguments.

7.4 Toy models

The unitary dynamics of most quantum systems is quite complicated and difficult to understand. Among the few exceptions are: trivial dynamics, in which $T(t', t) = I$ independent of t and t'; a spin-half particle in a constant magnetic field with Hamiltonian (7.16); and the harmonic oscillator, which has a simple time dependence because its energy levels have a uniform spacing, (7.22). Even a particle moving in one dimension in the potential $V(x) = 0$ represents a nontrivial dynamical problem in quantum theory. Though one can write down closed-form solutions, they tend to be a bit messy, especially in comparison with the simple trajectory $x = x_0 + (p_0/m)t$ and $p = p_0$ in the classical phase space.

In order to gain some intuitive understanding of quantum dynamics, it is important to have simple model systems whose properties can be worked out explicitly with very little effort "on the back of an envelope", but which allow more complicated behavior than occurs in the case of a spin-half particle or a harmonic oscillator. We want to be able to discuss interference effects, measurements, radioactive decay, and so forth. For this purpose toy models resembling the one introduced in Sec. 2.5, where a particle can be located at one of a finite number of discrete sites, turn out to be particularly useful. The key to obtaining simple dynamics in a toy model is to make *time* (like space) a *discrete variable*. Thus we shall assume that the time t takes on only integer values: $-1, 0, 1, 2$, etc. These could, in principle, be integer multiples of some very short interval of time, say 10^{-50} seconds, so discretization is not, by itself, much of a limitation (or simplification).

Though it is not essential, in many cases one can assume that $T(t, t')$ depends only on the time difference $t - t'$; this is the toy analog of a time-independent Hamiltonian. Then one can write

$$T(t, t') = T^{t-t'}, \tag{7.49}$$

where the symbol T *without any arguments* will represent a unitary operator on the (usually finite-dimensional) Hilbert space of the toy model. The strategy for constructing a useful toy model is to make T a very simple operator, as in the examples discussed below. Because t takes integer values, $T(t, t')$ is given by integer powers of the operator T, and can be calculated by applying T several times in a row. To be sure, these powers can be negative, but that is not so bad, because we will be able to choose T in such a way that its inverse $T^{-1} = T^{\dagger}$ is also a very simple operator.

As a first example, consider the model introduced in Sec. 2.5 with a particle located at one of $M = M_a + M_b + 1$ sites placed in a one-dimensional line and labeled with an integer m in the interval

$$-M_a \le m \le M_b, \tag{7.50}$$

where M_a and M_b are large integers. This becomes a *hopping model* if the time development operator T is set equal to the *shift operator* S defined by

$$S|m\rangle = |m + 1\rangle, \quad S|M_b\rangle = |-M_a\rangle. \tag{7.51}$$

That is, during a single time step the particle hops one space to the right, but when it comes to the maximum value of m it hops to the minimum value. Thus the dynamics has a "periodic boundary condition", and one may prefer to imagine the successive sites as located not on a line but on a large circle, so that the one labeled M_b is just to the left of the one labeled $-M_a$. One must check that $T = S$ is unitary, and this is easily done. The collection of kets $\{|m\rangle\}$ forms an orthonormal basis of the Hilbert space, and the collection $\{S|m\rangle\}$, since it consists of precisely the same elements, is also an orthonormal basis. Thus the criterion in the second definition in Sec. 7.2 is satisfied, and S is unitary.

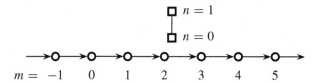

Fig. 7.1. Toy model of particle with detector.

To make the hopping model a bit more interesting, let us add a detector, a second particle which can be at only one of the two sites $n = 0$ or 1 indicated in Fig. 7.1. The Hilbert space $\mathcal{H}$ for this system is, as noted in Sec. 6.3, a tensor product $\mathcal{M} \otimes \mathcal{N}$ of an M-dimensional space $\mathcal{M}$ for the first particle and a two-dimensional space $\mathcal{N}$ for the detector, and the $2M$ kets $\{|m, n\rangle\}$ form an orthonormal basis. What makes the detector act like a detector is a choice for the unitary dynamics in which the time development operator is

$$T = SR, \tag{7.52}$$

where $S = S \otimes I$, using the notation of Sec. 6.4, is the extension to $\mathcal{M} \otimes \mathcal{N}$ of the shift operator defined earlier on $\mathcal{M}$ using (7.51), and R is defined by

$$R|m, n\rangle = |m, n\rangle \text{ for } m \neq 2,$$
$$R|2, n\rangle = |2, 1 - n\rangle. \tag{7.53}$$

Thus R does nothing at all unless the particle is at $m = 2$, in which case it "flips" the detector from $n = 0$ to $n = 1$ and vice versa. That R is unitary follows from the fact that the collection of kets $\{R|m, n\rangle\}$ is identical to the collection $\{|m, n\rangle\}$, as all that R does is interchange two of them, and is thus an orthonormal basis of $\mathcal{H}$. The extended operator $S \otimes I$ satisfies (7.28) when S satisfies this condition, so it is unitary. The unitarity of $T = SR$ is then a consequence of the fact that the product of unitary operators is unitary, as noted in Sec. 7.2. (While it is not hard to show directly that T is unitary, the strategy of writing it as a product of other unitary operators is useful in more complicated cases, which is why we have used it here.) The action of $T = SR$ on the combined system of particle plus detector is as follows. At each time step the particle hops from m to $m + 1$ (except when it makes the big jump from M_b to $-M_a$). The detector remains at $n = 0$ or at $n = 1$, wherever it happens to be, except during a time step in which the particle hops from 2 to 3, when the detector hops from n to $1 - n$, that is, from 0 to 1 or 1 to 0.

What justifies calling the detector a detector? Let us use a notation in which $\mapsto$ denotes the action of T in the sense that

$$|\psi\rangle \mapsto T|\psi\rangle \mapsto T^2|\psi\rangle \mapsto \cdots . \tag{7.54}$$

Suppose that the particle starts off at $m = 0$ and the detector is in the state $n = 0$, "ready to detect the particle", at $t = 0$. The initial state of the combined system of particle plus detector develops in time according to

$$|0, 0\rangle \mapsto |1, 0\rangle \mapsto |2, 0\rangle \mapsto |3, 1\rangle \mapsto |4, 1\rangle \mapsto \cdots . \tag{7.55}$$

That is to say, during the time step from $t = 2$ to $t = 3$, in which the particle hops from $m = 2$ to $m = 3$, the detector moves from $n = 0$ "ready" to $n = 1$, "have detected the particle," and it continues in the "have detected" state at later times. Not at all later times, since the particle will eventually hop from $M_b > 0$ to $-M_a < 0$, and then m will increase until, eventually, the particle will pass by the detector a second time and "untrigger" it. But by making M_a or M_b large compared with the times we are interested in, we can ignore this possibility. More sophisticated models of detectors are certainly possible, and some of these will be introduced in later chapters. However, the essential spirit of the toy model approach is to use the simplest possibility which provides some physical intuition. The detector in Fig. 7.1 is perfectly adequate for many purposes, and will be used repeatedly in later chapters.

It is worth noting that the measurement of the particle's position (or its passing the position of the detector) in this way does *not* influence the motion of the particle: in the absence of the detector one would have the same sequence of positions m as a function of time as those in (7.55). But is it not the case that *any* quantum measurement perturbs the measured system? One of the benefits of introducing toy models is that they make it possible to study this and other pieces of quantum folklore in specific situations. In later chapters we will explore the issue of perturbations produced by measurements in more detail. For the present it is enough to note that quantum measurement apparatus can be designed so that it does not perturb certain properties, even though it may perturb other properties.

Another example of a toy model is the one in Fig. 7.2, which can be used to illustrate the process of radioactive decay. Consider alpha decay, and adopt the picture in which an alpha particle is rattling around inside a nucleus until it eventually tunnels out through the Coulomb barrier. One knows that this is a fairly good description of the escape process, even though it is bad nuclear physics if

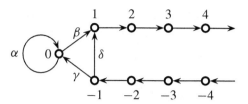

Fig. 7.2. Toy model for alpha decay.

taken too literally. However, the unitary time development of a particle tunneling through a potential barrier is not easy to compute; one needs WKB formulas and other approximations.

By contrast, the unitary time development of the toy model in Fig. 7.2, which is a slight modification of the hopping model (without a detector) introduced earlier, can be worked out very easily. The different sites represent possible locations of the alpha particle, with the $m = 0$ site inside the nucleus, and the other sites outside the nucleus. At each time step, the particle at $m = 0$ can either stay put, with amplitude α, or escape to $m = 1$, with amplitude β. Once it reaches $m = 1$, it hops to $m = 2$ at the next time step, and then on to $m = 3$, etc. Eventually it will hop from $m = +M_b$ to $m = -M_a$ and begin its journey back towards the nucleus, but we will assume that M_b is so large that we never have to consider the return process. (One could make M_a and M_b infinite, at the price of introducing an infinite-dimensional Hilbert space.) The time development operator is $T = S_a$, where

$$S_a|m\rangle = |m+1\rangle \text{ for } m \neq 0, -1, M_b, \quad S_a|M_b\rangle = |-M_a\rangle,$$
$$S_a|0\rangle = \alpha|0\rangle + \beta|1\rangle, \quad S_a|-1\rangle = \gamma|0\rangle + \delta|1\rangle. \tag{7.56}$$

Thus S_a is identical to the simple shift S of (7.51), except when applied to the two kets $|0\rangle$ and $|-1\rangle$.

The operator S_a is unitary if the complex constants $\alpha, \beta, \gamma, \delta$ form a unitary matrix

$$\begin{pmatrix} \alpha & \beta \\ \gamma & \delta \end{pmatrix}. \tag{7.57}$$

If we use the criterion, Sec. 7.2, that the row vectors are normalized and mutually orthogonal, the conditions for unitarity can be written in the form:

$$|\alpha|^2 + |\beta|^2 = 1 = |\gamma|^2 + |\delta|^2, \quad \alpha^*\gamma + \beta^*\delta = 0. \tag{7.58}$$

That S_a is unitary when (7.58) is satisfied can be seen from the fact that it maps the orthonormal basis $\{|m\rangle\}$ into an orthonormal collection of vectors, which, since the Hilbert space is finite, must itself be an orthonormal basis. In particular, S_a applied to $|0\rangle$ and to $|-1\rangle$ yields two normalized vectors which are mutually orthogonal to each other, a result ensured by (7.58).

Note how the requirement of unitarity leads to the nontrivial consequence that if the action of the shift operator S on $|0\rangle$ is modified so that the particle can either hop or remain in place during one time step, there must be an additional modification of S someplace else. In this example the other modification occurs at $|-1\rangle$, which is a fairly natural place to put it. The fact that $|\gamma| = |\beta|$ means that if there is an amplitude for the alpha particle to escape from the nucleus, there is also an amplitude for an alpha particle approaching the nucleus along the $m < 0$ sites to be captured at $m = 0$, rather than simply being scattered to $m = 1$. As one might expect, $|\beta|^2$ is the probability that the alpha particle will escape during a particular time step, and $|\alpha|^2$ the probability that it will remain in the nucleus. However, showing that this is so requires additional developments of the theory found in the following chapters; see Secs. 9.5 and 12.4.

The unitary time development of an initial state $|0\rangle$ at $t = 0$, corresponding to the alpha particle being inside the nucleus, is easily worked out. Using the $\mapsto$ notation of (7.54), one has:

$$
\begin{aligned}
|0\rangle &\mapsto \alpha|0\rangle + \beta|1\rangle \mapsto \alpha^2|0\rangle + \alpha\beta|1\rangle + \beta|2\rangle \\
&\mapsto \alpha^3|0\rangle + \alpha^2\beta|1\rangle + \alpha\beta|2\rangle + \beta|3\rangle \mapsto \cdots ,
\end{aligned}
\tag{7.59}
$$

so that for any time $t > 0$,

$$
|\psi_t\rangle = T^t|0\rangle = \alpha^t|0\rangle + \alpha^{t-1}\beta|1\rangle + \alpha^{t-2}\beta|2\rangle + \cdots \beta|t\rangle.
\tag{7.60}
$$

The magnitude of the coefficient of $|0\rangle$ decreases exponentially with time. The rest of the time development can be thought of in the following way. An "initial wave" reaches site m at $t = m$. Thereafter, the coefficient of $|m\rangle$ decreases exponentially. That is, the wave function is spreading out and, at the same time, its amplitude is decreasing. These features are physically correct in that they will also emerge from a more sophisticated model of the decay process. Even though not every detail of the toy model is realistic, it nonetheless provides a good beginning for understanding some of the quantum physics of radioactive and other decay processes.

8

Stochastic histories

8.1 Introduction

Despite the fact that classical mechanics employs deterministic dynamical laws, random dynamical processes often arise in classical physics, as well as in everyday life. A *stochastic* or *random* process is one in which states-of-affairs at successive times are not related to one another by deterministic laws, and instead probability theory is employed to describe whatever regularities exist. Tossing a coin or rolling a die several times in succession are examples of stochastic processes in which the previous history is of very little help in predicting what will happen in the future. The motion of a baseball is an example of a stochastic process which is to some degree predictable using classical equations of motion that relate its acceleration to the total force acting upon it. However, a lack of information about its initial state (e.g., whether it is spinning), its precise shape, and the condition and motion of the air through which it moves limits the precision with which one can predict its trajectory.

The Brownian motion of a small particle suspended in a fluid and subject to random bombardment by the surrounding molecules of fluid is a well-studied example of a stochastic process in classical physics. Whereas the instantaneous velocity of the particle is hard to predict, there is a probabilistic correlation between successive positions, which can be predicted using stochastic dynamics and checked by experimental measurements. In particular, given the particle's position at a time t, it is possible to compute the probability that it will have moved a certain distance by the time $t + \Delta t$. The stochastic description of the motion of a Brownian particle uses the deterministic law for the motion of an object in a viscous fluid, and assumes that there is, in addition, a random force or "noise" which is unpredictable, but whose statistical properties are known.

In classical physics the need to use stochastic rather than deterministic dynamical processes can be blamed on ignorance. If one knew the precise positions and

108

velocities of all the molecules making up the fluid in which the Brownian particle is suspended, along with the same quantities for the molecules in the walls of the container and inside the Brownian particle itself, it would in principle be possible to integrate the classical equations of motion and make precise predictions about the motion of the particle. Of course, integrating the classical equations of motion with infinite precision is not possible. Nonetheless, in classical physics one can, in principle, construct more and more refined descriptions of a mechanical system, and thereby continue to reduce the noise in the stochastic dynamics in order to come arbitrarily close to a deterministic description. Knowing the spin imparted to a baseball by the pitcher allows a more precise prediction of its future trajectory. Knowing the positions and velocities of the fluid molecules inside a sphere centered at a Brownian particle makes it possible to improve one's prediction of its motion, at least over a short time interval.

The situation in quantum physics is similar, up to a point. A quantum description can be made more precise by using smaller, that is, lower-dimensional subspaces of the Hilbert space. However, while the refinement of a classical description can go on indefinitely, one reaches a limit in the quantum case when the subspaces are one-dimensional, since no finer description is possible. However, at this level quantum dynamics is still stochastic: there is an irreducible "quantum noise" which cannot be eliminated, even in principle. To be sure, quantum theory allows for a deterministic (and thus noise free) unitary dynamics, as discussed in the previous chapter. But there are many processes in the real world which cannot be discussed in terms of purely unitary dynamics based upon Schrödinger's equation. Consequently, stochastic descriptions are a fundamental part of quantum mechanics in a sense which is not true in classical mechanics.

In this chapter we focus on the kinematical aspects of classical and quantum stochastic dynamics: how to construct sample spaces and the corresponding event algebras. As usual, classical dynamics is simpler and provides a valuable guide and useful analogies for the quantum case, so various classical examples are taken up in Sec. 8.2. Quantum dynamics is the subject of the remainder of the chapter.

8.2 Classical histories

Consider a coin which is tossed three times in a row. The eight possible outcomes of this experiment are HHH, HHT, HTH, ... TTT: heads on all three tosses, heads the first two times and tails the third, and so forth. These eight possibilities constitute a *sample space* as that term is used in probability theory, see Sec. 5.1, since the different possibilities are mutually exclusive, and one and only one of them will occur in any particular experiment in which a coin is tossed three times in a row. The *event algebra* (Sec. 5.1) consists of the 2^8 subsets of elements in the sample space: the empty set, HHH by itself, the pair $\{HHT, TTT\}$, and so forth. The elements of the sample space will be referred to as *histories*, where a history is to be thought of as a *sequence of events at successive times*. Members of the event algebra will also be called "histories" in a somewhat looser sense, or *compound histories* if they include two or more elements from the sample space.

As a second example, consider a die which is rolled f times in succession. The sample space consists of 6^f possibilities $\{s_1, s_2, \ldots s_f\}$, where each s_j takes some value between 1 and 6.

A third example is a Brownian particle moving in a fluid and observed under a microscope at successive times $t_1, t_2, \ldots t_f$. The sequence of positions $\mathbf{r}_1, \mathbf{r}_2, \ldots \mathbf{r}_f$ is an example of a history, and the sample space consists of all possible sequences of this type. Since any measuring instrument has finite resolution, one can, if one wants, suppose that for the purpose of recording the data the region inside the fluid is thought of as divided up into a collection of small cubical cells, with $\mathbf{r}_j$ the label of the cell containing the particle at time t_j.

A fourth example is a particle undergoing a random walk in one dimension, a sort of "toy model" of Brownian motion. Assume that the location of the particle or random walker, denoted by s, is an integer in the range

$$-M_a \leq s \leq M_b. \tag{8.1}$$

One could allow s to be any integer, but using the limited range (8.1) results in a finite sample space of $M = M_a + M_b + 1$ possibilities at any given time. At each time step the particle either remains where it is, or hops to the right or to the left. Hence a *history* of the particle's motion consists in giving its positions at a set of times $t = 0, 1, \ldots f$ as a sequence of integers

$$\mathbf{s} = (s_0, s_1, s_2, \ldots s_f), \tag{8.2}$$

where each s_j falls in the interval (8.1). The *sample space of histories* consists of the M^{f+1} different sequences $\mathbf{s}$. (Letting s_0 rather than s_1 be the initial position of the particle is of no importance; the convention used here agrees with that in the next chapter.) One could employ histories extending to $t = \infty$, but that would mean using an infinite sample space.

This sample space can be thought of as produced by successively refining an initial, coarse sample space in which s_0 takes one of M possible values, and nothing is said about what happens at later times. Histories involving the two times $t = 0$ and 1 are produced by taking a point in this initial sample space, say $s_0 = 3$, and "splitting it up" into two-time histories of the form $(3, s_1)$, where s_1 can take on any one of the M values in (8.1). Given a point, say $(3, 2)$, in this new sample space, it can again be split up into elements of the form $(3, 2, s_2)$, and so forth. Note that any history involving less than $n+1$ times can be thought of as a compound history on the full sample space. Thus (3, 2) consists of all sequences **s** for which $s_0 = 3$ and $s_1 = 2$. Rather than starting with a coarse sample space of events at $t = 0$, one could equally well begin with a later time, such as all the possibilities for s_2 at $t = 2$, and then refine this space by including additional details at both earlier and later times.

8.3 Quantum histories

A *quantum history* of a physical system is a sequence of *quantum events* at successive times, where a quantum event at a particular time can be any *quantum property* of the system in question. Thus given a set of times $t_1 < t_2 < \cdots t_f$, a quantum history is specified by a collection of projectors $(F_1, F_2, \ldots F_f)$, one projector for each time. It is convenient, both for technical and for conceptual reasons, to suppose that the number f of distinct times is finite, though it might be very large. It is always possible to add additional times to those in the list $t_1 < t_2 < \cdots t_f$ in the manner indicated in Sec. 8.4. Sometimes the initial time will be denoted by t_0 rather than t_1.

For a spin-half particle, $([z^+], [x^+])$ is an example of a history involving two times, while $([z^+], [x^+], [z^+])$ is an example involving three times.

As a second example, consider a harmonic oscillator. A possible history with three different times is the sequence of events

$$F_1 = [\phi_1] + [\phi_2], \quad F_2 = [\phi_1], \quad F_3 = X, \quad (8.3)$$

where $[\phi_n]$ is the projector on the energy eigenstate with energy $(n + 1/2)\hbar\omega$, and X is the projector defined in (4.20) corresponding to the position x lying in the interval $x_1 \le x \le x_2$. Note that the projectors making up a history do not have to project onto a one-dimensional subspace of the Hilbert space. In this example, F_1 projects onto a two-dimensional subspace, F_2 onto a one-dimensional subspace, and X onto an infinite-dimensional subspace.

As a third example, consider a coin tossed three times in a row. A physical coin is made up of atoms, so it has in principle a (rather complicated) quantum mechanical description. Thus a "classical" property such as "heads" will correspond to some quantum projector H onto a subspace of enormous dimension, and there will be another projector T for "tails". Then by using the projectors

$$F_1 = H, \quad F_2 = T, \quad F_3 = T \quad (8.4)$$

at successive times one obtains a quantum history HTT for the coin.

As a fourth example of a quantum history, consider a Brownian particle suspended in a fluid. Whereas this is usually described in classical terms, the particle and the surrounding fluid are, in reality, a quantum system. At time t_j let F_j be the projector, in an appropriate Hilbert space, for the property that the center of mass of the Brownian particle is inside a particular cubical cell. Then $(F_1, F_2, \ldots F_f)$ is the quantum counterpart of the classical history $\mathbf{r}_1, \mathbf{r}_2, \ldots \mathbf{r}_f$ introduced earlier, with $\mathbf{r}_j$ understood as a cell label, rather than a precise position.

One does not normally think of coin tossing in "quantum" terms, and there is really no advantage to doing so, since a classical description is simpler, and is perfectly adequate. Similarly, a classical description of the motion of a Brownian particle is usually quite adequate. However, these examples illustrate the fact that the concept of a quantum history is really quite general, and is by no means limited to processes and events at an atomic scale, even though that is where quantum histories are most useful, precisely because the corresponding classical descriptions are not adequate.

The sample space of a coin tossed f times in a row is formally the same as the sample space of f coins tossed simultaneously: each consists of 2^f mutually exclusive possibilities. Since in quantum theory the Hilbert space of a collection of f systems is the tensor product of the separate Hilbert spaces, Ch. 6, it seems reasonable to use a tensor product of f spaces for describing the different histories of a single quantum system at f successive times. Thus we define a *history Hilbert space* as a tensor product

$$\breve{\mathcal{H}} = \mathcal{H}_1 \odot \mathcal{H}_2 \odot \cdots \mathcal{H}_f, \tag{8.5}$$

where for each j, $\mathcal{H}_j$ is a copy of the Hilbert space $\mathcal{H}$ used to describe the system at a single time, and $\odot$ is a variant of the tensor product symbol $\otimes$. We could equally well write $\mathcal{H}_1 \otimes \mathcal{H}_2 \otimes \cdots$, but it is helpful to have a distinctive notation for a tensor product when the factors in it refer to different times, and reserve $\otimes$ for a tensor product of spaces at a single time. On the space $\breve{\mathcal{H}}$ the history $(F_1, F_2, \ldots F_f)$ is represented by the (tensor) product projector

$$Y = F_1 \odot F_2 \odot \cdots F_f. \tag{8.6}$$

That Y is a projector, that is, $Y^\dagger = Y = Y^2$, follows from the fact that each F_j is a projector, and from the rules for adjoints and products of operators on tensor products as discussed in Sec. 6.4.

8.4 Extensions and logical operations on histories

Suppose that $f = 3$ in (8.6), so that

$$Y = F_1 \odot F_2 \odot F_3. \tag{8.7}$$

This history can be *extended* to additional times by introducing the identity operator at the times not included in the initial set t_1, t_2, t_3. Suppose, for example, that we wish to add an additional time t_4 later than t_3. Then for times $t_1 < t_2 < t_3 < t_4$, (8.7) is equivalent to

$$Y = F_1 \odot F_2 \odot F_3 \odot I, \tag{8.8}$$

because the identity operator I represents the property which is always true, and therefore provides no additional information about the system at t_4. In the same way, one can introduce earlier and intermediate times, say t_0 and $t_{1.5}$, in which case (8.7) is equivalent to

$$Y = I \odot F_1 \odot I \odot F_2 \odot F_3 \odot I \tag{8.9}$$

on the history space $\breve{\mathcal{H}}$ for the times $t_0 < t_1 < t_{1.5} < t_2 < t_3 < t_4$. We shall always use a notation in which the events in a history are in temporal order, with time increasing from left to right.

The notational convention for extensions of operators introduced in Sec. 6.4 justifies using the same symbol Y in (8.7), (8.8), and (8.9). And its intuitive significance is precisely the same in all three cases: Y means "F_1 at t_1, F_2 at t_2, and F_3 at t_3", and tells us nothing at all about what is happening at any other time. Using the same symbol for F and $F \odot I$ can sometimes be confusing for the reason pointed out at the end of Sec. 6.4. For example, the projector for a two-time history of a spin-half particle can be written as an operator product

$$[z^+] \odot [x^+] = \big([z^+] \odot I\big) \cdot \big(I \odot [x^+]\big) \tag{8.10}$$

of two projectors. If on the right side we replace $\big([z^+] \odot I\big)$ with $[z^+]$ and $\big(I \odot [x^+]\big)$ with $[x^+]$, the result $[z^+] \cdot [x^+]$ is likely to be incorrectly interpreted as the product of two noncommuting operators on a single copy of the Hilbert space $\mathcal{H}$, rather than as the product of two commuting operators on the tensor product $\mathcal{H}_1 \odot \mathcal{H}_2$. Using the longer $\big([z^+] \odot I\big)$ avoids this confusion.

If histories are written as projectors on the history Hilbert space $\breve{\mathcal{H}}$, the rules for the logical operations of negation, conjunction, and disjunction are precisely the same as for quantum properties at a single time, as discussed in Secs. 4.4 and 4.5. In particular, the negation of the history Y, "Y did not occur", corresponds to a projector

$$\tilde{Y} = \breve{I} - Y, \tag{8.11}$$

where $\breve{I}$ is the identity on $\breve{\mathcal{H}}$. (Our notational convention allows us to write $\breve{I}$ as I, but $\breve{I}$ is clearer.)

Note that a history does not occur if any event in it fails to occur. Thus the negation of HH when a coin is tossed two times in a row is not TT, but instead the compound history consisting of HT, TH, and TT. Similarly, the negation of the quantum history

$$Y = F_1 \odot F_2 \tag{8.12}$$

given by (8.11) is a sum of three orthogonal projectors,

$$\tilde{Y} = F_1 \odot \tilde{F}_2 + \tilde{F}_1 \odot F_2 + \tilde{F}_1 \odot \tilde{F}_2, \tag{8.13}$$

where $\tilde{F}_j$ means $I - F_j$. Note that the compound history $\tilde{Y}$ in (8.13) cannot be written in the form $G_1 \odot G_2$, that is, as an event at t_1 followed by another event at t_2.

The conjunction Y AND Y', or $Y \wedge Y'$, of two histories is represented by the product YY' of the projectors, provided they commute with each other. If $YY' \neq Y'Y$, the conjunction is not defined. The situation is thus entirely analogous to the conjunction of two quantum properties at a single time, as discussed in Secs. 4.5 and 4.6. Let us suppose that the history

$$Y' = F_1' \odot F_2' \odot F_3' \tag{8.14}$$

is defined at the same three times as Y in (8.7). Their conjunction is represented by the projector

$$Y' \wedge Y = Y'Y = F_1'F_1 \odot F_2'F_2 \odot F_3'F_3, \tag{8.15}$$

which is equal to YY' provided that at each of the three times the projectors in the two histories commute:

$$F_j'F_j = F_jF_j' \text{ for } j = 1, 2, 3. \tag{8.16}$$

However, there is a case in which Y and Y' commute even if some of the conditions in (8.16) are not satisfied. It occurs when the product of the two projectors at one of the times is 0, for this means that $YY' = 0$ independent of what projectors occur at other times. Here is an example involving a spin-half particle:

$$\begin{aligned} Y &= [x^+] \odot [x^+] \odot [z^+], \\ Y' &= [y^+] \odot [z^+] \odot [z^-]. \end{aligned} \tag{8.17}$$

The two projectors at t_1, $[x^+]$ and $[y^+]$, clearly do not commute with each other, and the same is true at time t_2. However, the projectors at t_3 are orthogonal, and thus $YY' = 0 = Y'Y$.

A simple example of a nonvanishing conjunction is provided by a spin-half particle and two histories

$$Y = [z^+] \odot I, \quad Y' = I \odot [x^+], \tag{8.18}$$

defined at the times t_1 and t_2. The conjunction is

$$Y' \wedge Y = Y'Y = YY' = [z^+] \odot [x^+], \qquad (8.19)$$

and this is sensible, for the intuitive significance of (8.19) is "$S_z = +1/2$ at t_1 and $S_x = +1/2$ at t_2." Indeed, any history of the form (8.6) can be understood as "F_1 at t_1, and F_2 at t_2, and ... F_f at t_f." This example also shows how to generate the conjunction of two histories defined at different sets of times. First one must extend each history by including I at additional times until the extended histories are defined on a common set of times. If the extended projectors commute with each other, the operator product of the projectors, as in (8.15), is the projector for the conjunction of the two histories.

The disjunction "Y' or Y or both" of two histories is represented by a projector

$$Y' \vee Y = Y' + Y - Y'Y \qquad (8.20)$$

provided $Y'Y = YY'$; otherwise it is undefined. The intuitive significance of the disjunction of two (possibly compound) histories is what one would expect, though there is a subtlety associated with the quantum disjunction which does not arise in the case of classical histories, as has already been noted in Sec. 4.5 for the case of properties at a single time. It can best be illustrated by means of an explicit example. For a spin-half particle, define the two histories

$$Y = [z^+] \odot [x^+], \quad Y' = [z^+] \odot [x^-]. \qquad (8.21)$$

The projector for the disjunction is

$$Y \vee Y' = Y + Y' = [z^+] \odot I, \qquad (8.22)$$

since in this case $YY' = 0$. The *projector* $Y \vee Y'$ tells us nothing at all about the spin of the particle at the second time: in and of itself it does *not* imply that $S_x = +1/2$ or $S_x = -1/2$ at t_2, since the subspace of $\breve{\mathcal{H}}$ on which it projects contains, among others, the history $[z^+] \odot [y^+]$, which is incompatible with S_x having any value at all at t_2. On the other hand, when the projector $Y \vee Y'$ occurs in the context of a discussion in which both Y and Y' make sense, it can be safely interpreted as meaning (or implying) that at t_2 either $S_x = +1/2$ or $S_x = -1/2$, since any other possibility, such as $S_y = +1/2$, would be incompatible with Y and Y'.

This example illustrates an important principle of quantum reasoning: The *context*, that is, the sample space or event algebra used for constructing a quantum description or discussing the histories of a quantum system, can make a difference in how one understands or interprets various symbols. In quantum theory it is important to be clear about precisely what sample space is being used.

8.5 Sample spaces and families of histories

As discussed in Sec. 5.2, a sample space for a quantum system at a single time is a decomposition of the identity operator for the Hilbert space $\mathcal{H}$: a collection of mutually orthogonal projectors which sum to I. In the same way, a sample space of histories is a decomposition of the identity on the history Hilbert space $\breve{\mathcal{H}}$, a collection $\{Y^\alpha\}$ of mutually orthogonal projectors representing histories which sum to the history identity:

$$\breve{I} = \sum_\alpha Y^\alpha. \tag{8.23}$$

It is convenient to label the history projectors with a superscript in order to be able to reserve the subscript position for time. Since the square of a projector is equal to itself, we will not need to use superscripts on projectors as exponents.

Associated with a sample space of histories is a Boolean "event" algebra, called a *family of histories*, consisting of projectors of the form

$$Y = \sum_\alpha \pi^\alpha Y^\alpha, \tag{8.24}$$

with each π^α equal to 0 or 1, as in (5.12). Histories which are members of the sample space will be called *elementary* histories, whereas those of the form (8.24) with two or more π^α equal to 1 are *compound* histories. The term "family of histories" is also used to denote the sample space of histories which generates a particular Boolean algebra. Given the intimate connection between the sample space and the corresponding algebra, this double usage is unlikely to cause confusion.

The simplest way to introduce a history sample space is to use a *product of sample spaces* as that term was defined in Sec. 6.6. Assume that at each time t_j there is a decomposition of the identity I_j for the Hilbert space $\mathcal{H}_j$,

$$I_j = \sum_{\alpha_j} P_j^{\alpha_j}, \tag{8.25}$$

where the subscript j labels the time, and the superscript α_j labels the different projectors which occur in the decomposition at this time. The decompositions (8.25) for different values of j could be the same or they could be different; they need have no relationship to one another. (Note that the sample spaces for the different classical systems discussed in Sec. 8.2 have this sort of product structure.) Projectors of the form

$$Y^\alpha = P_1^{\alpha_1} \odot P_2^{\alpha_2} \odot \cdots P_f^{\alpha_f}, \tag{8.26}$$

where α is an f-component label

$$\alpha = (\alpha_1, \alpha_2, \ldots \alpha_f), \tag{8.27}$$

make up the sample space, and it is straightforward to check that (8.23) is satisfied.

Here is a simple example for a spin-half particle with $f = 2$:

$$I_1 = [z^+] + [z^-], \quad I_2 = [x^+] + [x^-]. \tag{8.28}$$

The product of sample spaces consists of the four histories

$$\begin{aligned} Y^{++} = [z^+] \odot [x^+], \quad Y^{+-} = [z^+] \odot [x^-], \\ Y^{-+} = [z^-] \odot [x^+], \quad Y^{--} = [z^-] \odot [x^-], \end{aligned} \tag{8.29}$$

in an obvious notation. The Boolean algebra or family of histories contains $2^4 = 16$ elementary and compound histories, including the null history 0 (which never occurs).

Another type of sample space that arises quite often in practice consists of histories which begin at an initial time t_0 with a specific state represented by a projector Ψ_0, but behave in different ways at later times. We shall refer to it as a *family based upon the initial state* Ψ_0. A relatively simple version is that in which the histories are of the form

$$Y^\alpha = \Psi_0 \odot P_1^{\alpha_1} \odot P_2^{\alpha_2} \odot \cdots P_f^{\alpha_f}, \tag{8.30}$$

with the projectors at times later than t_0 drawn from decompositions of the identity of the type (8.25). The sum over α of the projectors in (8.30) is equal to Ψ_0, so in order to complete the sample space one adds one more history

$$Z = (I - \Psi_0) \odot I \odot I \odot \cdots I \tag{8.31}$$

to the collection. If, as is usually the case, one is only interested in the histories which begin with the initial state Ψ_0, the history Z is assigned zero probability, after which it can be ignored. The procedure for assigning probabilities to the other histories will be discussed in later chapters. Note that histories of the form

$$(I - \Psi_0) \odot P_1^{\alpha_1} \odot P_2^{\alpha_2} \odot \cdots P_f^{\alpha_f} \tag{8.32}$$

are *not* present in the sample space, and for this reason the family of histories based upon an initial state Ψ_0 is distinct from a product of sample spaces in which (8.25) is supplemented with an additional decomposition

$$I_0 = \Psi_0 + (I - \Psi_0) \tag{8.33}$$

at time t_0. As a consequence, later events in a family based upon an initial state Ψ_0 are dependent upon the initial state in the technical sense discussed in Ch. 14.

Other examples of sample spaces which are not products of sample spaces are used in various applications of quantum theory, and some of them will be discussed in later chapters. In all cases the individual histories in the sample space correspond to product projectors on the history space $\breve{\mathcal{H}}$ regarded as a tensor product of Hilbert spaces at different times, (8.5). That is, they are of the form (8.6): a quantum property at t_1, another quantum property at t_2, and so forth. Since the history space $\breve{\mathcal{H}}$ is a Hilbert space, it also contains subspaces which are not of this form, but might be said to be "entangled in time". For example, in the case of a spin-half particle and two times t_1 and t_2, the ket

$$|\epsilon\rangle = \left(|z^+\rangle \odot |z^-\rangle - |z^-\rangle \odot |z^+\rangle\right)/\sqrt{2} \qquad (8.34)$$

is an element of $\breve{\mathcal{H}}$, and therefore $[\epsilon] = |\epsilon\rangle\langle\epsilon|$ is a projector on $\breve{\mathcal{H}}$. It seems difficult to find a physical interpretation for histories of this sort, or sample spaces containing such histories.

8.6 Refinements of histories

The process of refining a sample space in which coarse projectors are replaced with finer projectors on subspaces of lower dimensionality was discussed in Sec. 5.3. Refinement is often used to construct sample spaces of histories, as was noted in connection with the classical random walk in one dimension in Sec. 8.2. Here is a simple example to show how this process works for a quantum system. Consider a spin-half particle and a decomposition of the identity $\{[z^+], [z^-]\}$ at time t_1. Each projector corresponds to a single-time history which can be extended to a second time t_2 in the manner indicated in Sec. 8.3, to make a history sample space containing

$$[z^+] \odot I, \quad [z^-] \odot I. \qquad (8.35)$$

If one uses this sample space, there is nothing one can say about the spin of the particle at the second time t_2, since I is always true, and is thus completely uninformative. However, the first projector in (8.35) is the sum of $[z^+] \odot [z^+]$ and $[z^+] \odot [z^-]$, and if one replaces it with these two projectors, and the second projector in (8.35) with the corresponding pair $[z^-] \odot [z^+]$ and $[z^-] \odot [z^-]$, the result is a sample space

$$\begin{aligned} [z^+] \odot [z^+], \quad [z^+] \odot [z^-], \\ [z^-] \odot [z^+], \quad [z^-] \odot [z^-], \end{aligned} \qquad (8.36)$$

which is a refinement of (8.35), and permits one to say something about the spin at time t_2 as well as at t_1.

When it is possible to refine a sample space in this way, there are always a large number of ways of doing it. Thus the four histories in (8.29) also constitute a refinement of (8.35). However, the refinements (8.29) and (8.36) are mutually incompatible, since it makes no sense to talk about S_x at t_2 at the same time that one is ascribing values to S_z, and vice versa. Both (8.29) and (8.36) are products of sample spaces, but refinements of (8.35) which are not of this type are also possible; for example,

$$[z^+] \odot [z^+], \quad [z^+] \odot [z^-],$$
$$[z^-] \odot [x^+], \quad [z^-] \odot [x^-], \tag{8.37}$$

where the decomposition of the identity used at t_2 is different depending upon which event occurs at t_1.

The process of refinement can continue by first extending the histories in (8.36) or (8.37) to an additional time, either later than t_2 or earlier than t_1 or between t_1 and t_2, and then replacing the identity I at this additional time with two projectors onto pure states. Note that the process of extension does not by itself lead to a refinement of the sample space, since it leaves the number of histories and their intuitive interpretation unchanged; refinement occurs when I is replaced with projectors on lower-dimensional spaces.

It is important to notice that refinement is *not* some sort of *physical process* which occurs in the quantum system described by these histories. Instead, it is a conceptual process carried out by the quantum physicist in the process of constructing a suitable mathematical description of the time dependence of a quantum system. Unlike deterministic classical mechanics, in which the state of a system at a single time yields a unique description (orbit in the phase space) of what happens at other times, stochastic quantum mechanics allows for a large number of alternative descriptions, and the process of refinement is often a helpful way of selecting useful and interesting sample spaces from among them.

8.7 Unitary histories

Thus far we have discussed quantum histories without any reference to the dynamical laws of quantum mechanics. The dynamics of histories is not a trivial matter, and is the subject of the next two chapters. However, at this point it is convenient to introduce the notion of a *unitary history*. The simplest example of such a history is the sequence of kets $|\psi_{t_1}\rangle, |\psi_{t_2}\rangle, \ldots |\psi_{t_f}\rangle$, where $|\psi_t\rangle$ is a solution of Schrödinger's equation, Sec. 7.3, or, to be more precise, the corresponding sequence of projectors $[\psi_{t_1}], [\psi_{t_2}], \ldots$. The general definition is that a history of the form (8.6) is unitary provided

$$F_j = T(t_j, t_1) F_1 T(t_1, t_j) \tag{8.38}$$

is satisfied for $j = 1, 2, \ldots f$. That is to say, all the projectors in the history are generated from F_1 by means of the unitary time development operators introduced in Sec. 7.3, see (7.44). In fact, F_1 does not play a distinguished role in this definition and could be replaced by F_k for any k, because for a set of projectors given by (8.38), $T(t_j, t_k) F_k T(t_k, t_j)$ is equal to F_j whatever the value of k.

One can also define *unitary families* of histories. We shall limit ourselves to the case of a product of sample spaces, in the notation of Sec. 8.5, and assume that for each time t_j there is a decomposition of the identity of the form

$$I_j = \sum_a P_j^a. \tag{8.39}$$

The corresponding family is unitary if for each choice of a these projectors satisfy (8.38), that is,

$$P_j^a = T(t_j, t_1) P_1^a T(t_1, t_j) \tag{8.40}$$

for every j. In the simplest (interesting) family of this type each decomposition of the identity contains only two projectors; for example, $[\psi_{t_1}]$ and $I - [\psi_{t_1}]$. Notice that while a unitary family will contain unitary histories, such as

$$P_1^1 \odot P_2^1 \odot P_3^1 \odot \cdots P_f^1, \tag{8.41}$$

it will also contain other histories, such as

$$P_1^1 \odot P_2^2 \odot P_3^1 \odot \ \cdots P_f^1, \tag{8.42}$$

which are not unitary. We will have more to say about unitary histories and families of histories in Secs. 9.3, 9.6, and 10.3.

9

The Born rule

9.1 Classical random walk

The previous chapter showed how to construct sample spaces of histories for both classical and quantum systems. Now we shall see how to use dynamical laws in order to assign probabilities to these histories. It is useful to begin with a classical random walk of a particle in one dimension, as it provides a helpful guide for quantum systems, which are discussed beginning in Sec. 9.3, as well as in the next chapter. The sample space of random walks, Sec. 8.2, consists of all sequences of the form

$$\mathbf{s} = (s_0, s_1, s_2, \dots s_f), \tag{9.1}$$

where s_j, an integer in the range

$$-M_a \leq s_j \leq M_b, \tag{9.2}$$

is the position of the particle or random walker at time $t = j$.

We shall assume that the *dynamical law* for the particle's motion is that when the time changes from t to $t + 1$, the particle can take one step to the left, from s to $s - 1$, with probability p, remain where it is with probability q, or take one step to the right, from s to $s + 1$, with probability r, where

$$p + q + r = 1. \tag{9.3}$$

The probability for hops in which s changes by 2 or more is 0. The endpoints of the interval (9.1) are thought of as connected by a periodic boundary condition, so that M_b is one step to the left of $-M_a$, which in turn is one step to the right of M_b. The dynamical law can be used to generate a probability distribution on the sample space of histories in the following way. We begin by assigning to each history a

121

weight

$$W(\mathbf{s}) = \prod_{j=1}^{n} w(s_j - s_{j-1}), \tag{9.4}$$

where the hopping probabilities

$$w(-1) = p, \quad w(0) = q, \quad w(+1) = r \tag{9.5}$$

were introduced earlier, and $w(\Delta s) = 0$ for $|\Delta s| \geq 2$.

The weights by themselves do not determine a probability. Instead, they must be combined with other information, such as the starting point of the particle at $t = 0$, or a probability distribution for this starting point, or perhaps information about where the particle is located at some later time(s). This information is not contained in the dynamical laws themselves, so we shall refer to it as *contingent information* or *initial data*. The "initial" in initial data refers to the beginning of an argument or calculation, and not necessarily to the earliest time in the random walk. The single contingent piece of information "$s = 3$ at $t = 2$" can be the initial datum used to generate a probability distribution on the space of all histories of the form (9.2). Contingent information is also needed for deterministic processes. The orbit of the planet Mars can be calculated using the laws of classical mechanics, but to get the calculation started one needs to provide its position and velocity at some particular time. These data are contingent in the sense that they are not determined by the laws of mechanics, but must be obtained from observations. Once they are given, the position of Mars can be calculated at earlier as well as later times.

Contingent information in the case of a random walk is often expressed as a probability distribution $p_0(s_0)$ on the coarse sample space of positions at $t = 0$. (If the particle starts at a definite location, the distribution p_0 assigns the value 1 to this position and 0 to all others.) The probability distribution on the refined sample space of histories is then determined by a *refinement rule* that says, in essence, that for each s_0, the probability $p_0(s_0)$ is to be divided up among all the different histories which start at this point at $t = 0$, with history $\mathbf{s}$ assigned a fraction of $p_0(s_0)$ proportional to its weight $W(\mathbf{s})$. One could also use a refinement rule if the contingent data were in the form of a position or a probability distribution at some later time, say $t = 2$ or $t = f$, or if positions were given at two or more different times. Things are more complicated when probability distributions are specified at two or more times.

In order to turn the refinement rule for a probability distribution at $t = 0$ into a formula, let $J(s_0)$ be the set of all histories which begin at s_0, and

$$N(s_0) := \sum_{\mathbf{s} \in J(s_0)} W(\mathbf{s}) \tag{9.6}$$

the sum of their weights. The probability of a particular history is then given by the formula

$$\Pr(\mathbf{s}) = p_0(s_0)W(\mathbf{s})/N(s_0). \tag{9.7}$$

These probabilities sum to 1 because the initial probabilities $p_0(s_0)$ sum to 1, and because the weights have been suitably normalized by dividing by the normalization factor $N(s_0)$. In fact, for the weights defined by (9.4) and (9.5) using hopping probabilities which satisfy (9.3), it is not hard to show that the sum in (9.6) is equal to 1, so that in this particular case the normalization can be omitted from (9.7). However, it is sometimes convenient to work with weights which are not normalized, and then the factor of $1/N(s_0)$ is needed.

Suppose the particle starts at $s_0 = 2$, so that $p_0(2) = 1$. Then the histories (2, 1), (2, 2), (2, 3), and (2, 4), which are compound histories for $f \geq 2$, have probabilities p, q, r, and 0, respectively. Likewise, the histories (2, 2, 2), (2, 2, 3), (2, 3, 4), (2, 4, 3) have probabilities of q^2, qr, r^2, and 0. Any history in which the particle hops by a distance of 2 or more in a single time step has zero probability, that is, it is impossible. One could reduce the size of the sample space by eliminating impossible histories, but in practice it is more convenient to use the larger sample space.

As another example, suppose that $p_0(0) = p_0(1) = p_0(2) = 1/3$. What is the probability that $s_1 = 2$ at time $t = 1$? Think of $s_1 = 2$ as a compound history given by the collection of all histories which pass through $s = 2$ when $t = 1$, so that its probability is the sum of probabilities of the histories in this collection. Clearly histories with zero probability can be ignored, and this leaves only three two-time histories: (1, 2), (2, 2), and (3, 2). In the case $f = 1$, formula (9.7) assigns them probabilities $r/3$, $q/3$, and 0, so the answer to the question is $(q + r)/3$. This answer is also correct for $f \geq 2$, but then it is not quite so obvious. The reader may find it a useful exercise to work out the case $f = 2$, in which there are nine histories of nonzero weight passing through $s = 2$ at $t = 1$.

Once probabilities have been assigned on the sample space, one can answer questions such as: "What is the probability that the particle was at $s = 2$ at time $t = 3$, given that it arrived at $s = 4$ at time $t = 5$?" by means of conditional probabilities:

$$\Pr(s_3 = 2 \mid s_5 = 4) = \Pr\left[(s_3 = 2) \wedge (s_5 = 4)\right]/\Pr(s_5 = 4). \tag{9.8}$$

Here the event $(s_3 = 2) \wedge (s_5 = 4)$ is the compound history consisting of all elementary histories which pass through $s = 2$ at time $t = 3$ and $s = 4$ at time $t = 5$. Such conditional probabilities depend, in general, both on the initial data and the weights. However, *if a value of s_0 is one of the conditions*, then the conditional probability does not depend upon $p_0(s_0)$ (assuming $p_0(s_0) > 0$, so that the conditional probability is defined). In particular,

$$\Pr(\mathbf{s} \mid s_0 = m) = \delta_{ms_0} W(\mathbf{s})/N(s_0). \tag{9.9}$$

To obtain similar formulas in other cases, it is convenient to extend the definition of weights to include compound histories in the event algebra using the formula

$$W(E) = \sum_{\mathbf{s} \in E} W(\mathbf{s}). \tag{9.10}$$

Defining $W(E)$ for the compound event E in this way makes it an "additive set function" or "measure" in the sense that if E and F are disjoint (they have no elementary histories in common) members of the event algebra of histories, then

$$W(E \cup F) = W(E) + W(F). \tag{9.11}$$

Using this extended definition of W, one can, for example, write

$$\Pr\big[s_3 = 2 \mid (s_0 = 1) \wedge (s_5 = 4)\big] = \frac{W\big[(s_0 = 1) \wedge (s_3 = 2) \wedge (s_5 = 4)\big]}{W\big[(s_0 = 1) \wedge (s_5 = 4)\big]}. \tag{9.12}$$

That is, take the total weight of all the histories which satisfy the conditions $s_0 = 1$ and $s_5 = 4$, and find what fraction of it corresponds to histories which also have $s_3 = 2$.

9.2 Single-time probabilities

The probability that at time t the random walker of Sec. 9.1 will be located at s is given by the *single-time probability distribution*[*]

$$\rho_t(s) = \sum_{\mathbf{s} \in J_t(s)} \Pr(\mathbf{s}), \tag{9.13}$$

where the sum is over the collection $J_t(s)$ of all histories which pass through s at time t. Because the particle must be somewhere at time t, it follows that

$$\sum_s \rho_t(s) = 1. \tag{9.14}$$

It is easy to show that the dynamical law used in Sec. 9.1 implies that $\rho_t(s)$ satisfies the difference equation

$$\rho_{t+1}(s) = p\,\rho_t(s+1) + q\,\rho_t(s) + r\,\rho_t(s-1). \tag{9.15}$$

In particular, if the contingent information is given by a probability distribution at $t = 0$, so that $\rho_0(s) = p_0(s)$, (9.15) can be used to calculate $\rho_t(s)$ at any later

---

[*] The term *one-dimensional distribution* is often used, but in the present context "one-dimensional" would be misleading.

time t. For example, if the random walker starts off at $s = 0$ when $t = 0$, and $p = q = r = 1/3$, then $\rho_0(0) = 1$, while

$$
\rho_1(-1) = \rho_1(0) = \rho_1(1) = 1/3,
$$
$$
\rho_2(-2) = \rho_2(2) = 1/9, \quad \rho_2(-1) = \rho_2(1) = 2/9, \quad \rho_2(0) = 1/3
$$

(9.16)

are the nonzero values of $\rho_t(s)$ for $t = 1$ and 2.

The single-time distribution $\rho_t(s)$ is a marginal probability distribution and contains less information than the full probability distribution Pr($\mathbf{s}$) on the set of all random walks. This is so even if one knows $\rho_t(s)$ for every value of t. In particular, $\rho_t(s)$ does not tell one how the particle's position is correlated at successive times. For example, given Pr($\mathbf{s}$), one can show that the conditional probability Pr($s_{t+1} \mid s_t$) is zero whenever $|s_{t+1} - s_t|$ is larger than 1, whereas the values of $\rho_1(s)$ and $\rho_2(s)$ in (9.16) are consistent with the possibility of the particle hopping from $s = 1$ at $t = 1$ to $s = -2$ at $t = 2$. It is not a defect of $\rho_t(s)$ that it contains less information than the total probability distribution Pr($\mathbf{s}$). Less detailed descriptions are often very useful in helping one see the forest and not just the trees. But one needs to be aware of the fact that the single-time distribution as a function of time is far from being the full story.

For a Brownian particle the analog of $\rho_t(s)$ for the random walker is the *single-time probability distribution density* $\rho_t(\mathbf{r})$, defined in such a way that the integral

$$
\int_R \rho_t(\mathbf{r}) \, d\mathbf{r}
$$

(9.17)

over a region R in three-dimensional space is the probability that the particle will lie in this region at time t. In the simplest theory of Brownian motion, $\rho_t(\mathbf{r})$ satisfies a partial differential equation

$$
\partial \rho / \partial t = D \nabla^2 \rho,
$$

(9.18)

where D is the diffusion constant and ∇^2 is the Laplacian. If the particle starts off at $\mathbf{r} = 0$ when $t = 0$, the solution is

$$
\rho_t(\mathbf{r}) = (4\pi D t)^{-3/2} e^{-r^2/4Dt},
$$

(9.19)

where r is the magnitude of $\mathbf{r}$.

Just as for $\rho_t(s)$ in the case of a random walk, $\rho_t(\mathbf{r})$ lacks information about the correlation between positions of the Brownian particle at successive times. Suppose, for example, that a particle starting at $\mathbf{r} = 0$ at time $t = 0$ is at $\mathbf{r}_1$ at a time $t_1 > 0$. Then at a time $t_2 = t_1 + \epsilon$, where ϵ is small compared to t_1, there is a high probability that the particle will still be quite close to $\mathbf{r}_1$. This fact is not, however, reflected in $\rho_{t_2}(\mathbf{r})$, as (9.19) gives the probability density for the particle to be at $\mathbf{r}$ using no information beyond the fact that it was at the origin at $t = 0$.

9.3 The Born rule

As in Ch. 7, we shall consider an *isolated system* which does not interact with its environment, so that one can define unitary time development operators of the form $T(t', t)$. To describe its stochastic time development one must assign probabilities to histories forming a suitable sample space of the type discussed in Sec. 8.5. Just as in the case of the random walk considered in Sec. 9.1, these probabilities are determined both by the contingent information contained in initial data, and by a set of weights. The weights are given by the laws of quantum mechanics, and for an isolated system they can be computed using the time development operators.

In this section we consider a very simple situation in which the initial datum is a normalized state $|\psi_0\rangle$ at time t_0, and the histories involve only two times, t_0 and a later time t_1 at which there is a decomposition of the identity corresponding to an orthonormal basis $\{|\phi_1^k\rangle, k = 1, 2, \dots\}$. Histories of the form

$$Y^k = [\psi_0] \odot [\phi_1^k], \tag{9.20}$$

together with a history

$$Z = (I - [\psi_0]) \odot I \tag{9.21}$$

constitute a decomposition of the history identity $\breve{I}$, and thus a sample space of histories based upon the initial state $[\psi_0]$, to use the terminology of Sec. 8.5. We assign initial probabilities $p_0(I - [\psi_0]) = 0$ and $p_0([\psi_0]) = 1$, in the notation of Sec. 9.1.

The *Born rule* assigns a weight

$$W(Y^k) = |\langle\phi_1^k|T(t_1, t_0)|\psi_0\rangle|^2 \tag{9.22}$$

to the history Y^k. These weights sum to 1,

$$\sum_{k>0} W(Y^k) = \sum_k \langle\psi_0|T(t_0, t_1)|\phi_1^k\rangle\langle\phi_1^k|T(t_1, t_0)|\psi_0\rangle$$
$$= \langle\psi_0|T(t_0, t_1)T(t_1, t_0)|\psi_0\rangle = \langle\psi_0|I|\psi_0\rangle = \langle\psi_0|\psi_0\rangle = 1, \tag{9.23}$$

because $|\psi_0\rangle$ is normalized and the $\{|\phi_1^k\rangle\}$ are an orthonormal basis. It is important to notice that the Born rule does not follow from any other principle of quantum mechanics. It is a fundamental postulate or axiom, the same as Schrödinger's equation. The weights can be used to assign probabilities to histories using the obvious analog of (9.7), with the normalization N equal to 1 because of (9.23):

$$\Pr(\phi_1^k) = \Pr(Y^k) = W(Y^k) = |\langle\phi_1^k|T(t_1, t_0)|\psi_0\rangle|^2, \tag{9.24}$$

where $\Pr(\phi_1^k)$, which could also be written as $\Pr(\phi_1^k \mid \psi_0)$, is the probability of the event ϕ_1^k at time t_1. The square brackets around ϕ_1^k have been omitted where these dyads appear as arguments of probabilities, since this makes the notation less awkward, and there is no risk of confusion. Given an observable of the form

$$V = V^\dagger = \sum_k v_k[\phi_1^k] = \sum_k v_k|\phi_1^k\rangle\langle\phi_1^k|, \tag{9.25}$$

one can compute its average, see (5.42), at the time t_1 using the probability distribution $\Pr(\phi_1^k)$:

$$\langle V\rangle = \sum_k v_k \Pr(\phi_1^k) = \langle\psi_0|T(t_0, t_1)VT(t_1, t_0)|\psi_0\rangle. \tag{9.26}$$

The validity of the right side becomes obvious when V is replaced by the right side of (9.25).

Let us analyze two simple but instructive examples. Consider a spin-half particle in zero magnetic field, so that the spin dynamics is trivial: $H = 0$ and $T(t', t) = I$. Let the initial state be

$$|\psi_0\rangle = |z^+\rangle. \tag{9.27}$$

For the first example use

$$|\phi_1^1\rangle = |z^+\rangle, \quad |\phi_1^2\rangle = |z^-\rangle \tag{9.28}$$

as the orthonormal basis at t_1. Then (9.24) results in

$$\Pr(\phi_1^1) = \Pr(z^+) = 1, \quad \Pr(\phi_1^2) = \Pr(z^-) = 0. \tag{9.29}$$

We have here an example of a unitary family of histories as defined in Sec. 8.7. Since the ket $T(t_1, t_0)|\psi_0\rangle$ is equal to one of the basis vectors at t_1, it is necessarily orthogonal to the other basis vector. Thus the unitary history $[\psi_0] \odot [\phi_1^1]$ has probability 1, whereas the other history $[\psi_0] \odot [\phi_1^2]$ which begins with $[\psi_0]$ has probability 0. It follows from (9.29) that

$$\langle S_z\rangle = 1/2, \tag{9.30}$$

where $S_z = \frac{1}{2}([z^+] - [z^-])$ is the operator for the z-component of spin angular momentum in units of $\hbar$ — see (5.30).

The second example uses the same initial state (9.27), but at t_1 an orthonormal basis

$$|\bar{\phi}_1^1\rangle = |x^+\rangle, \quad |\bar{\phi}_1^2\rangle = |x^-\rangle, \tag{9.31}$$

where bars have been added to distinguish these kets from those in (9.28). A straightforward calculation yields

$$\Pr(x^+) = 1/2 = \Pr(x^-). \tag{9.32}$$

Stated in words, if $S_z = +1/2$ at t_0, the probability is 1/2 that $S_x = +1/2$ at t_1, and $1/2$ that $S_x = -1/2$. Consequently, the average of the x-component of angular momentum is

$$\langle S_x \rangle = 0. \tag{9.33}$$

The second example may seem counterintuitive for the following reason. The unitary quantum dynamics is trivial: nothing at all is happening to this spin-half particle. It is not in a magnetic field, and therefore there is no reason why the spin should precess. Nonetheless, it might seem as if the spin orientation has managed to "jump" from being along the positive z axis at time t_0 to an orientation either along or opposite to the positive x axis at t_1. However, the idea that something is "jumping" comes from a misleading mental picture of a spin-half particle. To better understand the situation, imagine a classical object spinning in free space and not subject to any torques, so that its angular momentum is conserved. Suppose we know the z-component of its angular momentum at t_0, and for some reason want to discuss the x-component at a later time t_1. The fact that two different components of angular momentum are considered at the two different times does not mean there has been a change in the angular momentum of the object between t_0 and t_1. This analogy, like all classical analogies, is far from perfect, but in the present context it is less misleading than thinking of $S_z = +1/2$ for a spin-half particle as corresponding to a classical object with its total angular momentum in the $+z$ direction. Applying this analogy to the quantum case, we see that the probabilities in (9.32) are not unreasonable, given that we have adopted a sample space in which values of S_x occur at t_1, rather than values of S_z, as in the first example.

The odd thing about quantum theory is the fact that one cannot combine the conclusions in (9.29) and (9.32) to form a single description of the time development of the particle, whereas it would be perfectly reasonable to do so for a classical spinning object. It is incorrect to conclude from (9.29) and (9.32) that at t_1 either it is the case that $S_z = +1/2$ AND $S_x = +1/2$, or else it is the case that $S_z = +1/2$ AND $S_x = -1/2$. Both of the statements connected by AND are quantum nonsense, as they do not correspond to anything in the quantum Hilbert space; see Sec. 4.6. For the same reason the two averages (9.30) and (9.33) cannot be thought of as applying simultaneously to the same system, since the observables S_z and S_x do not commute with each other, and hence correspond to incompatible sample spaces. It is always possible to apply the Born rule in a large number of different ways by using different orthonormal bases at t_1, but these different results cannot be combined in a single sensible quantum description of the system. Attempting to do so violates the single-framework rule (to be discussed in Sec. 16.1) and leads to confusion.

The Born rule is often discussed in the context of *measurements*, as a formula to compute the probabilities of various outcomes of a measurement carried out by an apparatus $\mathcal{A}$ on a system $\mathcal{S}$. Hence it is worth emphasizing that the probabilities in (9.24) refer to an *isolated* system $\mathcal{S}$ which is *not* interacting with a separate measurement device. Indeed, our discussion of the Born rule has made no reference whatsoever to measurements of any sort. Measurements will be taken up in Chs. 17 and 18, where the usual formulas for the probabilities of different measurement outcomes will be derived by applying general quantum principles to the combined apparatus and measured system thought of as constituting a single, isolated system.

9.4 Wave function as a pre-probability

The basic formula (9.24) which expresses the Born rule can be rewritten in various ways. One rather common form is the following. Let

$$|\psi_1\rangle = T(t_1, t_0)|\psi_0\rangle \tag{9.34}$$

be the wave function obtained by integrating Schrödinger's equation from t_0 to t_1. Then (9.24) can be written in the compact form

$$\Pr(\phi_1^k) = |\langle\phi_1^k|\psi_1\rangle|^2. \tag{9.35}$$

Note that $|\psi_1\rangle$ or $[\psi_1]$, regarded as a quantum property at time t_1, is *incompatible* with the collection of properties $\{[\phi_1^k]\}$ if at least two of the probabilities in (9.35) are nonzero, that is, if one is not dealing with a unitary family. Thus in the second spin-half example considered above, $|\psi_1\rangle = |z^+\rangle$ is incompatible with both $|x^+\rangle$ and $|x^-\rangle$. Therefore, in the context of the family based on (9.20) and (9.21) it does not make sense to suppose that at t_1 the system possesses the *physical property* $|\psi_1\rangle$. Instead, $|\psi_1\rangle$ must be thought of as a *mathematical* construct suitable for calculating certain probabilities. We shall refer to $|\psi_1\rangle$ understood in this way as a *pre-probability*, since it is (obviously) not a probability, nor a property of the physical system, but instead something which is used to calculate probabilities. In addition to wave functions obtained by unitary time development, density matrices are often employed in quantum theory as pre-probabilities; see Ch. 15. The pre-probability $|\psi_1\rangle$ is very convenient for calculations because it does not depend upon which orthonormal basis $\{|\phi_1^k\rangle\}$ is employed at t_1. The theoretical physicist may want to compute probabilities for various different bases, that is, for various different families of histories, and $|\psi_1\rangle$ is a convenient tool for doing this. There is no harm in carrying out such calculations as long as one does not try to combine the results for incompatible bases into a single description of the quantum system.

Another way to see that $|\psi_1\rangle$ on the right side of (9.35) is a calculational device and not a physical property is to note that these probabilities can be computed equally well by an alternative procedure. For each k, let

$$|\phi_0^k\rangle = T(t_0, t_1)|\phi_1^k\rangle \qquad (9.36)$$

be the ket obtained by integrating Schrödinger's equation backwards in time from the final state $|\phi_1^k\rangle$. It is then obvious, see (9.24), that

$$\Pr(\phi_1^k) = |\langle \phi_0^k | \psi_0 \rangle|^2. \qquad (9.37)$$

There is no reason in principle to prefer (9.35) to (9.37) as a method of calculating these probabilities, and in fact there are a lot of other methods of obtaining the same answer. For example, one can integrate $|\psi_0\rangle$ forwards in time and each $|\phi_1^k\rangle$ backwards in time until they meet at some intermediate time, and then evaluate the absolute square of the inner product. To be sure, the most efficient procedure for calculating $\Pr(\phi_1^k \mid \psi_0)$ for all values of k is likely to be (9.35): one only has to do one time integration, and then evaluate a number of inner products. But the fact that other procedures are equally valid, and can give very different "pictures" of what is going on at intermediate times if one takes them literally, is a warning that one has no more justification for identifying $|\psi_1\rangle$, as defined in (9.34), as "the real state of the system" at time t_1 than one has for identifying one or more of the $|\phi_0^k\rangle$, as defined in (9.36), with "the real state of the system" at time t_0. Instead, both $|\psi_1\rangle$ and the $|\phi_0^k\rangle$ are functioning as pre-probabilities.

It is evident from (9.26) and (9.34) that the average of an observable V at time t_1 can be written in the compact and convenient form

$$\langle V \rangle = \langle \psi_1 | V | \psi_1 \rangle, \qquad (9.38)$$

where $|\psi_1\rangle$ is again functioning as a pre-probability. A similar expression holds for any other observable W, and there is no harm in simultaneously calculating averages for $\langle V \rangle$, $\langle W \rangle$ provided one keeps in mind the fact that when V and W do not commute with each other, one cannot regard $\langle V \rangle$ and $\langle W \rangle$ as belonging to a single (stochastic) description of a quantum system, for the two averages are necessarily based on incompatible sample spaces that cannot be combined. See the comments towards the end of Sec. 9.3 in connection with the example of a spin-half particle. Any time the symbol $\langle V \rangle$ is used with reference to the physical properties of a quantum system there is an implicit reference to a sample space, and ignoring this fact can lead to serious misunderstanding.

It is important to remember when applying the Born formula that a family of histories involving two times *tells us nothing at all about what happens at intermediate times*. Such times can, of course, be introduced formally by extending the history, in the manner indicated in Sec. 8.4,

$$Y^k = [\psi_0] \odot [\phi_1^k] = [\psi_0] \odot I \odot I \odot \cdots I \odot [\phi_1^k], \qquad (9.39)$$

for as many intermediate times as one wants. But each I at the intermediate time tells us nothing at all about what actually happens at this time. Imagine being outdoors on a dark night during a thunder storm. Each time the lightning flashes you can see the world around you. Between flashes, you cannot tell what is going on. To be sure, if we are curious about what is going on at intermediate times in a quantum history of the form (9.39), we can *refine* the history in the manner indicated in Sec. 8.6, by writing the projector as a sum of history projectors which include nontrivial information about the intermediate times, and then compute probabilities for these different possibilities. That, however, *cannot* be done by means of the Born formula (9.22), and requires an extension of this formula which will be introduced in the next chapter.

A similar restriction applies to a wave function understood as a pre-probability. Even if

$$|\psi_t\rangle = T(t, t_0)|\psi_0\rangle \tag{9.40}$$

is known for all values of the time t, it can only be used to compute probabilities of histories involving just two times, t_0 and t. These probabilities are the quantum analogs of the single-time probabilities $\rho_t(\mathbf{r})$ for a classical Brownian particle which started off at a definite location at the initial time t_0. As discussed in Sec. 9.2, $\rho_t(\mathbf{r})$ does not contain probabilistic information about correlations between particle positions at intermediate times, and in the same way correlations between quantum properties at different times cannot be computed from $|\psi_t\rangle$. Instead, one must use the procedures discussed in the next chapter.

9.5 Application: Alpha decay

A toy model of alpha decay was introduced in Sec. 7.4, see Fig. 7.2, as an example of unitary time evolution. In this section we shall apply the Born formula in order to calculate some of the associated probabilities, but before doing so it will be convenient to add a toy detector of the sort shown in Fig. 7.1, in order to detect the alpha particle after it leaves the nucleus, see Fig. 9.1. Let $\mathcal{M}$ be the Hilbert space of the particle, and $\mathcal{N}$ that of the detector. For the combined system $\mathcal{M} \otimes \mathcal{N}$ we define the time development operator to be

$$T = S_a R, \tag{9.41}$$

where S_a is defined in (7.56) and R in (7.53). Note the similarity with (7.52), which means that the discussion of the operation of the detector found in Sec. 7.4, see Fig. 7.1, applies to the arrangement in Fig. 9.1, with a few obvious modifications.

Assume that at $t = 0$ the alpha particle is at $m = 0$ inside the nucleus, which has

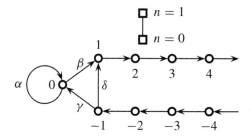

Fig. 9.1. Toy model of alpha decay (Fig. 7.2) plus a detector.

not yet decayed, and the detector is in its ready state $n = 0$, so the wave function for the total system is

$$|\Psi_0\rangle = |m = 0\rangle \otimes |n = 0\rangle = |0, 0\rangle. \tag{9.42}$$

Unitary time evolution using (9.41) results in

$$|\Psi_t\rangle = T^t |\Psi_0\rangle = |\chi_t\rangle \otimes |0\rangle + |\omega_t\rangle \otimes |1\rangle, \tag{9.43}$$

where

$$
\begin{aligned}
|\chi_1\rangle &= \alpha|0\rangle + \beta|1\rangle, & |\omega_1\rangle &= 0, \\
|\chi_2\rangle &= \alpha^2|0\rangle + \alpha\beta|1\rangle + \beta|2\rangle, & |\omega_2\rangle &= 0, \\
|\chi_3\rangle &= \alpha^3|0\rangle + \alpha^2\beta|1\rangle + \alpha\beta|2\rangle, & |\omega_3\rangle &= \beta|3\rangle,
\end{aligned}
\tag{9.44}
$$

and for $t \geq 4$

$$
\begin{aligned}
|\chi_t\rangle &= \alpha^t|0\rangle + \alpha^{t-1}\beta|1\rangle + \alpha^{t-2}\beta|2\rangle, \\
|\omega_t\rangle &= \alpha^{t-3}\beta|3\rangle + \alpha^{t-4}\beta|4\rangle + \cdots \beta|t\rangle.
\end{aligned}
\tag{9.45}
$$

Let us apply the Born rule with $t_0 = 0$, $t_1 = t$ for some integer $t > 0$, using $|\Psi_0\rangle$ as the initial state at time t_0, and at time t_1 the orthonormal basis $\{|m, n\rangle\}$, in which the alpha particle has a definite position m and the detector either has or has not detected the particle. The joint probability distribution of m and n at time t,

$$p_t(m, n) := \Pr([m, n]_t), \tag{9.46}$$

is easily computed by regarding $|\Psi_t\rangle$ in (9.43) as a pre-probability: $p_t(m, n)$ is the absolute square of the coefficient of $|m\rangle$ in $|\chi_t\rangle$ if $n = 0$, or in $|\omega_t\rangle$ if $n = 1$. These probabilities vanish except for the cases

$$p_t(0, 0) = e^{-t/\tau}, \tag{9.47}$$

$$p_t(m, 0) = \kappa e^{-(t-m)/\tau} \text{ for } m = 1, 2 \text{ and } m \leq t, \tag{9.48}$$

$$p_t(m, 1) = \kappa e^{-(t-m)/\tau} \text{ for } 3 \leq m \leq t. \tag{9.49}$$

The positive constants κ and τ are defined by

$$e^{-1/\tau} = |\alpha|^2, \quad \kappa = |\beta|^2 = 1 - |\alpha|^2. \tag{9.50}$$

The probabilities in (9.47)–(9.50) make good physical sense. The probability (9.47) that the alpha particle is still in the nucleus decreases exponentially with time, in agreement with the well-known exponential decay law for radioactive nuclei. That $p_t(m, n)$ vanishes for m larger than t reflects the fact that the alpha particle was (by assumption) inside the nucleus at $t = 0$ and, since it hops at most one step during any time interval, cannot arrive at m earlier than $t = m$. Finally, if the alpha particle is at $m = 0, 1$ or 2, the detector is still in its ready state $n = 0$, whereas for $m = 3$ or larger the detector will be in the state $n = 1$, indicating that it has detected the particle. This is just what one would expect for a detector designed to detect the particle as it hops from $m = 2$ to $m = 3$ (see the discussion in Sec. 7.4).

It is worth emphasizing once again that $p_t(m, n)$ is the quantum analog of the single-time probability $\rho_t(s)$ for the random walk discussed in Sec. 9.2. The reason is that the histories to which the Born rule applies involve only two times, t_0 and t_1 in the notation of Sec. 9.3, and thus no information is available as to what happens between these times. Consequently, just as $\rho_t(s)$ does not tell us all there is to be said about the stochastic behavior of a random walker, there is also more to the story of (toy) alpha decay and its detection than is contained in $p_t(m, n)$. However, providing a more detailed description of what is going on requires the additional mathematical tools introduced in the next chapter, and we shall return to the problem of alpha decay using more sophisticated methods (and a better detector) in Sec. 12.4.

It is not necessary to employ the basis $\{|m, n\rangle\}$ in order to apply the Born rule; one could use any other orthonormal basis of $\mathcal{M} \otimes \mathcal{N}$, and there are many possibilities. However, the physical properties which can be described by the resulting probabilities depend upon which basis is used, and not every choice of basis at time t (an example will be considered in the next section) allows one to say whether $n = 0$ or 1, that is, whether the detector has detected the particle. It is customary to use the term *pointer basis* for an orthonormal basis, or more generally a decomposition of the identity such as employed in the generalized Born rule defined in Sec. 10.3, that allows one to discuss the outcomes of a measurement in a sensible way. (The term arises from a mental picture of a measuring device equipped with a visible pointer whose position indicates the outcome after the measurement is over.) Thus $\{|m, n\rangle\}$ is a pointer basis, but so is any basis of the form $\{|\xi^j, n\rangle\}$, where $\{|\xi^j\rangle\}$, $j = 1, 2, \ldots$, is some orthonormal basis of $\mathcal{M}$. While quantum calculations which are to be compared with experiments usually employ a pointer basis for calculating probabilities, for obvious reasons, there is no fundamental principle of quantum theory which restricts the Born rule to bases of this type.

9.6 Schrödinger's cat

What is the physical significance of the state $|\Psi_t\rangle$ which evolves unitarily from $|\Psi_0\rangle$ in the toy model discussed in the preceding section? This is a difficult question to answer, because for $t \geq 3$ $|\Psi_t\rangle$ is of the form $|A\rangle + |B\rangle$, (9.43), where $|A\rangle = |\chi_t\rangle \otimes |0\rangle$ has the significance that the alpha particle is inside or very close to the nucleus and the detector is ready, whereas $|B\rangle = |\omega_t\rangle \otimes |1\rangle$ means that the detector has triggered and the nucleus has decayed. What can be the significance of a linear combination $|A\rangle + |B\rangle$ of states with quite distinct physical meanings? Could it signify that the detector both has and has not detected the particle?

The difficulty of interpreting such wave functions is often referred to as the problem or paradox of *Schrödinger's cat*. In a famous paper Schrödinger pointed out that in the case of alpha decay, unitary time evolution applied to the system consisting of a decaying nucleus plus a detector will quite generally lead to a superposition state $|S\rangle = |A\rangle + |B\rangle$, where the (macroscopic) detector either has, state $|B\rangle$, or has not, state $|A\rangle$, detected the alpha particle. To dramatize the conceptual difficulty Schrödinger imagined the detector hooked up to a device which would kill a live cat upon detection of an alpha particle, thus raising the problem of interpreting $|A\rangle + |B\rangle$ when $|A\rangle$ corresponds to an undecayed nucleus, untriggered detector, and live cat, and $|B\rangle$ to a nucleus which has decayed, a triggered detector, and a dead cat. We shall call $|A\rangle + |B\rangle$ a *macroscopic quantum superposition* or MQS state when $|A\rangle$ and $|B\rangle$ correspond to situations which are macroscopically distinct, and use the same terminology for a superposition of three or more macroscopically distinct states. In the literature MQS states are often called *Schrödinger cat states*.

Rather than addressing the general problem of MQS states, let us return to the toy model with its toy example of such a state and, to be specific, consider $|\Psi_5\rangle$ at $t = 5$ under the assumption that α and β have been chosen so that $\langle\chi_5|\chi_5\rangle$ and $\langle\omega_5|\omega_5\rangle$ are of the same order of magnitude, which will prevent us from escaping the problem of interpretation by supposing that either $|A\rangle$ or $|B\rangle$ is very small and can be ignored. It is easy to show that $[\Psi_5] = |\Psi_5\rangle\langle\Psi_5|$ does not commute with either of the projectors $[n = 0]$ or $[n = 1]$. Nor does it commute with a projector $[\hat{n}]$, where $|\hat{n}\rangle$ is some linear combination of $|n = 0\rangle$ and $|n = 1\rangle$. This means that it makes no sense to say that the combined system has the property $[\Psi_5]$, whatever its physical significance might be, while at the very same time the detector has or has not detected the particle (or has some other physical property). See the discussion in Sec. 4.6. Saying that the system is in the state $[\Psi_5]$ and then ascribing a property to the detector is no more meaningful than assigning simultaneous values to S_x and S_z for a spin-half particle. The converse is also true: if it makes sense (using an appropriate quantum description) to say that the detector is either ready or triggered at $t = 5$, then one cannot say that the combined system has the property $[\Psi_5]$, because that would be nonsense.

Note that these considerations cause no problem for the analysis in Sec. 9.5, because in applying the Born rule to the basis $\{|m, n\rangle\}$, $|\Psi_t\rangle$ is employed as a pre-probability, Sec. 9.4, a convenient mathematical tool for calculating probabilities which could also be computed by other methods. When it is used in this way there is obviously no need to ascribe some physical significance to $|\Psi_5\rangle$, nor is there any motivation for doing so, since $[\Psi_5]$ must in any case be excluded from any meaningful quantum description based upon $\{|m, n\rangle\}$.

Very similar considerations apply to the situation considered by Schrödinger, although analyzing it carefully requires a model of macroscopic measurement, see Secs. 17.3 and 17.4. The question of whether the cat is dead or alive can be addressed by using the Born rule with an appropriate pointer basis (as defined at the end of Sec. 9.5), and one never has to give a physical interpretation to Schrödinger's MQS state $|S\rangle$, since it only enters the calculations as a pre-probability. In any case, treating $[S]$ as a physical property is meaningless when one uses a pointer basis. To be sure, this does not prevent one from asking whether $|S\rangle$ by itself has some intuitive physical meaning. What the preceding discussion shows is that whatever that meaning may be, it cannot possibly have anything to do with whether the cat is dead or alive, as these properties will be incompatible with $[S]$. Indeed, it is probably the case that the very concept of a "cat" (small furry animal, etc.) cannot be meaningfully formulated in a way which is compatible with $[S]$.

Quite apart from MQS states, it is in general a mistake to associate a physical meaning with a linear combination $|C\rangle = |A\rangle + |B\rangle$ by referring to the properties of the separate states $|A\rangle$ and $|B\rangle$. For example, the state $|x^+\rangle$ for a spin-half particle is a linear combination of $|z^+\rangle$ and $|z^-\rangle$, but its physical signficance of $S_x = 1/2$ is unrelated to $S_z = \pm 1/2$. For another example, see the discussion of (2.27) in Sec. 2.5. In addition, there is the problem that for a given $|C\rangle = |A\rangle + |B\rangle$, the choice of $|A\rangle$ and $|B\rangle$ is far from unique. Think of an ordinary vector in three dimensions: there are lots of ways of writing it as the sum of two other vectors, even if one requires that these be mutually perpendicular, corresponding to the not-unreasonable orthogonality condition $\langle A|B\rangle = 0$. But if $|C\rangle$ is equal to $|A'\rangle + |B'\rangle$ as well as to $|A\rangle + |B\rangle$, why base a physical interpretation upon $|A\rangle$ rather than $|A'\rangle$? See the discussion of (2.28) in Sec. 2.5.

Returning once again to the toy model, it is worth emphasizing that $|\Psi_5\rangle$ is a perfectly good element of the Hilbert space, and enters fundamental quantum theory on precisely the same footing as all other states, despite our difficulty in assigning it a simple intuitive meaning. In particular, we can choose an orthonormal basis at

$t_1 = 5$ which contains $|\Psi_5\rangle$ as one of its members, and apply the Born rule. The result is that a weight of 1 is assigned to the unitary history $[\Psi_0] \odot [\Psi_5]$, and a weight of 0 to all other histories in the family with initial state $[\Psi_0]$. This means that the state $[\Psi_5]$ will certainly occur (probability 1) at $t = 5$ given the initial state $[\Psi_0]$ at $t = 0$.

But if $[\Psi_5]$ occurs with certainty, how is it possible for there to be a different quantum description in which $[n = 0]$ occurs with a finite probability, when we know the two properties $[\Psi_5]$ and $[n = 0]$ cannot consistently enter the same quantum description at the same time? The brief answer is that quantum probabilities only have meaning within specific families, and those from incompatible families — the term will be defined in Sec. 10.4, but we have here a particular instance — cannot be combined. Going beyond the brief answer to a more detailed discussion requires the material in the next chapter and its application to some additional examples. The general principle which emerges is called the *single-family* or *single-framework* rule, and is discussed in Sec. 16.1.

10

Consistent histories

10.1 Chain operators and weights

The previous chapter showed how the Born rule can be used to assign probabilities to a sample space of histories based upon an initial state $|\psi_0\rangle$ at t_0, and an orthonormal basis $\{|\phi_1^\alpha\rangle\}$ of the Hilbert space at a later time t_1. In this chapter we show how an extension of the Born rule can be used to assign probabilities to much more general families of histories, including histories defined at an arbitrary number of different times and using decompositions of the identity which are not limited to pure states, provided certain consistency conditions are satisfied.

We begin by rewriting the Born weight (9.22) for the history

$$Y^\alpha = [\psi_0] \odot [\phi_1^\alpha] \tag{10.1}$$

in the following way:

$$
\begin{aligned}
W(Y^\alpha) &= |\langle \phi_1^\alpha | T(t_1, t_0) |\psi_0\rangle|^2 = \langle \psi_0 | T(t_0, t_1) |\phi_1^\alpha\rangle \langle \phi_1^\alpha | T(t_1, t_0) |\psi_0\rangle \\
&= \mathrm{Tr}\Big([\psi_0] \, T(t_0, t_1) \, [\phi_1^\alpha] \, T(t_1, t_0) \Big) = \mathrm{Tr}\Big(K^\dagger(Y^\alpha) K(Y^\alpha) \Big), \quad (10.2)
\end{aligned}
$$

where the *chain operator* $K(Y)$ and its adjoint are given by the expressions

$$K(F_0 \odot F_1) = F_1 T(t_1, t_0) F_0, \quad K^\dagger(F_0 \odot F_1) = F_0 T(t_0, t_1) F_1 \tag{10.3}$$

in the case of a history $Y = F_0 \odot F_1$ involving just two times; recall that $T(t_0, t_1) = T^\dagger(t_1, t_0)$. The steps from the left side to the right side of (10.2) are straightforward but not trivial, and the reader may wish to work through them. Recall that if $|\psi\rangle$ is any normalized ket, $[\psi] = |\psi\rangle\langle\psi|$ is the projector onto the one-dimensional subspace containing $|\psi\rangle$, and $\langle\psi|A|\psi\rangle$ is equal to $\mathrm{Tr}(|\psi\rangle\langle\psi|A) = \mathrm{Tr}([\psi]A)$.

For a general history of the form

$$Y = F_0 \odot F_1 \odot F_2 \odot \cdots F_f \tag{10.4}$$

with events at times $t_0 < t_1 < t_2 < \cdots t_f$ the chain operator is defined as

$$K(Y) = F_f T(t_f, t_{f-1}) F_{f-1} T(t_{f-1}, t_{f-2}) \cdots T(t_1, t_0) F_0, \qquad (10.5)$$

and its adjoint is given by the expression

$$K^\dagger(Y) = F_0 T(t_0, t_1) F_1 T(t_1, t_2) \cdots T(t_{f-1}, t_f) F_f. \qquad (10.6)$$

Notice that the adjoint is formed by replacing each $\odot$ in (10.4) separating F_j from F_{j+1} by $T(t_j, t_{j+1})$. In both K and $K^\dagger$, each argument of any given T is adjacent to a projector representing an event at this particular time. One could just as well define $K(Y)$ using (10.6) and its adjoint $K^\dagger(Y)$ using (10.5). The definition used here is slightly more convenient for some purposes, but either convention yields exactly the same expressions for weights and consistency conditions, so there is no compelling reason to employ one rather than the other. Note that Y is an operator on the history Hilbert space $\breve{\mathcal{H}}$, and $K(Y)$ an operator on the original Hilbert space $\mathcal{H}$. Operators of these two types should not be confused with one another.

The definition of $K(Y)$ in (10.5) makes good sense when the F_j in (10.4) are any operators on the Hilbert space, not just projectors. In addition, K can be extended by linearity to sums of tensor product operators of the type (10.4):

$$K(Y' + Y'' + Y''' + \cdots) = K(Y') + K(Y'') + K(Y''') + \cdots. \qquad (10.7)$$

In this way, K becomes a linear map of operators on the history space $\breve{\mathcal{H}}$ to operators on the Hilbert space $\mathcal{H}$ of a system at a single time, and it is sometimes useful to employ this extended definition.

The sequence of operators which make up the "chain" on the right side of (10.6) is in the same order as the sequence of times $t_0 < t_1 < \cdots t_f$. This is important; one does *not* get the same answer (in general) if the order is different. Thus for $f = 2$, with $t_0 < t_1 < t_2$, the operator defined by (10.6) is different from

$$L^\dagger(Y) = F_0 T(t_0, t_2) F_2 T(t_2, t_1) F_1, \qquad (10.8)$$

and it is $K(Y)$ not $L(Y)$ which yields physically sensible results.

When all the projectors in a history are onto pure states, the chain operator has a particularly simple form when written in terms of dyads. For example, if

$$Y = |\psi_0\rangle\langle\psi_0| \odot |\phi_1\rangle\langle\phi_1| \odot |\omega_2\rangle\langle\omega_2|, \qquad (10.9)$$

then the chain operator

$$K(Y) = \langle\omega_2|T(t_2, t_1)|\phi_1\rangle \cdot \langle\phi_1|T(t_1, t_0)|\psi_0\rangle \cdot |\omega_2\rangle\langle\psi_0| \qquad (10.10)$$

is a product of complex numbers, often called *transition amplitudes*, with a dyad $|\omega_2\rangle\langle\psi_0|$ formed in an obvious way from the first and last projectors in the history.

Given any projector Y on the history space $\breve{\mathcal{H}}$, we assign to it a nonnegative weight

$$W(Y) = \text{Tr}[K^\dagger(Y)K(Y)] = \langle K(Y), K(Y) \rangle, \qquad (10.11)$$

where the angular brackets on the right side denote an *operator inner product* whose general definition is

$$\langle A, B \rangle := \text{Tr}[A^\dagger B], \qquad (10.12)$$

with A and B any two operators on $\mathcal{H}$. In an infinite-dimensional space the formula (10.12) does not always make sense, since the trace of an operator is not defined if one cannot write it as a convergent sum. Technical issues can be avoided by restricting oneself to a finite-dimensional Hilbert space, where the trace is always defined, or to operators on infinite-dimensional spaces for which (10.12) does makes sense.

Operators on a Hilbert space $\mathcal{H}$ form a linear vector space under addition and multiplication by (complex) scalars. If $\mathcal{H}$ is n-dimensional, its operators form an n^2-dimensional Hilbert space if one uses (10.12) to define the inner product. This inner product has all the usual properties: it is antilinear in its left argument, linear in its right argument, and satisfies:

$$\langle A, B \rangle^* = \langle B, A \rangle, \quad \langle A, A \rangle \geq 0, \qquad (10.13)$$

with $\langle A, A \rangle = 0$ only if $A = 0$; see (3.92). Consequently, the weight $W(Y)$ defined by (10.11) is a nonnegative real number, and it is zero if and only if the chain operator $K(Y)$ is zero. If one writes the operators as matrices using some fixed orthonormal basis of $\mathcal{H}$, one can think of them as n^2-component vectors, where each matrix element is one of the components of the vector. Addition of operators and multiplying an operator by a scalar then follow the same rules as for vectors, and the same is true of inner products. In particular, $\langle A, A \rangle$ is the sum of the absolute squares of the n^2 matrix elements of A.

If $\langle A, B \rangle = 0$, we shall say that the operators A and B are (mutually) *orthogonal*. Just as in the case of vectors in the Hilbert space, $\langle A, B \rangle = 0$ implies $\langle B, A \rangle = 0$, so orthogonality is a symmetrical relationship between A and B. Earlier we introduced a different definition of operator orthogonality by saying that two projectors P and Q are orthogonal if and only if $PQ = 0$. Fortunately, the new definition of orthogonality is an extension of the earlier one: if P and Q are projectors, they are also positive operators, and the argument following (3.93) in Sec. 3.9 shows that $\text{Tr}[PQ] = 0$ if and only if $PQ = 0$.

It is possible to have a history with a nonvanishing projector Y for which $K(Y) = 0$. These histories (and only these histories) have zero weight. We shall say that they are *dynamically impossible*. They never occur, because they have probability zero. For example, $[z^+] \odot [z^-]$ for a spin-half particle is dynamically impossible in the case of trivial dynamics, $T(t', t) = I$.

10.2 Consistency conditions and consistent families

Classical weights of the sort used to assign probabilities in stochastic processes such as a random walk, see Sec. 9.1, have the property that they are *additive* functions on the sample space: if E and F are two disjoint collections of histories from the sample space, then, as in (9.11),

$$W(E \cup F) = W(E) + W(F). \tag{10.14}$$

If quantum weights are to function the same way as classical weights, they too must satisfy (10.14), or its quantum analog. Suppose that our sample space of histories is a decomposition $\{Y^\alpha\}$ of the history identity. Any projector Y in the corresponding Boolean algebra can be written as

$$Y = \sum_\alpha \pi^\alpha Y^\alpha, \tag{10.15}$$

where each π^α is 0 or 1. Additivity of W then corresponds to

$$W(Y) = \sum_\alpha \pi^\alpha W(Y^\alpha). \tag{10.16}$$

However, the weights defined using (10.11) do not, in general, satisfy (10.16). Since the chain operator is a linear map, (10.15) implies that

$$K(Y) = \sum_\alpha \pi^\alpha K(Y^\alpha). \tag{10.17}$$

If we insert this in (10.11), and use the (anti)linearity of the operator inner product (note that the π^α are real), the result is

$$W(Y) = \sum_\alpha \sum_\beta \pi^\alpha \pi^\beta \langle K(Y^\alpha), K(Y^\beta) \rangle, \tag{10.18}$$

whereas the right side of (10.16) is given by

$$\sum_\alpha \pi^\alpha W(Y^\alpha) = \sum_\alpha \pi^\alpha \langle K(Y^\alpha), K(Y^\alpha) \rangle. \tag{10.19}$$

In general, (10.18) and (10.19) will be different. However, in the case in which

$$\langle K(Y^\alpha), K(Y^\beta) \rangle = 0 \text{ for } \alpha \neq \beta, \tag{10.20}$$

only the diagonal terms $\alpha = \beta$ remain in the sum (10.18), so it is the same as (10.19), and the additivity condition (10.16) will be satisfied. Thus a sufficient condition for the quantum weights to be additive is that the chain operators associated with the different histories in the sample space be *mutually orthogonal* in terms of the inner product defined in (10.12). The approach we shall adopt is to limit ourselves to sample spaces of quantum histories for which the equalities in (10.20), known as *consistency conditions*, are fulfilled. Such sample spaces, or the corresponding Boolean algebras, will be referred to as *consistent families* of histories, or *frameworks*.

The consistency conditions in (10.20) are also called "decoherence conditions", and the terms "decoherent family", "consistent set", and "decoherent set" are sometimes used to denote a consistent family or framework. The adjective "consistent", as we have defined it, applies to families of histories, and not to individual histories. However, a single history Y can be said to be inconsistent if there is no consistent family which contains it as one of the members of its Boolean algebra. For an example, see Sec. 11.8.

A consequence of the consistency conditions is the following. Let Y and $\bar{Y}$ be any two history projectors belonging to the Boolean algebra generated by the decomposition $\{Y^\alpha\}$. Then

$$Y\bar{Y} = 0 \text{ implies } \langle K(Y), K(\bar{Y}) \rangle = 0. \tag{10.21}$$

To see that this is true, write Y and $\bar{Y}$ in the form (10.15), using coefficients $\bar{\pi}^\alpha$ for $\bar{Y}$. Then

$$Y\bar{Y} = \sum_\alpha \pi^\alpha \bar{\pi}^\alpha Y^\alpha, \tag{10.22}$$

so $Y\bar{Y} = 0$ implies that

$$\pi^\alpha \bar{\pi}^\alpha = 0 \tag{10.23}$$

for every α. Next use the expansion (10.17) for both $K(Y)$ and $K(\bar{Y})$, so that

$$\langle K(Y), K(\bar{Y}) \rangle = \sum_{\alpha\beta} \pi^\alpha \bar{\pi}^\beta \langle K(Y^\alpha), K(Y^\beta) \rangle. \tag{10.24}$$

The consistency conditions (10.20) eliminate the terms with $\alpha \neq \beta$ from the sum, and (10.23) eliminates those with $\alpha = \beta$, so one arrives at (10.21). On the other hand, (10.21) implies (10.20) as a special case, since two different projectors in the decomposition $\{Y^\alpha\}$ are necessarily orthogonal to each other. Consequently, (10.20) and (10.21) are equivalent, and either one can serve as a definition of a consistent family.

While (10.20) is sufficient to ensure the additivity of W, (10.16), it is by no means necessary. It suffices to have

$$\text{Re}[\langle K(Y^\alpha), K(Y^\beta) \rangle] = 0 \text{ for } \alpha \neq \beta, \tag{10.25}$$

where Re denotes the real part. We shall refer to these as "weak consistency conditions". Even weaker conditions may work in certain cases. The subject has not been exhaustively studied. However, the conditions in (10.20) are easier to apply in actual calculations than are any of the weaker conditions, and seem adequate to cover all situations of physical interest which have been studied up till now. Consequently, we shall refer to them from now on as "the consistency conditions", while leaving open the possibility that further study may indicate the need to use a weaker condition that enlarges the class of consistent families.

What about sample spaces for which the consistency condition (10.20) is *not* satisfied? What shall be our attitude towards *inconsistent* families of histories? Within the consistent history approach to quantum theory such families are "meaningless" in the sense that there is no way to assign them probabilities within the context of a stochastic time development governed by the laws of quantum dynamics. This is not the first time we have encountered something which is "meaningless" within a quantum formalism. In the usual Hilbert space formulation of quantum theory, it makes sense to describe a spin-half particle as having its angular momentum along the $+z$ axis, or along the $+x$ axis, but trying to combine these two descriptions using "and" leads to something which lacks any meaning, because it does not correspond to any subspace in the quantum Hilbert space. See the discussion in Sec. 4.6. Consistency, on the other hand, is a more stringent condition, because a family of histories corresponding to an acceptable Boolean algebra of projectors on the history Hilbert space may still fail to satisfy the consistency conditions.

Consistency is always something which is *relative to dynamical laws*. As will be seen in an example in Sec. 10.3, changing the dynamics can render a consistent family inconsistent, or vice versa. Note that the conditions in (10.20) refer to an *isolated* quantum system. If a system is not isolated and is interacting with its environment, one must apply the consistency conditions to the system together with its environment, regarding the combination as an isolated system. A consistent family of histories for a system isolated from its environment may turn out to be inconsistent if interactions with the environment are "turned on." Conversely, interactions with the environment can sometimes ensure the consistency of a family of histories which would be inconsistent were the system isolated. Environmental effects go under the general heading of *decoherence*. (The term does not refer to the same thing as "decoherent" in "decoherent histories", though the two are related, and this sometimes causes confusion.) Decoherence is an active field of research, and while there has been considerable progress, there is much that is still not well understood. A brief introduction to the subject will be found in Ch. 26.

Must the orthogonality conditions in (10.20) be satisfied exactly, or should one allow small deviations from consistency? Inasmuch as the consistency conditions form part of the axiomatic structure of quantum theory, in the same sense as the

Born rule discussed in the previous chapter, it is natural to require that they be satisfied exactly. On the other hand, as first pointed out by Dowker and Kent, it is plausible that when the off-diagonal terms $\langle K(Y^\alpha), K(Y^\beta) \rangle$ in (10.24) are small compared with the diagonal terms $\langle K(Y^\alpha), K(Y^\alpha) \rangle$, one can find a "nearby" family of histories in which the consistency conditions are satisfied exactly. A nearby family is one in which the original projectors used to define the events (properties at a particular time) making up the histories in the family are replaced by projectors onto nearby subspaces of the same dimension. For example, consider a projector $[\phi]$ onto the subspace spanned by a normalized ket $|\phi\rangle$. The subspace $[\chi]$ spanned by a second normalized ket $|\chi\rangle$ can be said to be near to $[\phi]$ provided $|\langle\chi|\phi\rangle|^2$ is close to 1; that is, if the angle ϵ defined by

$$\sin^2(\epsilon) = 1 - |\langle\chi|\phi\rangle|^2 = \tfrac{1}{2}\operatorname{Tr}\left[([\chi] - [\phi])^2\right] \qquad (10.26)$$

is small. Notice that this measure is left unchanged by unitary time evolution: if $|\chi\rangle$ is near to $|\phi\rangle$ then $T(t', t)|\chi\rangle$ is near to $T(t', t)|\phi\rangle$. For example, if $|\phi\rangle$ corresponds to $S_z = +1/2$ for a spin-half particle, then a nearby $|\chi\rangle$ would correspond to $S_w = +1/2$ for a direction w close to the positive z axis. Or if $\phi(x)$ is a wave packet in one dimension, $\chi(x)$ might be the wave packet with its tails cut off, and then normalized. Of course, the histories in the nearby family are not the same as those in the original family. Nonetheless, since the subspaces which define the events are close to the original subspaces, their physical interpretation will be rather similar. In that case one would not commit a serious error by ignoring a small lack of consistency in the original family.

10.3 Examples of consistent and inconsistent families

As a first example, consider the family of two-time histories

$$Y^k = [\psi_0] \odot [\phi_1^k], \quad Z = (I - [\psi_0]) \odot I \qquad (10.27)$$

used in Sec. 9.3 when discussing the Born rule. The chain operators

$$K(Y^k) = [\phi_1^k]T(t_1, t_0)[\psi_0] = \langle\phi_1^k|T(t_1, t_0)|\psi_0\rangle \cdot |\phi_1^k\rangle\langle\psi_0| \qquad (10.28)$$

are mutually orthogonal because

$$\langle K(Y^k), K(Y^l) \rangle \propto \operatorname{Tr}\left(|\psi_0\rangle\langle\phi_1^k|\phi_1^l\rangle\langle\psi_0|\right) = \langle\psi_0|\psi_0\rangle\langle\phi_1^k|\phi_1^l\rangle \qquad (10.29)$$

is zero for $k \neq l$. To complete the argument, note that

$$\langle K(Y^k), K(Z) \rangle = \operatorname{Tr}\left([\psi_0]T(t_0, t_1)[\phi_1^k]T(t_1, t_0)(I - [\psi_0])\right) \qquad (10.30)$$

is zero, because one can cycle $[\psi_0]$ from the beginning to the end of the trace, and its product with $(I - [\psi_0])$ vanishes. Consequently, all the histories discussed in

Ch. 9 are consistent, which justifies our having omitted any discussion of consistency when introducing the Born rule.

The same argument works if we consider a more general situation in which the initial state is a projector Ψ_0 onto a subspace which could have a dimension greater than 1, and instead of an orthonormal basis we consider a general decomposition of the identity in projectors

$$I = \sum_k P^k \tag{10.31}$$

at time t_1. The family of two-time histories

$$Y^k = \Psi_0 \odot P^k, \quad Z = (I - \Psi_0) \odot I \tag{10.32}$$

is again consistent, since for $k \neq l$

$$\langle K(Y^k), K(Y^l) \rangle \propto \mathrm{Tr}\left(\Psi_0 T(t_0, t_1) P^k P^l T(t_1, t_0)\Psi_0\right) = 0 \tag{10.33}$$

because $P^k P^l = 0$, while $\langle K(Y^k), Z \rangle = 0$ follows, as in (10.30), from cycling operators inside the trace. (This argument is a special case of the general result in Sec. 11.3 that any family based on just two times is automatically consistent.) The probability of Y^k is given by

$$\mathrm{Pr}(Y^k) = \mathrm{Tr}\left(P^k T(t_1, t_0)\Psi_0 T(t_0, t_1)\right) / \mathrm{Tr}\left(\Psi_0\right), \tag{10.34}$$

which we shall refer to as the *generalized Born rule*. The factor of $1/\mathrm{Tr}(\Psi_0)$ is needed to normalize the probability when Ψ_0 projects onto a space of dimension greater than 1.

Another situation in which the consistency conditions are automatically satisfied is that of a unitary family as defined in Sec. 8.7. For a given initial state such a family contains one unitary history, (8.41), obtained by unitary time development of this initial state, and various nonunitary histories, such as (8.42). It is straightforward to show that the chain operator for any nonunitary history in such a family is zero, and that the chain operators for unitary histories with different initial states (belonging to the same decomposition of the identity) are orthogonal to one another. Thus the consistency conditions are satisfied. If the initial condition assigns probability 1 to a particular initial state, the corresponding unitary history occurs with probability 1, and zero probability is assigned to every other history in the family.

To find an example of an inconsistent family, one must look at histories defined at three or more times. Here is a fairly simple example for a spin-half particle. The five history projectors

$$
\begin{aligned}
Y^0 &= [z^-] \odot \ I \ \odot \ I, \\
Y^1 &= [z^+] \odot [x^+] \odot [z^+], \\
Y^2 &= [z^+] \odot [x^+] \odot [z^-], \\
Y^3 &= [z^+] \odot [x^-] \odot [z^+], \\
Y^4 &= [z^+] \odot [x^-] \odot [z^-]
\end{aligned}
\tag{10.35}
$$

defined at the three times $t_0 < t_1 < t_2$ form a decomposition of the history identity, and thus a sample space of histories. However, for trivial dynamics, $T = I$, the family is inconsistent. To show this it suffices to compute the chain operators using (10.10). In particular,

$$
\begin{aligned}
K(Y^1) &= |\langle z^+|x^+\rangle|^2 \cdot |z^+\rangle\langle z^+|, \\
K(Y^3) &= |\langle z^+|x^-\rangle|^2 \cdot |z^+\rangle\langle z^+|
\end{aligned}
\tag{10.36}
$$

are not orthogonal, since $|\langle z^+|x^+\rangle|^2$ and $|\langle z^+|x^-\rangle|^2$ are both equal to $1/2$; indeed,

$$
\langle K(Y^1), K(Y^3)\rangle = 1/4.
\tag{10.37}
$$

Similarly, $K(Y^2)$ and $K(Y^4)$ are not orthogonal, whereas $K(Y^1)$ is orthogonal to $K(Y^2)$, and $K(Y^3)$ to $K(Y^4)$. In addition, $K(Y^0)$ is orthogonal to the chain operators of the other histories. Since consistency requires that all pairs of chain operators for distinct histories in the sample space be orthogonal, this is not a consistent family.

On the other hand, the same five histories in (10.35) can form a consistent family if one uses a suitable dynamics. Suppose that there is a magnetic field along the y axis, and the time intervals $t_1 - t_0$ and $t_2 - t_1$ are chosen in such a way that

$$
T(t_1, t_0) = T(t_2, t_1) = R,
\tag{10.38}
$$

where R is the unitary operator such that

$$
\begin{aligned}
R|z^+\rangle &= |x^+\rangle, \quad R|z^-\rangle = |x^-\rangle, \\
R|x^+\rangle &= |z^-\rangle, \quad R|x^-\rangle = -|z^+\rangle,
\end{aligned}
\tag{10.39}
$$

where the second line is a consequence of the first when one uses the definitions in (4.14). With this dynamics, Y^2 is a unitary history whose chain operator is orthogonal to that of Y^0, because of the orthogonal initial states, while the chain operators for Y^1, Y^3, and Y^4 vanish. Thus the consistency conditions are satisfied. That the family (10.35) is consistent for one choice of dynamics and inconsistent for another serves to emphasize the important fact, noted earlier, that consistency depends upon the dynamical law of time evolution. This is not surprising given that the probabilities assigned to histories depend upon the dynamical law.

A number of additional examples of consistent and inconsistent histories will be discussed in Chs. 12 and 13. However, checking consistency by the process of finding chain operators for every history in a sample space is rather tedious and inefficient. Some general principles and various tricks explained in the next chapter make this process a lot easier. However, the reader may prefer to move on to the examples, and only refer back to Ch. 11 as needed.

10.4 Refinement and compatibility

The refinement of a sample space of histories was discussed in Sec. 8.6. In essence, the idea is the same as for any other quantum sample space: some or perhaps all of the projectors in a decomposition of the identity are replaced by two or more finer projectors whose sum is the coarser projector. It is important to note that even if the coarser family one starts with is consistent, the finer family need not be consistent.

Suppose that $\mathcal{Z} = \{Z^\beta\}$ is a consistent sample space of histories, $\mathcal{Y} = \{Y^\alpha\}$ is a refinement of $\mathcal{Z}$, and that

$$Z^1 = Y^1 + Y^2. \tag{10.40}$$

Then, by linearity,

$$K(Z^1) = K(Y^1) + K(Y^2). \tag{10.41}$$

When a vector is written as a sum of two other vectors, the latter need not be perpendicular to each other, and, by analogy, there is no reason to suppose that the terms on the right side of (10.3) are mutually orthogonal, $\langle K(Y^1), K(Y^2) \rangle = 0$, as is necessary if $\mathcal{Y}$ is to be a consistent family. Another way in which $\mathcal{Y}$ may fail to be consistent is the following. Since $\mathcal{Y}$ is a refinement of $\mathcal{Z}$, any projector in the sample space $\mathcal{Z}$, for example Z^3, belongs to the Boolean algebra generated by $\mathcal{Y}$. Because they represent mutually exclusive events, $Z^3 Z^1 = 0$, and because Y^1 and Y^2 in (10.40) are projectors, this means that

$$Z^3 Y^1 = 0 = Z^3 Y^2. \tag{10.42}$$

In addition, the consistency of $\mathcal{Z}$ implies that

$$\langle K(Z^3), K(Z^1) \rangle = \langle K(Z^3), K(Y^1) \rangle + \langle K(Z^3), K(Y^2) \rangle = 0. \tag{10.43}$$

However, this does not mean that either $\langle K(Z^3), K(Y^1) \rangle$ or $\langle K(Z^3), K(Y^2) \rangle$ is zero, whereas (10.21) implies, given (10.42), that both must vanish in order for $\mathcal{Y}$ to be consistent.

An example which illustrates these principles is the family $\mathcal{Y}$ whose sample space is (10.35), regarded as a refinement of the coarser family $\mathcal{Z}$ whose sample space consists of the three projectors Y^0, $Y^1 + Y^2$, and $Y^3 + Y^4$. As the histories in

$\mathcal{Z}$ depend (effectively) on only two times, t_0 and t_1, the consistency of this family is a consequence of the general argument for the first example in Sec. 10.3. However, the family $\mathcal{Y}$ is inconsistent for $T(t', t) = I$.

In light of these considerations, we shall say that two or more consistent families are *compatible* if and only if they have a common refinement which is itself a *consistent* family. In order for two consistent families $\mathcal{Y}$ and $\mathcal{Z}$ to be compatible, two conditions, taken together, are necessary and sufficient. First, the projectors for the two sample spaces, or decompositions of the history identity, $\{Y^\alpha\}$ and $\{Z^\beta\}$ must commute with each other:

$$Y^\alpha Z^\beta = Z^\beta Y^\alpha \text{ for all } \alpha, \beta. \tag{10.44}$$

Second, the chain operators associated with distinct projectors of the form $Y^\alpha Z^\beta$ must be mutually orthogonal:

$$\langle K(Y^\alpha Z^\beta), K(Y^{\hat\alpha} Z^{\hat\beta})\rangle = 0 \text{ if } \alpha \neq \hat\alpha \text{ or } \beta \neq \hat\beta. \tag{10.45}$$

Note that (10.45) is automatically satisfied when $Y^\alpha Z^\beta = 0$, so one only needs to check this condition for nonzero products. Similar considerations apply in an obvious way to three or more families. Consistent families that are not compatible are said to be (mutually) *incompatible*.

11

Checking consistency

11.1 Introduction

The conditions which define a consistent family of histories were stated in Ch. 10. The sample space must consist of a collection of mutually orthogonal projectors that add up to the history identity, and the chain operators for different members of the sample space must be mutually orthogonal, (10.20). Checking these conditions is in principle straightforward. In practice it can be rather tedious. Thus if there are n histories in the sample space, checking orthogonality involves computing n chain operators and then taking $n(n-1)/2$ operator inner products to check that they are mutually orthogonal. There are a number of simple observations, some definitions, and several "tricks" which can simplify the task of constructing a sample space of a consistent family, or checking that a given sample space is consistent. These form the subject matter of the present chapter. It is probably not worthwhile trying to read through this chapter as a unit. The reader will find it easier to learn the tricks by working through examples in Ch. 12 and later chapters, and referring back to this chapter as needed.

The discussion is limited to families in which all the histories in the sample space are of the product form, that is, represented by a projector on the history space which is a tensor product of quantum properties at different times, as in (8.7). As in the remainder of this book, the "strong" consistency conditions (10.20) are used rather than the weaker (10.25).

11.2 Support of a consistent family

A sample space of histories and the corresponding Boolean algebra it generates will be called *complete* if the sum of the projectors for the different histories in the sample space is the identity operator $\breve{I}$ on the history Hilbert space, (8.23). As noted at the end of Sec. 10.1, it is possible for the chain operator $K(Y)$ to be zero even if the history projector Y is not zero. The weight $W(Y) = \langle K(Y), K(Y) \rangle$ of such a history is obviously 0, so the history is dynamically impossible. Conversely, if $W(Y) = 0$, then $K(Y) = 0$; see the discussion in connection with (10.13). The *support* of a consistent family of histories is defined to be the set of all the histories in the sample space whose weight is strictly positive, that is, whose chain operators do not vanish. In other words, the support is what remains in the sample space if the histories of zero weight are removed. In general the support of a family is not complete, as that term was defined above, but one can say that it is *dynamically complete*.

When checking consistency, only histories lying in the support need be considered, because a chain operator which is zero is (trivially) orthogonal to all other chain operators. Using this fact can simplify the task of checking consistency in certain cases, such as the families considered in Ch. 12. Zero-weight histories are nonetheless of some importance, for they help to determine which histories, including histories of finite weight, are included in the Boolean event algebra. See the comments in Sec. 11.5.

11.3 Initial and final projectors

Checking consistency is often simplified by paying attention to the initial and final projectors of the histories in the sample space. Thus suppose that two histories

$$Y = F_0 \odot F_1 \odot \cdots F_f,$$
$$Y' = F_0' \odot F_1' \odot \cdots F_f' \tag{11.1}$$

are defined for the same set of times $t_0 < t_1 < \cdots t_f$. If either $F_0 F_0' = 0$ or $F_f F_f' = 0$, then one can easily show, by writing out the corresponding trace and cycling operators around the trace, that $\langle K(Y), K(Y') \rangle = 0$. Consequently, one can sometimes tell by inspection that two chain operators will be orthogonal, without actually computing what they are.

If the sample space consists of histories with just two times $t_0 < t_1$, then the family is automatically consistent. The reason is that the product of the history projectors for two different histories in the sample space is 0 (as the sample space consists of mutually exclusive possibilities). But in order that

$$(F_0 \odot F_1) \cdot (F_0' \odot F_1') = F_0 F_0' \odot F_1 F_1' \tag{11.2}$$

be 0, it is necessary that either $F_0 F_0'$ or $F_1 F_1'$ vanish. As we have just seen, either possibility implies that the chain operators for the two histories are orthogonal. As this holds for any pair of histories in the sample space, the consistency conditions are satisfied.

For families of histories involving three or more times, looking at the initial and final projectors does not settle the problem of consistency, but it does make

checking consistency somewhat simpler. Suppose, for example, we are considering
a family of histories based upon a fixed initial state Ψ_0 (see Sec. 11.5), with two
possible projectors at the final time based upon the decomposition

$$t_f : \quad I = P + \tilde{P}. \tag{11.3}$$

Then the sample space will consist of various histories, some of whose projectors
will have P at the final time, and some $\tilde{P}$. The chain operator of a projector with
a final P will be orthogonal to one with a final $\tilde{P}$. Thus we only need to check
whether the chain operators for the histories ending in P are mutually orthogonal
among themselves, and, similarly, the mutual orthogonality of the chain operators
for histories ending in $\tilde{P}$. If the decomposition of the identity at the final time t_f
involves more than two projectors, one need only check the orthogonality of chain
operators for histories which end in the same projector, as it is automatic when the
final projectors are different.

Yet another way of reducing the work involved in checking consistency can also
be illustrated using (11.3). Suppose that at t_{f-1} there is a decomposition of the
identity of the form

$$t_{f-1} : \quad I = \sum_m Q_m, \tag{11.4}$$

and suppose that we have already checked that the chain operators for the different
histories *ending in* P are all mutually orthogonal. In that case we can be sure that
the chain operators for two histories with projectors

$$
\begin{aligned}
Y &= \Psi_0 \odot \cdots Q_m \odot \tilde{P}, \\
Y' &= \Psi_0 \odot \cdots Q_{m'} \odot \tilde{P}
\end{aligned}
\tag{11.5}
$$

ending in $\tilde{P}$ will also be orthogonal to each other, provided $m' \neq m$. The reason
is that by cycling operators around the trace in a suitable fashion one obtains an
expression for the inner product of the chain operators in the form

$$
\begin{aligned}
\langle K(Y), K(Y') \rangle &= \text{Tr}\left(\cdots Q_m T(t_{f-1}, t_f) \tilde{P} T(t_f, t_{f-1}) Q_{m'} \right) \\
&= \text{Tr}\left(\cdots Q_m Q_{m'} \right) - \text{Tr}\left(\cdots Q_m T(t_{f-1}, t_f) P T(t_f, t_{f-1}) Q_{m'} \right),
\end{aligned}
\tag{11.6}
$$

where $\cdots$ refers to the same product of operators in each case. The second line of
the equation is obtained from the first by replacing $\tilde{P}$ by $(I - P)$, using the linearity
of the trace, and noting that $T(t_{f-1}, t_f) T(t_f, t_{f-1})$ is the identity operator, see
(7.40). The trace of the product which contains $Q_m Q_{m'}$ vanishes, because $m' \neq m$
means that $Q_m Q_{m'} = 0$. The final trace vanishes because it is the inner product of
the chain operators for the histories obtained from Y and Y' in (11.5) by replacing
$\tilde{P}$ at the final position with P; by assumption, the orthogonality of these has already

been checked. Thus the right side of (11.6) vanishes, so the chain operators for Y and Y' are orthogonal.

If the decomposition of the identity at t_f is into $n > 2$ projectors, the trick just discussed can still be used; however, it is necessary to check the mutual orthogonality of the chain operators for histories corresponding to each of $n - 1$ final projectors before one can obtain a certain number of results for those ending in the nth projector "for free". If, rather than a fixed initial state Ψ_0, one is interested in a decomposition of the identity at t_0 involving several projectors, there is an analogous trick in which the projectors at t_1 play the role of the Q_m in the preceding discussion.

11.4 Heisenberg representation

It is sometimes convenient to use the *Heisenberg representation* for the projectors and the chain operators, in place of the ordinary or *Schrödinger representation* which we have been using up to now. Suppose F_j is a projector representing an event thought of as happening at time t_j. We define the corresponding *Heisenberg projector* $\hat{F}_j$ using the formula

$$\hat{F}_j = T(t_r, t_j) F_j T(t_j, t_r), \tag{11.7}$$

where the *reference time* t_r is arbitrary, but must be kept fixed while analyzing a given family of histories. In particular, t_r cannot depend upon j. One can, for example, use $t_r = t_0$, but there are other possibilities as well. Given a history

$$Y = F_0 \odot F_1 \odot \cdots F_f \tag{11.8}$$

of events at the times $t_0 < t_1 < \cdots t_f$, the *Heisenberg chain operator* is defined by:

$$\hat{K}(Y) = \hat{F}_f \hat{F}_{f-1} \cdots \hat{F}_0 = T(t_r, t_f) K(Y) T(t_0, t_r), \tag{11.9}$$

where the second equality is easily verified using the definition of $K(Y)$ in (10.5) along with (11.7). Note that $\hat{K}(Y)$, like $K(Y)$, is a linear function of its argument.

Now let Y' be a history similar to Y, except that each F_j in (11.8) is replaced by an event F'_j (which may or may not be the same as F_j). Then it is easy to show that

$$\langle K(Y'), K(Y) \rangle = \langle \hat{K}(Y'), \hat{K}(Y) \rangle = \mathrm{Tr}\left(\hat{F}'_0 \hat{F}'_1 \cdots \hat{F}'_f \hat{F}_f \hat{F}_{f-1} \cdots \hat{F}_0 \right). \tag{11.10}$$

(Note that the inner product of the Heisenberg chain operators does not depend upon the choice of the reference time t_r.) Thus one obtains quite simple expressions for weights of histories and inner products of chain operators by using the Heisenberg representation. While this is not necessarily an advantage when doing an explicit calculation — time dependence has disappeared from (11.9), but one still has to use it to calculate the $\hat{F}_j$ in terms of the F_j, (11.7) — it does make some of the formulas simpler, and therefore more transparent. One disadvantage of using Heisenberg projectors is that, unlike ordinary (Schrödinger) projectors, they do not have a direct physical interpretation: what they signify in physical terms depends both on the form of the operator and on the dynamics of the quantum system.

11.5 Fixed initial state

A family of histories for the times $t_0 < t_1 < \cdots t_f$ based on an initial state Ψ_0 was introduced in Sec. 8.5, see (8.30). Let us write the elements of the sample space in the form

$$Y^\alpha = \Psi_0 \odot X^\alpha, \tag{11.11}$$

where for each α, X^α is a projector on the space $\bar{\mathcal{H}}$ of histories at times $t_1 < t_2 < \cdots < t_f$, with identity $\bar{I}$, and

$$\sum_\alpha X^\alpha = \bar{I}, \tag{11.12}$$

so that

$$\sum_\alpha Y^\alpha = \Psi_0 \odot \bar{I}. \tag{11.13}$$

The index α may have many components, as in the case of the product of sample spaces considered in Sec. 8.5. Since the Y^α do not add up to $\breve{I}$, we complete the sample space by adding another history

$$Z = (I - \Psi_0) \odot \bar{I}, \tag{11.14}$$

as in (8.31).

The chain operator $K(Z)$ is automatically orthogonal to the chain operators of all of the histories of the form (11.11) because the initial projectors are orthogonal, see Sec. 11.3. Consequently, the necessary and sufficient condition that the consistency conditions are satisfied for this sample space is that

$$\langle K(\Psi_0 \odot X^\alpha), K(\Psi_0 \odot X^\beta) \rangle = 0 \text{ for } \alpha \neq \beta. \tag{11.15}$$

As one normally assigns Ψ_0 probability 1 and $\tilde{\Psi}_0$ probability 0, the history Z can be ignored, and we shall henceforth assume that our sample space consists of the histories of the form (11.11).

One consequence of (11.12) and the fact that the chain operator $K(Y)$ is a linear function of Y, (10.7), is that

$$\sum_\alpha K(Y^\alpha) = K(\Psi_0 \odot \bar{I}) = T(t_f, t_0)\Psi_0. \qquad (11.16)$$

Of course, this is still true if we omit all the zero terms from the sum on the left side, that is to say, if we sum only over histories in the support S of the sample space (as defined in Sec. 11.2):

$$\sum_{\alpha \in S} K(Y^\alpha) = K(\Psi_0 \odot \bar{I}) = T(t_f, t_0)\Psi_0. \qquad (11.17)$$

One can sometimes make use of the result (11.17) in the following way. Suppose that we have found a certain collection S of histories of the form (11.11), represented by mutually orthogonal history projectors (that is, $X^\alpha X^\beta = 0$ if $\alpha \neq \beta$) with nonzero weights. Suppose that, in addition, (11.17) is satisfied, but the X^α for $\alpha \in S$ do not add up to $\bar{I}$, (11.12). Can we be sure of finding a set of zero-weight histories of the form (11.11) so that we can complete our sample space in the sense that (11.13) is satisfied? Generally there are several ways of completing a sample space with histories of zero weight. One way is to define

$$X' = \bar{I} - \sum_{\alpha \in S} X^\alpha, \quad Y' = \Psi_0 \odot X'. \qquad (11.18)$$

Then, since

$$Y' + \sum_{\alpha \in S} Y^\alpha = \Psi_0 \odot \bar{I}, \qquad (11.19)$$

it follows from the linearity of K, see (11.17), that

$$K(Y') = 0. \qquad (11.20)$$

Consequently, Y' as defined in (11.18) is a zero-weight history of the correct type, showing that there is at least one solution to our problem.

However, Y' might not be the sort of solution we are looking for. The point is that while zero-weight histories never occur, and thus in some sense they can be ignored, nonetheless they help to determine what constitutes the Boolean algebra of histories, since this depends upon the sample space. Sometimes one wants to discuss a particular item in the Boolean algebra which occurs with finite probability, but whose very presence in the algebra depends upon the existence of certain zero-weight histories in the sample space. In such a case one might need to use a collection of zero-weight history projectors adding up to Y' rather than Y' by itself.

The argument which begins at (11.16) looks a bit simpler if one uses the Heisenberg representation for the projectors and the chain operators. In particular, since

$$\hat{K}(\Psi_0 \odot \bar{I}) = \hat{\Psi}_0, \qquad (11.21)$$

we can write (11.17) as

$$\sum_{\alpha \in S} \hat{K}(Y^\alpha) = \hat{\Psi}_0, \tag{11.22}$$

and since $\hat{K}(Y)$ is, like $K(Y)$, a linear function of its argument Y, the argument leading to $\hat{K}(Y') = 0$, obviously equivalent to $K(Y') = 0$, is somewhat more transparent.

11.6 Initial pure state. Chain kets

If the initial projector of Sec. 11.5 projects onto a pure state,

$$\Psi_0 = [\psi_0] = |\psi_0\rangle\langle\psi_0|, \tag{11.23}$$

where we will assume that $|\psi_0\rangle$ is normalized, there is an alternative route for calculating weights and checking consistency which involves using *chain kets* rather than chain operators. Since it is usually easier to manipulate kets than it is to carry out the corresponding tasks on operators, using chain kets has advantages in terms of both speed and simplicity. Suppose that Y^α in (11.11) has the form given in (8.30),

$$Y^\alpha = [\psi_0] \odot P_1^{\alpha_1} \odot P_2^{\alpha_2} \odot \cdots P_f^{\alpha_f}, \tag{11.24}$$

with projectors at t_1, t_2, etc. drawn from decompositions of the identity of the type (8.25). Then it is easy to see that the corresponding chain operator is of the form

$$K(Y^\alpha) = |\alpha\rangle\langle\psi_0|, \tag{11.25}$$

where the *chain ket* $|\alpha\rangle$ is given by the expression

$$|\alpha\rangle = P_f^{\alpha_f} T(t_f, t_{f-1}) \cdots P_2^{\alpha_2} T(t_2, t_1) P_1^{\alpha_1} T(t_1, t_0)|\psi_0\rangle. \tag{11.26}$$

That is, start with $|\psi_0\rangle$, integrate Schrödinger's equation from t_0 to t_1, and apply the projector $P_1^{\alpha_1}$ to the result in order to obtain

$$|\phi_1\rangle = P_1^{\alpha_1} T(t_1, t_0)|\psi_0\rangle. \tag{11.27}$$

Next use $|\phi_1\rangle$ as the starting state, integrate Schrödinger's equation from t_1 to t_2, and apply $P_2^{\alpha_2}$. Continuing in this way will eventually yield $|\alpha\rangle$, where the symbol α stands for $(\alpha_1, \alpha_2, \ldots \alpha_f)$.

The inner product of two chain operators of the form (11.22) is the same as the inner product of the corresponding chain kets:

$$\begin{aligned} \langle K(Y^\alpha), K(Y^\beta)\rangle &= \mathrm{Tr}\left(K^\dagger(Y^\alpha)K(Y^\beta)\right) \\ &= \mathrm{Tr}\left(|\psi_0\rangle\langle\alpha|\beta\rangle\langle\psi_0|\right) = \langle\alpha|\beta\rangle. \end{aligned} \tag{11.28}$$

Consequently, the consistency condition becomes

$$\langle\alpha|\beta\rangle = 0 \text{ for } \alpha \neq \beta, \tag{11.29}$$

while the weight of a history is

$$W(Y^\alpha) = \langle\alpha|\alpha\rangle. \tag{11.30}$$

In the special case in which one of the projectors at time t_f projects onto a pure state $|\alpha_f\rangle$, the chain ket will be a complex constant, which could be 0, times $|\alpha_f\rangle$. If two or more histories in the sample space have the same final projector onto a pure state $|\alpha_f\rangle$, then consistency requires that at most one of these chain kets can be nonzero.

The analog of the argument in Sec. 11.5 following (11.17) leads to the following conclusion. Suppose one has a collection S of nonzero chain kets of the form (11.26) with the property that

$$\sum_{\alpha\in S} |\alpha\rangle = T(t_f, t_0)|\psi_0\rangle. \tag{11.31}$$

That is, they add up to the state produced by the unitary time evolution of $|\psi_0\rangle$ from t_0 to t_f. Suppose also that for the collection S the consistency conditions (11.29) are satisfied. Then one knows that the collection of histories $\{Y^\alpha : \alpha \in S\}$ is the support of a consistent family: there is at least one way (and usually there are many different ways) to add histories of zero weight to the support S in order to have a sample space satisfying (11.13), with $\Psi_0 = [\psi_0]$. Nonetheless, for the reasons discussed towards the end of Sec. 11.5, it is sometimes a good idea to go ahead and construct the zero-weight histories explicitly, in order to have a Boolean algebra of history projectors with certain specific properties, rather than relying on a general existence proof.

11.7 Unitary extensions

For the following discussion it is convenient to use the Heisenberg representation introduced in Sec. 11.4, even though the concept of unitary extensions works equally well for the ordinary (Schrödinger) representation. Unitary histories were introduced in Sec. 8.7 and defined by (8.38). An equivalent definition is that the corresponding Heisenberg operators be identical,

$$\hat{F}_0 = \hat{F}_1 = \cdots \hat{F}_f, \tag{11.32}$$

where we have used t_0 as the initial time rather than t_1 as in Sec. 8.7. It is obvious from (11.9) that the Heisenberg chain operator $\hat{K}$ for a unitary history is the projector $\hat{F}_0$.

Next suppose that in place of (11.32) we have

$$\hat{F}_0 = \hat{F}_1 = \cdots \hat{F}_{m-1} \neq \hat{F}_m = \hat{F}_{m+1} = \cdots \hat{F}_f, \qquad (11.33)$$

where m is some integer in the interval $1 \leq m \leq f$. We shall call this a "one-jump history", because the Heisenberg projectors are not all equal; there is a change, or "jump" between t_{m-1} and t_m. In a one-jump history there are precisely two types of Heisenberg projectors, with all the projectors of one type occurring at times which are earlier than the first occurrence of a projector of the other type. The chain operator for a history with one jump is $\hat{K} = \hat{F}_f \hat{F}_0$. (If, as is usually the case, $\hat{F}_0$ and $\hat{F}_f$ do not commute, $\hat{K}$ is not a projector.) Similarly, a history with two jumps is characterized by

$$\hat{F}_0 = \cdots \hat{F}_{m-1} \neq \hat{F}_m = \cdots \hat{F}_{m'-1} \neq \hat{F}_{m'} = \cdots \hat{F}_f, \qquad (11.34)$$

with m and m' two integers in the range $1 \leq m < m' \leq f$, and its chain operator is $\hat{K} = \hat{F}_f \hat{F}_m \hat{F}_0$. (It could be the case that $\hat{F}_f = \hat{F}_0$.) Histories with three or more jumps are defined in a similar way.

A *unitary extension of a unitary history* (11.32) is obtained by adding some additional times, which may be earlier than t_0 or between t_0 and t_f or later than t_f; the only restriction is that the new times do not appear in the original list $t_0, t_1, \ldots t_f$. At each new time the projector for the event is chosen so that the corresponding Heisenberg projector is identical to those in the original history, (11.32). Hence, a unitary extension of a unitary history is itself a unitary history, and its Heisenberg chain operator is $\hat{F}_0$, the same as for the original history.

A *unitary extension of a history with one jump* is obtained by including additional times, and requiring that the corresponding Heisenberg projectors are such that the new history has one jump. This means that if a new time t' precedes t_{m-1} in (11.33), the corresponding Heisenberg projector $\hat{F}'$ will be $\hat{F}_0$, whereas if it follows t_m, $\hat{F}'$ will be $\hat{F}_m$. If additional times are introduced between t_{m-1} and t_m, then the Heisenberg projectors corresponding to these times must all be $\hat{F}_0$, or all $\hat{F}_m$, or if some are $\hat{F}_0$ and some are $\hat{F}_m$, then all the times associated with the former must precede the earliest time associated with the latter. The Heisenberg chain operator of the extension is the same as for the original history, $\hat{F}_f \hat{F}_0$.

Unitary extensions of histories with two or more jumps follow the same pattern. One or more additional times are introduced, and the corresponding Heisenberg projectors must be such that the number of jumps in the new history is the same as in the original history. As a consequence, the Heisenberg chain operator is left unchanged. By using a limiting process it is possible to produce a unitary extension of a history in which events are defined on a continuous time interval. However, it is not clear that there is any advantage to doing so.

The fact that the Heisenberg chain operator is not altered in forming a unitary extension means that the weight W of an extended history is the same as that of the original history. Likewise, if the chain operators for a collection of histories are mutually orthogonal, the same is true for the chain operators of the unitary extensions. These results can be used to extend a consistent family of histories to include additional times without having to recheck the consistency conditions or recalculate the weights.

There is a slight complication in that while the histories obtained by unitary extension of the histories in the original sample space form the support of the new sample space, one needs additional zero-weight histories so that the projectors will add up to the history identity (or the projector for an initial state). The argument which follows shows that such zero-weight histories will always exist. Imagine that some history is being extended in steps, adding one more time at each step. Suppose that t' has just been added to the set of times, with $\hat{F}'$ the corresponding Heisenberg projector. We now define a zero-weight history which has the same set of times as the newly extended history, and the same projectors at these times, except that at t' the projector $\hat{F}'$ is replaced with its complement

$$\hat{F}'' = I - \hat{F}'. \tag{11.35}$$

What is $\hat{K}''$ for the history containing $\hat{F}''$? Since the unitary extension had the same number of jumps as the original history, $\hat{F}''$ must occur next to an $\hat{F}'$ in the product which defines $\hat{K}''$, and this means that $\hat{K}'' = 0$, since $\hat{F}'\hat{F}'' = 0$. Thus we have produced a zero-weight history whose history projector when added to that of the newly extended history yields the projector for the history before the extension, since $\hat{F}' + \hat{F}'' = I$. Consequently, by carrying out unitary extensions in successive steps, at each step we generate zero-weight histories of the form needed to produce a final sample space in which all the history projectors add up to the desired answer. While the procedure just described can always be applied to generate a sample space, there will usually be other ways to add zero-weight histories, and since the choice of zero-weight histories can determine what events occur in the final Boolean algebra, as noted towards the end of Sec. 11.5, one may prefer to use some alternative to the procedure just described.

11.8 Intrinsically inconsistent histories

A single history is said to be *intrinsically inconsistent*, or simply *inconsistent*, if there is no consistent family which contains it as one of the elements of the Boolean algebra. The smallest Boolean algebra of histories which contains a history projector Y consists of 0, Y, $\tilde{Y} = \breve{I} - Y$, and the history identity $\breve{I}$. Since $Y\tilde{Y} = 0$,

$$\langle K(Y), K(\tilde{Y})\rangle \neq 0, \tag{11.36}$$

see (10.21), is a necessary and sufficient condition that Y be intrinsically inconsistent.

If one restricts attention to histories which are product projectors, (8.6), no history involving just two times can be intrinsically inconsistent, so the simplest possibility is a three-time history of the form

$$Y = A \odot B \odot C. \tag{11.37}$$

Given Y, define the three histories

$$\begin{aligned} Y' &= A \odot \tilde{B} \odot C, \\ Y'' &= A \odot I \odot \tilde{C}, \\ Y''' &= \tilde{A} \odot I \odot I, \end{aligned} \tag{11.38}$$

where, as usual, $\tilde{P}$ stands for $I - P$. Then it is evident that

$$Y + Y' + Y'' + Y''' = I \odot I \odot I = \breve{I}, \tag{11.39}$$

so that

$$\tilde{Y} = Y' + Y'' + Y''', \tag{11.40}$$

and thus

$$K(\tilde{Y}) = K(Y') + K(Y'') + K(Y'''). \tag{11.41}$$

By considering initial and final projectors, Sec. 11.3, it is at once evident that $K(Y'')$ and $K(Y''')$ are orthogonal to $K(Y)$. Consequently,

$$\langle K(Y), K(\tilde{Y})\rangle = \langle K(Y), K(Y')\rangle, \tag{11.42}$$

so that Y is an inconsistent history if the right side of this equation is nonzero.

As an example, consider the histories in (10.35), and let $Y = Y^1$. Then $Y' = Y^3$, and (10.37), which was used to show that (10.35) is an inconsistent family, also shows that Y^1 is intrinsically inconsistent. The same is true of Y^2, Y^3, and Y^4. The same basic strategy can be applied in certain cases which are at first sight more complicated; e.g., the histories in (13.19).

12

Examples of consistent families

12.1 Toy beam splitter

Beam splitters are employed in optics, in devices such as the Michelson and Mach–Zehnder interferometers, to split an incoming beam of light into two separate beams propagating perpendicularly to each other. The analogous situation in a neutron interferometer is achieved using a single crystal of silicon as a beam splitter. The toy beam splitter in Fig. 12.1 can be thought of as a model of either an optical or a neutron beam splitter. It has two entrance channels (or ports) a and b, and two exit channels c and d. The sites are labeled by a pair mz, where m is an integer, and z is one of the four letters a, b, c, or d, indicating the channel in which the site is located.

The unitary time development operator is $T = S_b$, where the action of the operator S_b is given by

$$S_b|mz\rangle = |(m+1)z\rangle, \tag{12.1}$$

with the exceptions:

$$
\begin{aligned}
S_b|0a\rangle &= \big(+|1c\rangle + |1d\rangle\big)/\sqrt{2}, \\
S_b|0b\rangle &= \big(-|1c\rangle + |1d\rangle\big)/\sqrt{2}.
\end{aligned}
\tag{12.2}
$$

The physical significance of the states $|0a\rangle$, $|1c\rangle$, etc., is not altered if they are multiplied by arbitrary phase factors, see Sec. 2.2, and this means that (12.2) is not the only possible way of representing the action of the beam splitter. One could equally well replace the states on the right side with

$$\big(i|1c\rangle + |1d\rangle\big)/\sqrt{2}, \quad \big(|1c\rangle + i|1d\rangle\big)/\sqrt{2}, \tag{12.3}$$

or make other choices for the phases. There are two other exceptions to (12.1) that are needed to supply the model with periodic boundary conditions which connect the c channel back into the a channel and the d channel back into the b channel (or

159

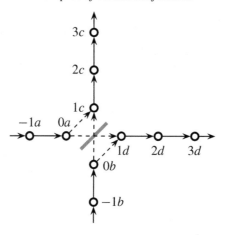

Fig. 12.1. Toy beam splitter.

c into b and d into a if one prefers). It is not necessary to write down a formula, since we shall only be interested in short time intervals during which the particle will not pass across the periodic boundaries and come back to the beam splitter. That S_b is unitary follows from the fact that it maps an orthonormal basis of the Hilbert space, namely the collection of all kets of the form $|mz\rangle$, onto another orthonormal basis of the same space; see Sec. 7.2.

Suppose that at $t = 0$ the particle starts off in the state

$$|\psi_0\rangle = |0a\rangle, \tag{12.4}$$

that is, it is in the a channel and about to enter the beam splitter. Unitary time development up to a time $t > 0$ results in

$$|\psi_t\rangle = S_b^t|\psi_0\rangle = \big(|tc\rangle + |td\rangle\big)/\sqrt{2} = |t\bar{a}\rangle, \tag{12.5}$$

where

$$|m\bar{a}\rangle := \big(|mc\rangle + |md\rangle\big)/\sqrt{2}, \quad |m\bar{b}\rangle := \big(-|mc\rangle + |md\rangle\big)/\sqrt{2} \tag{12.6}$$

are the states resulting from unitary time evolution when the particle starts off in $|0a\rangle$ or $|0b\rangle$, respectively.

Let us consider histories involving just two times, with an initial state $|\psi_0\rangle = |0a\rangle$ at $t = 0$, and a basis at some time $t > 0$ consisting of the states $\{|mz\rangle\}$, $z = a$, b, c, or d, corresponding to a decomposition of the identity

$$I = \sum_{m,z}[mz]. \tag{12.7}$$

By treating $|\psi_t\rangle$ as a pre-probability, see Sec. 9.4, one finds that

$$\Pr([mc]_t) = (1/2)\delta_{tm} = \Pr([md]_t), \tag{12.8}$$

while all other probabilities vanish; that is, at time t the particle will be either in the c output channel at the site tc, or in the d channel at td. Here $[mc]$ is a projector onto the ray which contains $|mc\rangle$, and the subscript indicates the time at which the event occurs.

If, on the other hand, one employs a unitary history, Sec. 8.7, in which at time t the particle is in the state $|t\bar{a}\rangle$, one cannot say that it is in either the c or the d channel. The situation is analogous to the case of a spin-half particle with an initial state $|z^+\rangle$ and trivial dynamics, discussed in Sec. 9.3. In a unitary history with $S_z = +1/2$ at a later time it is not meaningful to ascribe a value to S_x, whereas by using a sample space in which S_x at the later time makes sense, one concludes that $S_x = +1/2$ or $S_x = -1/2$, each with probability $1/2$.

The toy beam splitter is a bit more complicated than a spin-half particle, because when we say that "the particle is in the c channel", we are not committed to saying that it is at a *particular* site in the c channel. Instead, being in the c channel or being in the d channel is represented by means of projectors

$$C = \sum_m |mc\rangle\langle mc| = \sum_m [mc], \quad D = \sum_m [md]. \tag{12.9}$$

Neither of these projectors commutes with a projector $[m\bar{a}]$ corresponding to the state $|m\bar{a}\rangle$ defined in (12.6), so if we use a unitary history, we cannot say that the particle is in channel c or channel d. Note that whenever it *is* sensible to speak of a particle being in channel c or channel d, it cannot possibly be in both channels, since

$$CD = 0; \tag{12.10}$$

that is, these properties are mutually exclusive. A quantum particle can lack a definite location, as in the state $|m\bar{a}\rangle$, but, as already pointed out in Sec. 4.5, it cannot be in two places at the same time.

The fact that the particle is at the site tc with probability $1/2$ and at the site td with probability $1/2$ at a time $t > 0$, (12.8), might suggest that with probability $1/2$ the particle is moving out the c channel through a succession of sites $1c$, $2c$, $3c$, and so forth, and with probability $1/2$ out the d channel through $1d$, $2d$, etc. But this is *not* something one can infer by considering histories defined at only two times, for it would be equally consistent to suppose that the particle hops from $2c$ to $3d$ during the time step from $t = 2$ to $t = 3$, and from $2d$ to $3c$ if it happens to be in the d channel at $t = 2$. In order to rule out unphysical possibilities of this sort we need to consider histories involving more than just two times.

Consider a family of histories based upon the initial state $[0a]$ and at each time $t > 0$ the decomposition of the identity (12.7), so that the particle has a definite location. The histories are then of the form, for a set of times $t = 0, 1, 2, \ldots f$,

$$Y = [0a] \odot [mz] \odot [m'z'] \odot \cdots [m''z''], \tag{12.11}$$

with a chain operator of the form $K(Y) = |\phi\rangle\langle 0a|$, Sec. 11.6, where the chain ket is

$$|\phi\rangle = |m''z''\rangle \cdots \langle m'z'|S_b|mz\rangle\langle mz|S_b|0a\rangle. \tag{12.12}$$

From (12.2) it is obvious that the term $\langle mz|S_b|0a\rangle$ is 0 unless $m = 1$ and $z = c$ or d, and given $m = 1$, it follows from (12.1) that $\langle m'z'|S_b|mz\rangle$ vanishes unless $m' = 2$ and $z' = z$. By continuing this argument one sees that $|\phi\rangle$, and therefore $K(Y)$, will vanish for all but two histories, which in the case $f = 4$ are

$$\begin{aligned} Y^c &= [0a] \odot [1c] \odot [2c] \odot [3c] \odot [4c], \\ Y^d &= [0a] \odot [1d] \odot [2d] \odot [3d] \odot [4d]. \end{aligned} \tag{12.13}$$

The fact that the final projectors $[4c]$ and $[4d]$ in (12.13) are orthogonal to each other means that the chain operators $K(Y^c)$ and $K(Y^d)$ are orthogonal, in accordance with a general principle noted in Sec. 11.3. Since the chain operators of all the other histories are zero, it follows that Y^c and Y^d form the support, as defined in Sec. 11.2, of a consistent family. It is straightforward to show, either by means of chain kets as discussed in Sec. 11.6 or by a direct use of $W(Y) = \langle K(Y), K(Y)\rangle$, that

$$W(Y^c) = 1/2 = W(Y^d), \tag{12.14}$$

and hence, assuming an initial state of $[0a]$ with probability 1, the two histories Y^c and Y^d each have probability 1/2, while all other histories in this family have probability 0.

The fact that the only histories with finite probability are Y^c and Y^d means that if the particle arrives at the site $1c$ at $t = 1$, it continues to move out along the c channel, and does not hop to the d channel, and if the particle is at $1d$ at time $t = 1$, it moves out along the d channel. Thus by using multiple-time histories one can eliminate the possibility that the particle hops back and forth between channels c and d, something which cannot be excluded by considering only two-time histories, as noted earlier. A formal argument confirming what is rather obvious from looking at (12.13) can be constructed by calculating the probability

$$\Pr(D_t \,|\, [1c]_1) = \Pr(D_t \wedge [1c]_1)/\Pr([1c]_1) \tag{12.15}$$

that the particle will be in the d channel at some time $t > 0$, given that it was at the site $[1c]$ at $t = 1$. Here D_t is a projector on the history space for the particle to be in channel d at time t. For example, for $t = 2$,

$$D_2 = I \odot I \odot D \odot I \odot I, \tag{12.16}$$

and thus

$$D_2 \wedge [1c]_1 = I \odot [1c] \odot D \odot I \odot I. \tag{12.17}$$

This projector gives 0 when applied to either Y^c or Y^d, the only two histories with positive probability, and therefore the numerator on the right side of (12.15) is 0. Thus if the particle is at $1c$ at $t = 1$, it will not be in the d channel at $t = 2$. The same argument works equally well for other values of t, and analogous results are obtained if the particle is initially in the d channel. Thus one has

$$\begin{aligned} \Pr(D_t \mid [1c]_1) &= 0 = \Pr(C_t \mid [1d]_1), \\ \Pr(C_t \mid [1c]_1) &= 1 = \Pr(D_t \mid [1d]_1) \end{aligned} \tag{12.18}$$

for any $t \geq 1$, where C_t is defined in the same manner as D_t, with C in place of D.

(Since we are considering a family which is based on the initial state $[0a]$, the preceding discussion runs into the technical difficulty that C_t and D_t do not belong to the corresponding Boolean algebra of histories when the latter is constructed in the manner indicated in Sec. 8.5. One can get around this problem by replacing C_t and D_t with the operators $C_t \wedge [a0]_0$ and $D_t \wedge [a0]_0$, and remembering that the probabilities in (12.15) and (12.18) always contain the initial state $[a0]$ at $t = 0$ as an (implicit) condition. Also see the remarks in Sec. 14.4.)

Another family of consistent histories can be constructed in the following way. At the times $t = 1, 2$ use, in place of (12.7), a three-projector decomposition of the identity

$$I = [t\bar{a}] + [t\bar{b}] + J_t, \tag{12.19}$$

where the states $|t\bar{a}\rangle$, $|t\bar{b}\rangle$ are defined in (12.6), and

$$J_t = I - [t\bar{a}] - [t\bar{b}] = I - [tc] - [td] \tag{12.20}$$

is a projector for the particle to be someplace other than the two sites tc or td. At later times $t \geq 3$ use the decomposition (12.7). It is easy to show that in the case $f = 4$ the two histories

$$\begin{aligned} \bar{Y}^c &= [0a] \odot [1\bar{a}] \odot [2\bar{a}] \odot [3c] \odot [4c], \\ \bar{Y}^d &= [0a] \odot [1\bar{a}] \odot [2\bar{a}] \odot [3d] \odot [4d], \end{aligned} \tag{12.21}$$

each with weight $1/2$, form the support of the sample space of a consistent family; all other histories have zero weight.

The histories $\bar{Y}^c$ and $\bar{Y}^d$ in (12.21) have the physical significance that at $t =$

1 and $t = 2$ the particle is in a coherent superposition of states in both output channels. After $t = 2$ a "split" occurs, and at later times the two histories are no longer identical: one represents the particle as traveling out the c channel, and the other the particle traveling out the d channel. What causes this split? To think of a physical cause for it is to look at the problem in the wrong way. Recall the case of a spin-half particle with trivial dynamics, discussed in Sec. 9.3, with $S_z = 1/2$ initially and then $S_x = \pm 1/2$ at a later time. There is no physical transformation of the particle, since the dynamics is trivial. Instead, different aspects of the particle's spin angular momentum are being described at two successive times. In the same way, the histories in (12.21) allow us to describe a property at times $t = 1$ and $t = 2$, corresponding to the linear superposition $|m\bar{a}\rangle$, which cannot be described if we use the histories in (12.13). Conversely, using (12.21) makes it impossible to discuss whether the particle is in the c or in the d channel when $t = 1$ or 2, because these properties are incompatible with the projectors employed in $\bar{Y}^c$ and $\bar{Y}^d$. There is a similar split in the case of the histories Y^c and Y^d: they start with the same initial state $[0a]$, and the split occurs when t changes from 0 to 1. In this situation one may be tempted to suppose that the beam splitter causes the split, but that surely cannot be the case, for the very same beam splitter does not cause a split in the case of $\bar{Y}^c$ and $\bar{Y}^d$.

We have one family of histories based upon Y^c and Y^d, and a distinct family based upon $\bar{Y}^c$ and $\bar{Y}^d$. The two families are incompatible, as they have no common refinement. Which one provides the *correct* description of the physical system? Consider two histories of Great Britain: one a political history which discusses the monarchs, the other an intellectual history focusing upon developments in British science. Which is the *correct* history of Great Britain? That is not the proper way to compare them. Instead, there are certain questions which can be answered by one history rather than the other. For certain purposes one history is more useful, for other purposes the other is to be preferred. In the same way, both the Y^c, Y^d family and the $\bar{Y}^c$, $\bar{Y}^d$ family provide correct (stochastic) descriptions of the physical system, descriptions which are useful for answering different sorts of questions. There are, to be sure, certain questions which can be answered using either family, such as "Will the particle be in the c or the d channel at $t = 4$ if it was at $3c$ at $t = 3$?" For such questions, both families give precisely the same answer, in agreement with a general principle of consistency discussed in Sec. 16.3.

Next consider a family in which the histories start off like $\bar{Y}^c$ and $\bar{Y}^d$ in (12.21), but later on revert back to the coherent superposition states corresponding to (12.19); for example

$$
\begin{aligned}
Y' &= [0a] \odot [1\bar{a}] \odot [2\bar{a}] \odot [3c] \odot [4\bar{a}], \\
Y'' &= [0a] \odot [1\bar{a}] \odot [2\bar{a}] \odot [3d] \odot [4\bar{a}],
\end{aligned}
\qquad (12.22)
$$

plus other histories needed to make up a sample space. This family is not consistent. The reason is that the chain kets $|y'\rangle$ and $|y''\rangle$ corresponding to $K(Y')$ and $K(Y'')$ are nonzero multiples of $|4\bar{a}\rangle$, so $\langle y'|y''\rangle \neq 0$, and hence $K(Y')$ and $K(Y'')$ are not orthogonal to each other, see (11.28). There is a certain analogy between (12.22) and the inconsistent family for a spin-half particle involving three times discussed in Sec. 10.3. The precise time at which the split and the rejoining occur is not important; for example, the chain operators associated with the histories

$$
\begin{aligned}
X' &= [0a] \odot [1c] \odot [2c] \odot [3c] \odot [4\bar{a}], \\
X'' &= [0a] \odot [1d] \odot [2d] \odot [3d] \odot [4\bar{a}]
\end{aligned}
\tag{12.23}
$$

are also not mutually orthogonal, so the corresponding family is inconsistent. Inconsistency does not require a perfect rejoining; even a partial one can cause trouble! But why might someone want to consider families of histories of the form (12.22) or (12.23)? We will see in Ch. 13 that in the case of a simple interferometer the analogous histories look rather "natural", and it will be of some importance that they are not part of a consistent family.

12.2 Beam splitter with detector

Let us now add a detector of the sort described in Sec. 7.4 to the c output channel of the beam splitter, Fig. 12.2. The detector has two states: $|0\hat{c}\rangle$ "ready", and $|1\hat{c}\rangle$ "triggered", which span a Hilbert space $\mathcal{C}$. The Hilbert space of the total quantum system is

$$
\mathcal{H} = \mathcal{M} \otimes \mathcal{C},
\tag{12.24}
$$

where $\mathcal{M}$ is the Hilbert space of the particle passing through the beam splitter, and the collection $\{|mz, n\hat{c}\rangle\}$ for different values of m, z, and n is an orthonormal basis of $\mathcal{H}$.

The unitary time development operator takes the form

$$
T = S_b R_c,
\tag{12.25}
$$

where S_b is the unitary transformation defined in (12.1) and (12.2), extended in the usual way to the operator $S_b \otimes I$ on $\mathcal{M} \otimes \mathcal{C}$, and R_c (the subscript indicates that this detector is attached to the c channel) is defined in analogy with (7.53) as

$$
R_c|mz, n\hat{c}\rangle = |mz, n\hat{c}\rangle,
\tag{12.26}
$$

with the exception that

$$
R_c|2c, n\hat{c}\rangle = |2c, (1-n)\hat{c}\rangle.
\tag{12.27}
$$

That is, R_c is the identity operator unless the particle is at the site $2c$, in which case

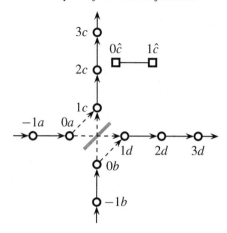

Fig. 12.2. Toy beam splitter with detector.

the detector flips from $0\hat{c}$ to $1\hat{c}$, or $1\hat{c}$ to $0\hat{c}$. As noted in Sec. 7.4, such a detector does not perturb the motion of the particle, in the sense that the particle moves from $1c$ to $2c$ to $3c$, etc. at successive time steps whether or not the detector is present.

We shall assume an initial state

$$|\Psi_0\rangle = |0a, 0\hat{c}\rangle \tag{12.28}$$

at $t = 0$: the particle is at $0a$, and is about to enter the beam splitter, and the detector is ready. Unitary time development of this initial state leads to

$$|\Psi_t\rangle = T^t|\Psi_0\rangle = \begin{cases} \big(|tc\rangle + |td\rangle\big) \otimes |0\hat{c}\rangle/\sqrt{2} & \text{for } t = 1, 2, \\ \big(|tc, 1\hat{c}\rangle + |td, 0\hat{c}\rangle\big)/\sqrt{2} & \text{for } t \geq 3. \end{cases} \tag{12.29}$$

If one regards $|\Psi_t\rangle$ for $t \geq 3$ as representing a physical state or physical property of the combined particle and detector, then the detector is not in a definite state. Instead one has a toy counterpart of a *macroscopic quantum superposition* (MQS) or *Schrödinger's cat* state. See the discussion in Sec. 9.6. It is impossible to say whether or not the detector has detected something at times $t \geq 3$ if one uses a unitary family based upon the initial state $|\Psi_0\rangle$.

A useful family of histories for studying the process of detection is based on the initial state $|\Psi_0\rangle$ and a decomposition of the identity in pure states

$$I = \sum_{m,z,n} [mz, n\hat{c}], \tag{12.30}$$

in which the particle has a definite location and the detector is in one of its pointer states at every time $t > 0$. The histories

$$Z^c = [0a, 0\hat{c}] \odot [1c, 0\hat{c}] \odot [2c, 0\hat{c}] \odot [3c, 1\hat{c}] \odot [4c, 1\hat{c}] \odot \cdots ,$$
$$Z^d = [0a, 0\hat{c}] \odot [1d, 0\hat{c}] \odot [2d, 0\hat{c}] \odot [3d, 0\hat{c}] \odot [4d, 0\hat{c}] \odot \cdots ,$$

(12.31)

continuing for as long a sequence of times as one wants to consider, are the obvious counterparts of Y^c and Y^d in (12.13). Because the final projectors are orthogonal, $K(Z^c)$ and $K(Z^d)$ are orthogonal, and it is not hard to show that Z^c and Z^d constitute the support of a consistent family $\mathcal{F}$ based on the initial state $|\Psi_0\rangle$. The physical interpretation of these histories is straightforward. In Z^c the particle moves out the c channel and triggers the detector, changing $0\hat{c}$ to $1\hat{c}$ as it moves from $2c$ to $3c$. In Z^d the particle moves out the d channel, and the detector remains in its untriggered or ready state $0\hat{c}$.

We can use the property that the detector has (or has not) detected the particle at some time $t' \geq 3$ to determine which channel the particle is in, by computing a conditional probability. Thus one finds — see the discussion following (12.15) — that

$$\Pr(C_t \,|\, [1\hat{c}]_{t'}) = 1, \quad \Pr(D_t \,|\, [1\hat{c}]_{t'}) = 0,$$
$$\Pr(C_t \,|\, [0\hat{c}]_{t'}) = 0, \quad \Pr(D_t \,|\, [0\hat{c}]_{t'}) = 1,$$

(12.32)

for $t' \geq 3$ and $t \geq 1$. That is, if at some time $t' \geq 3$, the detector has detected the particle, then at time t, the particle is (or was) in the c and not in the d channel, while if the detector has not detected the particle, the particle is (or was) in the d and not in the c channel.

Note that the conditional probabilities in (12.32) are valid not simply for $t \geq 3$; they also hold for $t = 1$ and 2. That is, if the detector is triggered at time $t' = 3$, then the particle was in the c channel at $t = 1$ and 2, and if the detector is not triggered at $t' = 3$, then at these earlier times the particle was in the d channel. These results are perfectly reasonable from a physical point of view. How could the particle have triggered the detector unless it was already moving out along the c channel? And if it did not trigger the detector, where could it have been except in the d channel? As long as the particle does not hop from one channel to the other in some magical way, the results in (12.32) are just what one would expect.

Another family in which the detector is always in one of its pointer states is the counterpart of (12.21), modified by the addition of a detector:

$$\bar{Z}^c = [0a, 0\hat{c}] \odot [1\bar{a}, 0\hat{c}] \odot [2\bar{a}, 0\hat{c}] \odot [3c, 1\hat{c}] \odot [4c, 1\hat{c}] \odot \cdots ,$$
$$\bar{Z}^d = [0a, 0\hat{c}] \odot [1\bar{a}, 0\hat{c}] \odot [2\bar{a}, 0\hat{c}] \odot [3d, 0\hat{c}] \odot [4d, 0\hat{c}] \odot \cdots .$$

(12.33)

The chain operators for $\bar{Z}^c$ and $\bar{Z}^d$ are orthogonal, and it is easy to find zero-weight histories to complete the sample space, so that (12.33) is the support of a consistent family $\mathcal{G}$. It differs from $\mathcal{F}$, (12.31), in that at $t = 1$ and 2 the particle is in the superposition state $|t\bar{a}\rangle$ rather than in the c or the d channel, but for times after $t = 2$ $\mathcal{F}$ and $\mathcal{G}$ are identical.

Both families $\mathcal{F}$, (12.31), and $\mathcal{G}$, (12.33), represent equally good quantum descriptions. The only difference is that they allow one to discuss somewhat different properties of the particle at a time after it has passed through the beam splitter and before it has been detected. In particular, if one is interested in knowing the location of the particle before the measurement occurred (or could have occurred), it is necessary to employ a consistent family in which questions about its location are meaningful, so $\mathcal{F}$ must be used, not $\mathcal{G}$. On the other hand, if one is interested in whether the particle was in the superposition $|1\bar{a}\rangle$ at $t = 1$ rather than in $|1\bar{b}\rangle$ — see the definitions in (12.6) — then it is necessary to use $\mathcal{G}$, for questions related to such superpositions are meaningless in $\mathcal{F}$.

The family $\mathcal{G}$, (12.33), is useful for understanding the idea, which goes back to von Neumann, that a measurement produces a "collapse" or "reduction" of the wave function. As applied to our toy model, a measurement which serves to detect the presence of the particle in the c channel is thought of as collapsing the superposition wave function $|2\bar{a}\rangle$ produced by unitary time evolution into a state $|3c\rangle$ located in the c channel. This is the step from $[2\bar{a}, 0\hat{c}]$ to $[3c, 1\hat{c}]$ in the history $\bar{Z}^c$. Similarly, if the detector does not detect the particle, $|2\bar{a}\rangle$ collapses to a state $|3d\rangle$ in the d channel, as represented by the step from $t = 2$ to $t = 3$ in the history $\bar{Z}^d$.

The approach to measurements based on wave function collapse is the subject of Sec. 18.2. While it can often be employed in a way which gives correct results, wave function collapse is not really needed, since the same results can always be obtained by straightforward use of conditional probabilities. On the other hand, it has given rise to a lot of confusion, principally because the collapse tends to be thought of as a physical effect produced by the measuring apparatus. With reference to our toy model, this might be a reasonable point of view when the particle is detected to be in the c channel, but it seems very odd that a *failure* to detect the particle in the c channel has the effect of collapsing its wave function into the d channel, which might be a long way away from the c detector. That the collapse is not any sort of physical effect is clear from the fact that it occurs in the family (12.21) in the absence of a detector, and in $\mathcal{F}$, (12.31), it occurs prior to detection. To be sure, in $\mathcal{F}$ one might suppose that the collapse is caused by the beam splitter. However, one could modify (12.31) in an obvious way to produce a consistent family in which the collapse takes place between $t = 1$ and $t = 2$, and thus has nothing to do with either the beam splitter or detector.

Another way in which the collapse approach to quantum measurements is somewhat unsatisfactory is that it does not provide a connection between the outcome of a measurement and a corresponding property of the measured system before the measurement took place. For example, if at $t \geq 3$ the detector is in the state $1\hat{c}$, there is no way to infer that the particle was earlier in the c channel if one uses the family (12.33) rather than (12.31). The connection between measurements and what they measure will be discussed in Ch. 17.

12.3 Time-elapse detector

A simple two-state toy detector is useful for thinking about a number of situations in quantum theory involving detection and measurement. However, it has its limitations. In particular, unlike real detectors, it does not have sufficient complexity to allow the *time* at which an event occurs to be recorded by the detector. While it is certainly possible to include a clock as part of a toy detector, a slightly simpler solution to the timing problem is to use a *time-elapse detector*: when an event is detected, a clock is started, and reading this clock tells how much time has elapsed since the detection occurred. As in Sec. 7.4, the Hilbert space $\mathcal{H}$ is a tensor product $\mathcal{M} \otimes \mathcal{N}$ of the space $\mathcal{M}$ of the particle, spanned by kets $|m\rangle$ with $-M_a \leq m \leq M_b$, and the space $\mathcal{N}$ of the detector, with kets $|n\rangle$ labeled by n in the range

$$-N \leq n \leq N. \tag{12.34}$$

In effect, one can think of the detector as a second particle that moves according to an appropriate dynamics. However, to avoid confusion the term *particle* will be reserved for the toy particle whose position is labeled by m, and which the detector is designed to detect, while n will be the position of the detector's *pointer* (see the remarks at the end of Sec. 9.5). We shall suppose that M_a, M_b, and N are sufficiently large that we do not have to worry about either the particle or the pointer "coming around the cycle" during the time period of interest.

The unitary time development operator is

$$T = SRS_d, \tag{12.35}$$

where S is the shift operator on $\mathcal{M}$,

$$S|m\rangle = |m + 1\rangle, \tag{12.36}$$

with a periodic boundary condition $S|M_b\rangle = |-M_a\rangle$, and S_d acts on $\mathcal{N}$,

$$S_d|n\rangle = |n + 1\rangle, \tag{12.37}$$

with the exceptions

$$S_d|0\rangle = |0\rangle, \quad S_d|-1\rangle = |1\rangle, \tag{12.38}$$

and $S_d|N\rangle = |-N\rangle$ to take care of the periodic boundary condition. The unitary operator R which couples the pointer to the particle is the identity,

$$R|m, n\rangle = |m, n\rangle, \tag{12.39}$$

except for

$$R|2, 0\rangle = |2, 1\rangle, \quad R|2, 1\rangle = |2, 0\rangle. \tag{12.40}$$

That is, when the particle is at $m = 2$, R moves the pointer from $n = 0$ to $n = 1$, or from $n = 1$ to $n = 0$, while if the pointer is someplace else, R has no effect on it. The unitarity of T in (12.35) follows from that of S, R, and S_d.

When its pointer is at $n = 0$, the detector is in its "ready" state, where it remains until the particle reaches $m = 2$, at which point the "detection event" (12.40) occurs, and the pointer hops to $n = 1$ at the same time as the particle hops to $m = 3$, since T includes the shift operator S for the particle, (12.35). This is identical to the operation of the two-state detector of Sec. 7.4. But once the detector pointer is at $n = 1$ it keeps going, (12.37), so a typical unitary time development of $|m, n\rangle$ is of the form

$$|0, 0\rangle \mapsto |1, 0\rangle \mapsto |2, 0\rangle \mapsto |3, 1\rangle \mapsto |4, 2\rangle \mapsto |5, 3\rangle \mapsto \cdots . \tag{12.41}$$

Thus the pointer reading n (assumed to be less than N) tells how much time has elapsed since the detection event occurred.

As an example of the operation of this detector in a stochastic context, suppose that at $t = 0$ there is an initial state

$$|\Psi_0\rangle = |\psi_0\rangle \otimes |0\rangle, \tag{12.42}$$

where the particle wave packet

$$|\psi_0\rangle = a|0\rangle + b|1\rangle + c|2\rangle \tag{12.43}$$

has three nonzero coefficients a, b, c. Consider histories which for $t > 0$ employ a decomposition of the identity corresponding to the orthonormal basis $\{|m, n\rangle\}$. The chain operators for the three histories

$$Z^0 = [\Psi_0] \odot [1, 0] \odot [2, 0] \odot [3, 1],$$

$$Z^1 = [\Psi_0] \odot [2, 0] \odot [3, 1] \odot [4, 2], \tag{12.44}$$

$$Z^2 = [\Psi_0] \odot [3, 1] \odot [4, 2] \odot [5, 3],$$

involving the four times $t = 0, 1, 2, 3$, are obviously orthogonal to one another (because of the final projectors, Sec. 11.3). The corresponding weights are $|a|^2$, $|b|^2$, and $|c|^2$, while all other histories beginning with $[\Psi_0]$ have zero weight. Hence (12.44) is the support of a consistent family with initial state $|\Psi_0\rangle$.

Suppose that the pointer is located at $n = 2$ when $t = 3$. Since the pointer position indicates the time that has elapsed since the particle was detected, we should be able to infer that the detection event $[2, 0]$ occurred at $t = 3 - 2 = 1$. Indeed, one can show that

$$\Pr([2, 0] \text{ at } t = 1 \mid n = 2 \text{ at } t = 3) = 1, \tag{12.45}$$

using the fact that the condition $n = 2$ when $t = 3$ is only true for Z^1. If the pointer is at $n = 1$ when $t = 3$, one can use the family (12.44) to show not only that the detection event $[2, 0]$ occurred at $t = 2$, but also that at $t = 1$ the particle was at $m = 1$, one site to the left of the detector. Being able to infer where the particle was before it was detected is intuitively reasonable, and is the sort of inference often employed when analyzing data from real detectors in the laboratory. Such inferences depend, of course, on using an appropriate consistent family, as discussed in Sec. 12.2.

12.4 Toy alpha decay

A toy model of alpha decay was introduced in Sec. 7.4, and discussed using the Born rule in Sec. 9.5. We assume the sites are labeled as in Fig. 7.2 on page 106, and will employ the same $T = S_a$ dynamics used previously, (7.56). That is,

$$S_a|m\rangle = |m + 1\rangle, \tag{12.46}$$

with the exceptions

$$S_a|0\rangle = \alpha|0\rangle + \beta|1\rangle, \quad S_a|-1\rangle = \gamma|0\rangle + \delta|1\rangle, \tag{12.47}$$

together with a periodic boundary condition. The coefficients α, β, γ, and δ satisfy (7.58).

Consider histories which begin with the initial state

$$|\psi_0\rangle = |0\rangle, \tag{12.48}$$

the alpha particle inside the nucleus, and employ a decomposition of the identity based upon particle position states $|m\rangle$ at all later times. That such a family of histories, thought of as extending from the initial state at $t = 0$ till a later time $t = f$, is consistent can be seen by working out what happens when f is small. In particular, when $f = 1$, there are two histories with nonzero weight:

$$\begin{aligned} &[0] \odot [0], \\ &[0] \odot [1]. \end{aligned} \tag{12.49}$$

The chain operators are orthogonal because the projectors at the final time are mutually orthogonal (Sec. 11.3). With $f = 2$, there are three histories with nonzero weight:

$$[0] \odot [0] \odot [0],$$
$$[0] \odot [0] \odot [1], \qquad (12.50)$$
$$[0] \odot [1] \odot [2],$$

and again it is obvious that the chain operators are orthogonal, so that the corresponding family is consistent.

These examples suggest the general pattern, valid for any f. The support of the consistent family contains a history in which $m = 0$ at all times, together with histories with a decay time $t = \tau$, with τ in the range $0 \le \tau \le f - 1$, of the form

$$[0]_0 \odot [0]_1 \odot \cdots [0]_\tau \odot [1]_{\tau+1} \odot [2]_{\tau+2} \odot \cdots . \qquad (12.51)$$

That is, the alpha particle remains in the nucleus, $m = 0$, until the time $t = \tau$, then hops to $m = 1$ at $t = \tau + 1$, and after that it keeps going. If one uses this particular family of histories, the quantum problem is much the same as that of a classical particle which hops out of a well with a certain probability at each time step, and once out of the well moves away from it at a constant speed. This is not surprising, since as long as one employs a single consistent family the mathematics of a quantum stochastic process is formally identical to that of a classical stochastic process.

In Sec. 9.5 a simple two-state detector was used in analyzing toy alpha decay by means of the Born rule. Additional insight can be gained by replacing the two-state detector in Fig. 9.1 with the time-elapse detector of Sec. 12.3 to detect the alpha particle as it hops from $m = 2$ to $m = 3$ after leaving the nucleus. On the Hilbert space $\mathcal{M} \otimes \mathcal{N}$ of the alpha particle and detector pointer, the unitary time development operator is

$$T = S_a R S_d, \qquad (12.52)$$

where S_d and R are defined in (12.37)–(12.40).

Suppose that at the time $t = \bar{t}$ the detector pointer is at $\bar{n}$. Then the detection event should have occurred at the time $\bar{t} - \bar{n}$. And since the particle was detected at the site $m = 2$, the actual decay time τ when it left the nucleus would have been a bit earlier,

$$\tau = \bar{t} - \bar{n} - 2, \qquad (12.53)$$

because of the finite travel time from the nucleus to the detector. This line of reasoning can be confirmed by a straightforward calculation of the conditional probabilities

$$\Pr(m = 0 \text{ at } t = \bar{t} - \bar{n} - 2 \,|\, n = \bar{n} \text{ at } t = \bar{t}) = 1,$$
$$\Pr(m = 0 \text{ at } t = \bar{t} - \bar{n} - 1 \,|\, n = \bar{n} \text{ at } t = \bar{t}) = 0.$$

(12.54)

That is, at the time τ given in (12.53), the particle was still in the nucleus, while one time step later it was no longer there. (Of course this only makes sense if $\bar{t}$ and $\bar{n}$ are such that $\Pr(n = \bar{n} \text{ at } t = \bar{t})$ is positive.) Note once again that by adopting an appropriate family of histories one can make physically reasonable inferences about events prior to the detection of the alpha particle.

Does the fact that we can assign a decay time in the case of our toy model mean that the same thing is possible for real alpha decay? The answer is presumably "yes", provided one does not require that the decay time be defined too precisely. However, finding a suitable criterion for the nucleus to have or have not decayed and checking consistency conditions for an appropriate family pose nontrivial technical issues, and the matter does not seem to have been studied in detail. Note that even in the toy model the decay time is not precisely defined, because time is discretized, and $\tau + 1$ has as much justification for being identified with the decay time as does τ. This uncertainty can, however, be much shorter than the half life of the nucleus, which is of the order of $|\beta|^{-2}$.

13

Quantum interference

13.1 Two-slit and Mach–Zehnder interferometers

Interference effects involving quantum particles reflect both the wave-like and particle-like properties of quantum entities. One of the best-known examples is the interference pattern produced by a double slit. Quantum particles — photons or neutrons or electrons — are sent one at a time through the slit system shown in Fig. 13.1, and later arrive at a series of detectors located in the diffraction zone far from the slits. The detectors are triggered at random, with each particle triggering just one detector. After enough particles have been detected, an interference pattern can be discerned in the relative counting rates of the different detectors, indicated by the length of the horizontal bars in the figure. Lots of particles arrive at some detectors, very few particles at others.

Fig. 13.1. Interference pattern for a wave arriving from the left and passing through the two slits. Each circle on the right side represents a detector, and the black bar to its right indicates the relative counting rate.

The relative number of particles arriving at each detector depends on the *difference* of the distances between the detector and the two slits, in units of the particle's de Broglie wavelength. Furthermore, this interference pattern persists even at very low intensities, say one particle per second passing through the slit system. Hence it seems very unlikely that it arises from a sort of cooperative phenomenon in which a particle going through one slit compares notes with a particle going through the other slit. Instead, each particle must somehow pass through both slits, for how else can one understand the interference effect?

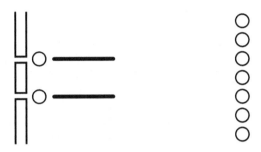

Fig. 13.2. Detectors directly behind the two slits. The black bars are again proportional to the counting rates.

However, if detectors are placed directly behind the two slits, Fig. 13.2, then either one or the other detector detects a particle, and it is never the case that both detectors simultaneously detect a particle. Furthermore, the total counting rate for the arrangement in Fig. 13.2 is the same as that in Fig. 13.1, suggesting that if a particle had not been detected just behind one of the slits, it would have continued on into the diffraction zone and arrived at one of the detectors located there. Thus it seems plausible that the particles which do arrive in the diffraction zone in Fig. 13.1 have earlier passed through one or the other of the two slits, and not both. But this is difficult to reconcile with the interference effect seen in the diffraction zone, which seems to require that each particle pass through both slits. Could a particle passing through one slit somehow sense the presence of the other slit, and take this into account when it arrives in the diffraction zone?

In Feynman's discussion of two-slit interference (see bibliography), he considers what happens if there is a nondestructive measurement of which slit the particle passes through, a measurement that allows the particle to continue on its way and later be detected in the diffraction zone. His quantum particles are electrons, and he places a light source just behind the slits, Fig. 13.3. By scattering a photon off the electron one can "see" which slit it has just passed through. Illuminating the slits in this way washes out the interference effect, and the intensities in the diffraction

Fig. 13.3. A light source L between the slits washes out the electron interference pattern.

zone can be explained as sums of intensities due to electrons coming through each of the two slits.

Feynman then imagines reducing the intensity of the light source to such a degree that sometimes an electron scatters a photon, revealing which slit it passed through, and sometimes it does not. Data for electrons arriving in the diffraction zone are then segregated into two sets: one set for "visible" electrons which earlier scattered a photon, and the other for electrons which were "invisible" as they passed through the slit system. When the set of data for the "visible" electrons is examined it shows no interference effects, whereas that for the "invisible" electrons indicates that they arrive in the diffraction zone with the same interference pattern as when there is no source of light behind the slits. Can the behavior of an electron really depend upon whether or not it has been seen?

In this chapter we explore these paradoxes using a toy Mach–Zehnder interferometer, which exhibits the same sorts of paradoxes as a double slit, but is easier to discuss. A Mach–Zehnder interferometer, Fig. 13.4, consists of a beam splitter followed by two mirrors which bring the split beams back together again, and a second beam splitter placed where the reflected beams intersect. Detectors can be placed on the output channels. We assume that light from a monochromatic source enters the first beam splitter through the a channel. The intensity of light emerging in the two output channels e and f depends on the *difference* in path length, measured in units of the wavelength of the light, in the c and d arms of the interferometer. (The classical wave theory of light suffices for calculating these intensities; one does not need quantum theory.) We shall assume that this difference has been adjusted so that after the second beam splitter all the light which enters through the a channel emerges in the f channel and none in the e channel. Rather than changing the physical path lengths, it is possible to alter the final

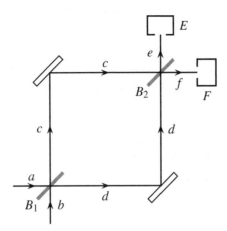

Fig. 13.4. Mach–Zehnder interferometer with detectors. The beam splitters are labeled B_1 and B_2.

intensities by inserting *phase shifters* in one or both arms of the interferometer. (A phase shifter is a piece of dielectric material which, when placed in the light beam, alters the optical path length (number of wavelengths) between the two beam splitters.)

An interferometer for neutrons which is analogous to a Mach–Zehnder interferometer for photons can be constructed using a single crystal of silicon. For our purposes the difference between these two types of interferometer is not important, since neutrons are quantum particles that behave like waves, and photons are light waves that behave like particles. Thus while we shall continue to think of photons going through a Mach–Zehnder interferometer, the toy model introduced in Sec. 13.2 could equally well describe the interference of neutrons.

The analogy between a Mach–Zehnder interferometer and double-slit interference is the following. Each photon on its way through the interferometer must pass through the c arm or the d arm in much the same way that a particle (photon or something else) must pass through one of the two slits on its way to a detector in the diffraction zone. The first beam splitter provides a source of coherent light (that is, the relative phase is well defined) for the two arms of the interferometer, just as one needs a coherent source of particles illuminating the two slits. (This coherent source can be a single slit a long distance to the left of the double slit.) The second beam splitter in the interferometer combines beams from the separate arms and makes them interfere in a way which is analogous to the interference of the beams emerging from the two slits when they reach the diffraction zone.

13.2 Toy Mach–Zehnder interferometer

We shall set up a stochastic or probabilistic model of a toy Mach–Zehnder interferometer, Fig. 13.5, and discuss what happens when a *single* particle or photon passes through the instrument. The model will supply us with probabilities for different possible histories of this single particle. If one imagines, as in a real experiment, lots of particles going through the apparatus, one after another, then each particle represents an "independent trial" in the sense of probability theory. That is, each particle will follow (or undergo) a particular history chosen randomly from the collection of all possible histories. If a large number of particles are used, then the number which follow some given history will be proportional to the probability, computed by the laws of quantum theory, that a single particle will follow that history.

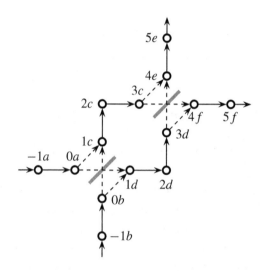

Fig. 13.5. Toy Mach–Zehnder interferometer constructed from two beam splitters of the sort shown in Fig. 12.1.

The toy Mach–Zehnder interferometer consists of two toy beam splitters, of the type shown in Fig. 12.1, in series. The arms and the entrance and output channels are labeled in a way which corresponds to Fig. 13.4. The unitary time transformation for the toy model is $T = S_i$, where the operator S_i is defined by

$$S_i|mz\rangle = |(m+1)z\rangle \tag{13.1}$$

for m an integer, and $z = a, b, c, d, e$ or f, with the exceptions

$$S_i|0a\rangle = (+|1c\rangle + |1d\rangle)/\sqrt{2}, \quad S_i|0b\rangle = (-|1c\rangle + |1d\rangle)/\sqrt{2},$$
$$S_i|3c\rangle = (+|4e\rangle + |4f\rangle)/\sqrt{2}, \quad S_i|3d\rangle = (-|4e\rangle + |4f\rangle)/\sqrt{2}. \tag{13.2}$$

(See the comment following (12.2) on the choice of phases.) In addition, the usual provision must be made for periodic boundary conditions, but (as usual) these will not play any role in the discussion which follows; see the remarks in Sec. 12.1. The transformation S_i is unitary because it maps an orthonormal basis, the collection of states $\{|mz\rangle\}$, onto an orthonormal basis of the Hilbert space. A particle (photon) which enters the a channel undergoes a unitary time evolution of the form

$$|0a\rangle \mapsto |1\bar{a}\rangle \mapsto |2\bar{a}\rangle \mapsto |3\bar{a}\rangle \mapsto |4f\rangle \mapsto |5f\rangle \mapsto \cdots, \tag{13.3}$$

where, as in (12.6),

$$|m\bar{a}\rangle = (+|mc\rangle + |md\rangle)/\sqrt{2}, \quad |m\bar{b}\rangle = (-|mc\rangle + |md\rangle)/\sqrt{2} \tag{13.4}$$

are superpositions of states of the particle in the c and d arms of the interferometer, with phases chosen to correspond to unitary evolution under S_i starting with $|0a\rangle$, and $|0b\rangle$, respectively.

The probability that the particle emerges in the e or in the f channel is influenced by what happens in *both* arms of the interferometer, as can be seen in the following way. Let us introduce *toy phase shifters* in the c and d arms by using in place of S_i a unitary time transformation S_i' identical to S_i, (13.1) and (13.2), except that

$$S_i'\,|2c\rangle = e^{i\phi_c}\,|3c\rangle, \quad S_i'\,|2d\rangle = e^{i\phi_d}\,|3d\rangle, \tag{13.5}$$

where ϕ_c and ϕ_d are phase shifts. Obviously S_i' is unitary, and it is the same as S_i when ϕ_c and ϕ_d are zero. If we use S_i' in place of S_i, the unitary time evolution in (13.3) becomes

$$\begin{aligned}|0a\rangle \mapsto |1\bar{a}\rangle \mapsto |2\bar{a}\rangle &= \left(|2c\rangle + |2d\rangle\right)/\sqrt{2} \mapsto \left(e^{i\phi_c}|3c\rangle + e^{i\phi_d}|3d\rangle\right)/\sqrt{2}\\ &\mapsto \tfrac{1}{2}\Big[\left(e^{i\phi_c} - e^{i\phi_d}\right)|4e\rangle + \left(e^{i\phi_c} + e^{i\phi_d}\right)|4f\rangle\Big] \mapsto \cdots,\end{aligned} \tag{13.6}$$

where the result at $t = 5$ is obtained by replacing $|4e\rangle$ by $|5e\rangle$, and $|4f\rangle$ by $|5f\rangle$.

Consider a consistent family of histories based upon an initial state $|0a\rangle$ at $t = 0$ and a decomposition of the identity corresponding to the orthonormal basis $\{|mz\rangle\}$ at a second time $t = 4$. There are two histories with positive weight,

$$Y = [0a]_0 \odot [4e]_4, \quad Y' = [0a]_0 \odot [4f]_4, \tag{13.7}$$

where, as usual, subscripts indicate the time. The probabilities can be read off from the $t = 4$ term in (13.6), treating it as a pre-probability, by taking the absolute squares of the coefficients of $|4e\rangle$ and $|4f\rangle$:

$$\begin{aligned}\Pr([4e]_4) = \Pr(Y) &= |e^{i\phi_c} - e^{i\phi_d}|^2/4 = [\sin(\Delta\phi/2)]^2,\\ \Pr([4f]_4) = \Pr(Y') &= |e^{i\phi_c} + e^{i\phi_d}|^2/4 = [\cos(\Delta\phi/2)]^2,\end{aligned} \tag{13.8}$$

where

$$\Delta\phi = \phi_c - \phi_d \tag{13.9}$$

is the difference between the two phase shifts. Since these probabilities depend upon $\Delta\phi$, and thus upon what is happening in *both* arms of the interferometer, the quantum particle must in some sense be delocalized as it passes through the interferometer, rather than localized in arm c or in arm d. On the other hand, it is a mistake to think of it as simultaneously present in both arms in the sense that "it is in c *and at the same time* it is in d." See the remarks in Sec. 4.5: a quantum particle cannot be in two places at the same time.

Similarly, if we want to understand double-slit interference using this analogy, we would like to say that the particle "goes through both slits," without meaning that it is present in the upper slit at the same time as it is present in the lower slit, or that it went through one slit or the other and we do not know which. See the discussion of the localization of quantum particles in Secs. 2.3 and 4.5. Speaking of the particle as "passing through the slit system" conveys roughly the right meaning. In the double-slit experiment, one could introduce phase shifters behind each slit, and thereby shift the positions of the peaks and valleys of the interference pattern in the diffraction zone. Again, it is the *difference* of the phase shifts which is important, and this shows that one somehow has to think of the quantum particle as a coherent entity as it passes through the slit system.

Very similar results are obtained if instead of $|0a\rangle$ one uses a wave packet

$$|\psi_0\rangle = c\,|-2a\rangle + c'\,|-1a\rangle + c''|0a\rangle \tag{13.10}$$

in the a channel as the initial state at $t = 0$, where c, c', and c'' are numerical coefficients. For such an initial state it is convenient to use histories

$$X = [\psi_0] \odot E_t, \quad X' = [\psi_0] \odot F_t \tag{13.11}$$

rather than Y and Y' in (13.7), where

$$E = \sum_m [me], \quad F = \sum_m [mf] \tag{13.12}$$

are projectors for the particle to be someplace in the e and f channels, respectively, and E_t means the particle is in the e channel at time t; see the analogous (12.16). As long as $t \geq 6$, so that the entire wave packet corresponding to $|\psi_0\rangle$ has a chance to emerge from the interferometer, one finds that the corresponding probabilities are

$$\begin{aligned} \Pr(E_t) &= \Pr(X) = [\sin(\Delta\phi/2)]^2, \\ \Pr(F_t) &= \Pr(X') = [\cos(\Delta\phi/2)]^2, \end{aligned} \tag{13.13}$$

precisely the same as in (13.8). Since the philosophy behind toy models is simplicity and physical insight, not generality, we shall use only the simple initial state $|0a\rangle$ in what follows, even though a good part of the discussion would hold (with some fairly obvious modifications) for a more general initial state representing a wave packet entering the interferometer in the a channel.

What can we say about the particle while it is *inside* the interferometer, during the time interval for which the histories in (13.7) provide no information? There are various ways of refining these histories by inserting additional events at times between $t = 0$ and 4. For example, one can employ unitary extensions, Sec. 11.7, of Y and Y' by using the unitary time development of the initial $|0a\rangle$ at intermediate times to obtain two histories

$$Y^e = [0a] \odot [1\bar{a}] \odot [2\bar{a}] \odot [3\bar{q}] \odot [4e],$$
$$Y^f = [0a] \odot [1\bar{a}] \odot [2\bar{a}] \odot [3\bar{q}] \odot [4f], \tag{13.14}$$

defined at $t = 0, 1, 2, 3, 4$, which form the support of a consistent family with initial state $[0a]$. The projector $[3\bar{q}]$ is onto the state

$$|3\bar{q}\rangle := \left(e^{i\phi_c}|3c\rangle + e^{i\phi_d}|3d\rangle\right)/\sqrt{2}. \tag{13.15}$$

The histories in (13.14) are identical up to $t = 3$, and then split. One can place the split earlier, between $t = 2$ and $t = 3$, by mapping $[4e]$ and $[4f]$ unitarily backwards in time to $t = 3$:

$$\bar{Y}^e = [0a] \odot [1\bar{a}] \odot [2\bar{a}] \odot [3\bar{b}] \odot [4e],$$
$$\bar{Y}^f = [0a] \odot [1\bar{a}] \odot [2\bar{a}] \odot [3\bar{a}] \odot [4f]. \tag{13.16}$$

Note that Y, Y^e, and $\bar{Y}^e$ all have exactly the same chain operator, for reasons discussed in Sec. 11.7, and the same is true of Y', Y^f, and $\bar{Y}^f$. The consistency of the family (13.7) is automatic, as only two times are involved, Sec. 11.3. As a consequence the unitary extensions (13.14) and (13.16) of that family are supports of consistent families; see Sec. 11.7.

The families in (13.14) and (13.16) can be used to discuss some aspects of the particle's behavior while inside the interferometer, but cannot tell us whether it was in the c or in the d arm, because the projectors C and D, (12.9), do not commute with projectors onto superposition states, such as $[1\bar{a}]$, $[3\bar{q}]$, or $[3\bar{b}]$. Instead, we must look for alternative families in which events of the form $[mc]$ or $[md]$ appear at intermediate times. It will simplify the discussion if we assume that $\phi_c = 0 = \phi_d$, that is, use S_i for time development rather than the more general S'_i.

One consistent family of this type has for its support the two elementary histories

$$Y^c = [0a] \odot [1c] \odot [2c] \odot [3c] \odot [4\bar{c}] \odot [5\bar{c}] \odot \cdots [\tau\bar{c}],$$
$$Y^d = [0a] \odot [1d] \odot [2d] \odot [3d] \odot [4\bar{d}] \odot [5\bar{d}] \odot \cdots [\tau\bar{d}], \tag{13.17}$$

where

$$|m\bar{c}\rangle = (+|me\rangle + |mf\rangle)/\sqrt{2}, \quad |m\bar{d}\rangle = (-|me\rangle + |mf\rangle)/\sqrt{2} \qquad (13.18)$$

for $m \geq 4$ correspond to unitary time evolution starting with $|3c\rangle$ and $|3d\rangle$, respectively. The final time τ can be as large as one wants, consistent with the particle not having passed out of the e or f channels due to the periodic boundary condition. The histories in (13.17) are unitary extensions of $[0a] \odot [1c]$ and $[0a] \odot [1d]$, and consistency follows from the general arguments given in Sec. 11.7. Note that if we use Y^c and Y^d, we cannot say whether the particle emerges in the e or f channel of the second beam splitter, whereas if we use Y^e and Y^f in (13.14), with $\phi_c = 0 = \phi_d$, we can say that the particle leaves this beam splitter in a definite channel, but we cannot discuss the channel in which it arrives at the beam splitter.

In order to describe the particle as being in a definite arm of the interferometer *and* emerging in a definite channel from the second beam splitter, one might try a family which includes

$$
\begin{aligned}
Y^{ce} &= [0a] \odot [1c] \odot [2c] \odot [3c] \odot [4e] \odot [5e] \odot \cdots [\tau e], \\
Y^{cf} &= [0a] \odot [1c] \odot [2c] \odot [3c] \odot [4f] \odot [5f] \odot \cdots [\tau f], \\
Y^{de} &= [0a] \odot [1d] \odot [2d] \odot [3d] \odot [4e] \odot [5e] \odot \cdots [\tau e], \\
Y^{df} &= [0a] \odot [1d] \odot [2d] \odot [3d] \odot [4f] \odot [5f] \odot \cdots [\tau f],
\end{aligned}
\qquad (13.19)
$$

continuing till some final time τ. Alas, this will not work. The family is inconsistent, because

$$\langle K(Y^{ce}), K(Y^{de}) \rangle \neq 0, \quad \langle K(Y^{cf}), K(Y^{df}) \rangle \neq 0, \qquad (13.20)$$

as is easily shown using the corresponding chain kets (Sec. 11.6). In fact, each of the histories in (13.19) is *intrinsically inconsistent* in the sense that there is no way of making it part of some consistent family. See the discussion of intrinsic inconsistency in Sec. 11.8; the strategy used there for histories involving three times is easily extended to cover the somewhat more complicated situation represented in (13.19).

The analog of (13.14) for two-slit interference is a consistent family $\mathcal{F}$ in which the particle passes through the slit system in a delocalized state, but arrives at a definite location in the diffraction zone. It is $\mathcal{F}$ which lies behind conventional discussions of two-slit interference, which emphasize (correctly) that in such circumstances it is meaningless to discuss which slit the particle passed through. However, there is also another consistent family $\mathcal{G}$, the analog of (13.17), in which the particle passes through one or the other of the two slits, and is described in the diffraction zone by one of two delocalized wave packets, the counterparts of the $\bar{c}$ and $\bar{d}$ superpositions defined in (13.18). Although these wave packets overlap in space, they are orthogonal to each other and thus represent distinct quantum states. The families $\mathcal{F}$ and $\mathcal{G}$ are incompatible, and hence the descriptions they provide cannot be combined. Attempting to do so by assuming that the particle goes through a definite slit *and* arrives at a definite location in the diffraction zone gives rise to inconsistencies analogous to those noted in connection with (13.19).

From the perspective of fundamental quantum theory there is no reason to prefer one of these two families to the other. Each has its use for addressing certain types of physical question. If one wants to know the location of the particle when it reaches the diffraction zone, $\mathcal{F}$ must be used in preference to $\mathcal{G}$, because it is only in $\mathcal{F}$ that this location makes sense. On the other hand, if one wants to know which slit the particle passed through, $\mathcal{G}$ must be employed, for in $\mathcal{F}$ the concept of passing through a particular slit makes no sense. Experiments can be carried out to check the predictions of either family, and the Mach–Zehnder analogs of these two kinds of experiments are discussed in the next two sections.

13.3 Detector in output of interferometer

Let us add to the e output channel of our toy Mach–Zehnder interferometer a simple two-state detector of the type introduced in Sec. 7.4 and used in Sec. 12.2, see Fig. 12.2. The detector states are $|0\hat{e}\rangle$, "ready", and $|1\hat{e}\rangle$, "triggered", and the unitary time development operator is

$$T = S_i' R_e, \tag{13.21}$$

where R_e is the identity on the Hilbert space $\mathcal{M} \otimes \mathcal{E}$ of particle-plus-detector, except for

$$R_e|4e, n\hat{e}\rangle = |4e, (1-n)\hat{e}\rangle, \tag{13.22}$$

with $n = 0$ or 1, which is the analog of (12.27). Thus, in particular,

$$T|4e, 0\hat{e}\rangle = |5e, 1\hat{e}\rangle, \quad T|4f, 0\hat{e}\rangle = |5f, 0\hat{e}\rangle, \tag{13.23}$$

so the detector is triggered by the particle emerging in the e channel as it hops from $4e$ to $5e$, but is not triggered if the particle emerges in the f channel. We could add a second detector for the f channel, but that is not necessary: if the e channel detector remains in its ready state after a certain time, that will tell us that the particle emerged in the f channel. See the discussion in Sec. 12.2.

Assume an initial state

$$|\Psi_0\rangle = |0a, 0\hat{e}\rangle, \tag{13.24}$$

and consider histories that are the obvious counterparts of those in (13.14),

$$Z^e = Z^i \odot [4e, 0\hat{e}] \odot [5e, 1\hat{e}] \odot [6e, 1\hat{e}] \odot \cdots [\tau e, 1\hat{e}],$$
$$Z^f = Z^i \odot [4f, 0\hat{e}] \odot [5f, 0\hat{e}] \odot [6f, 0\hat{e}] \odot \cdots [\tau f, 0\hat{e}], \tag{13.25}$$

but which continue on to some final time τ. The initial unitary portion

$$Z^i = [\Psi_0] \odot [1\bar{a}, 0\hat{e}] \odot [2\bar{a}, 0\hat{e}] \odot [3\bar{q}, 0\hat{e}] \tag{13.26}$$

is the same for both Z^e and Z^f. The histories in (13.25) are the support of a consistent family with initial state $|\Psi_0\rangle$, and they contain no surprises. If the particle passes through the interferometer in a coherent superposition and emerges in channel e, it triggers the detector and keeps going. If it emerges in f it does not trigger the detector, and continues to move out that channel. The probability that the detector will be in its triggered state at $t = 5$ or later is $\sin^2(\Delta\phi/2)$, the same as the probability calculated earlier, (13.8), that the particle will emerge in the e channel when no detector is present.

As a second example, suppose that $\phi_c = 0 = \phi_d$, and consider the consistent family whose support consists of the two histories

$$Z^c = [\Psi_0] \odot [1c] \odot [2c] \odot [3c] \odot [4\bar{c}, 0\hat{e}] \odot [5r] \odot [6r] \odot \cdots [\tau r],$$
$$Z^d = [\Psi_0] \odot [1d] \odot [2d] \odot [3d] \odot [4\bar{d}, 0\hat{e}] \odot [5s] \odot [6s] \odot \cdots [\tau s], \tag{13.27}$$

where the detector state $[0\hat{e}]$ has been omitted for times earlier than $t = 4$ (it could be included at all these times in both histories), and

$$|mr\rangle = \frac{|me, 1\hat{e}\rangle + |mf, 0\hat{e}\rangle}{\sqrt{2}}, \quad |ms\rangle = \frac{-|me, 1\hat{e}\rangle + |mf, 0\hat{e}\rangle}{\sqrt{2}} \tag{13.28}$$

are superpositions of states in which the detector has and has not been triggered, so they are toy MQS (macroscopic quantum superposition) states, as in (12.29). The histories in (13.27) are obvious counterparts of those in (13.17), and they are unitary extensions (Sec. 11.7) to later times of $[\Psi_0] \odot [1c, 0\hat{e}]$, and $[\Psi_0] \odot [1d, 0\hat{e}]$.

The toy MQS states at time $t \geq 5$ in (13.27) are hard to interpret, and their grown-up counterparts for a real Mach–Zehnder or neutron interferometer are impossible to observe in the laboratory. Can we get around this manifestation of Schrödinger's cat (Sec. 9.6) by the same method we used in Sec. 12.2: using histories in which the detector is in its pointer basis (see the definition at the end of Sec. 9.5) rather than in some MQS state? The obvious choice would be something like

$$Z^{ce} = [\Psi_0] \odot [1c] \odot [2c] \odot [3c] \odot [4e, 0\hat{e}] \odot [5e, 1\hat{e}] \odot \cdots ,$$
$$Z^{cf} = [\Psi_0] \odot [1c] \odot [2c] \odot [3c] \odot [4f, 0\hat{e}] \odot [5f, 0\hat{e}] \odot \cdots ,$$
$$Z^{de} = [\Psi_0] \odot [1d] \odot [2d] \odot [3d] \odot [4e, 0\hat{e}] \odot [5e, 1\hat{e}] \odot \cdots ,$$
$$Z^{df} = [\Psi_0] \odot [1d] \odot [2d] \odot [3d] \odot [4f, 0\hat{e}] \odot [5f, 0\hat{e}] \odot \cdots ,$$

$$(13.29)$$

where, once again, we have omitted the detector state $[0\hat{e}]$ at times earlier than $t = 4$. However, this family is inconsistent: (13.20) holds with Y replaced by Z, and one can even show that the individual histories in (13.29), like those in (13.19), are intrinsically inconsistent. Indeed, the history

$$[\Psi_0]_0 \odot C_t \odot [1\hat{e}]_{t'},$$

$$(13.30)$$

in which the initial state is followed by a particle in the c arm at some time in the interval $1 \le t \le 3$, and then the detector in its triggered state at a later time $t' \ge 5$, is intrinsically inconsistent, and the same is true if C_t is replaced by D_t, or $[1\hat{e}]_{t'}$ by $[0\hat{e}]_{t'}$. (For the meaning of C_t or D_t, see the discussion following (12.15).)

A similar analysis can be applied to the analogous situation of two-slit interference in which a detector is located at some point in the diffraction zone. By using a family in which the particle passes through the slit system in a delocalized state corresponding to unitary time evolution, the analog of (13.25), one can show that the probability of detection is the same as the probability of the particle arriving at the corresponding region in space in the absence of a detector. There is also a family, the analog of (13.27), in which the particle passes through a definite slit, and later on the detector is described by an MQS state, the counterpart of one of the states defined in (13.28). There is no way of "collapsing" these MQS states into pointer states of the detector — this is the lesson to be drawn from the inconsistent family (13.29) — as long as one insists upon assigning a definite slit to the particle.

This example shows that it is possible to construct families of histories using events at earlier times which are "normal" (non-MQS), but which have the consequence that at later times one is "forced" to employ MQS states. If one does not want to use MQS states at a later time, it is necessary to change the events in the histories at earlier times, or alter the initial states. Note that consistency depends upon *all* the events which occur in a history, because the chain operator depends upon all the events, so one cannot say that inconsistency is "caused" by a particular event in the history, unless one has decided that other events shall, by definition, not share in the blame.

13.4 Detector in internal arm of interferometer

Let us see what happens if a detector is placed in the c arm inside the toy interferometer. (A detector could also be placed in the d arm, but this would not lead to anything new, since if the particle is not detected in the c arm one can conclude that it passed through the d arm.) The detector states are $|0\hat{c}\rangle$ "ready" and $|1\hat{c}\rangle$ "triggered". The unitary time operator is

$$T = S_i' R_c, \qquad (13.31)$$

where S_i' is defined in (13.5), and R_c is the identity on the space $\mathcal{M} \otimes \mathcal{C}$ of particle and detector, except for

$$R_c|2c, n\hat{c}\rangle = |2c, (1-n)\hat{c}\rangle. \qquad (13.32)$$

In particular,

$$T|2c, 0\hat{c}\rangle = e^{i\phi_c}|3c, 1\hat{c}\rangle, \quad T|2d, 0\hat{c}\rangle = e^{i\phi_d}|3d, 0\hat{c}\rangle, \qquad (13.33)$$

so the detector is triggered as the particle hops from $2c$ to $3c$ when passing through the c arm, but is not triggered if the particle passes through the d arm.

Consider the unitary time development,

$$|\Phi_t\rangle = T^t|\Phi_0\rangle, \quad |\Phi_0\rangle = |0a, 0\hat{c}\rangle, \qquad (13.34)$$

of an initial state in which the particle is in the a channel, and the c channel detector is in its ready state. At $t = 4$ we have

$$|\Phi_4\rangle = \tfrac{1}{2}\big(e^{i\phi_c}|4e, 1\hat{c}\rangle - e^{i\phi_d}|4e, 0\hat{c}\rangle + e^{i\phi_c}|4f, 1\hat{c}\rangle + e^{i\phi_d}|4f, 0\hat{c}\rangle\big), \qquad (13.35)$$

where all four states in the sum on the right side are mutually orthogonal.

One can use (13.35) as a pre-probability to compute the probabilities of two-time histories beginning with the initial state $|\Phi_0\rangle$ at $t = 0$, and with the particle in either the e or the f channel at $t = 4$. Thus consider a family in which the four histories with nonzero weight are of the form $\Phi_0 \odot [\phi_j]$, where $|\phi_j\rangle$ is one of the four kets on the right side of (13.35). Each will occur with probability $1/4$, and thus

$$\Pr([4e]_4) = 1/4 + 1/4 = 1/2 = \Pr([4f]_4). \qquad (13.36)$$

Upon comparing these with (13.8) when no detector is present, one sees that inserting a detector in one arm of the interferometer has a drastic effect: there is no longer any dependence of these probabilities upon the phase difference $\Delta\phi$. Thus a measurement of which arm the particle passes through wipes out all the interference effects which would otherwise be apparent in the output intensities following the second beam splitter. Note the analogy with Feynman's discussion of the double slit: determining which slit the electron goes through, by scattering light off of it, destroys the interference pattern in the diffraction zone.

Now let us consider various possible histories describing what the particle does while it is inside the interferometer, assuming $\phi_c = 0 = \phi_d$ in order to simplify the discussion. Straightforward unitary time evolution will result in a family in which every $[\Phi_t]$ for $t \geq 3$ is a toy MQS state involving both $|0\hat{c}\rangle$ and the triggered state $|1\hat{c}\rangle$ of the detector. In order to obtain a consistent family without MQS states, we can let unitary time development continue up until the measurement occurs, and then have a split (or collapse) to produce the analog of (12.33) in the previous chapter: a family whose support consists of the two histories

$$
\begin{aligned}
V^c &= [0a, 0\hat{c}] \odot [1\bar{a}, 0\hat{c}] \odot [2\bar{a}, 0\hat{c}] \odot [3c, 1\hat{c}] \odot [4\bar{c}, 1\hat{c}] \odot \cdots, \\
V^d &= [0a, 0\hat{c}] \odot [1\bar{a}, 0\hat{c}] \odot [2\bar{a}, 0\hat{c}] \odot [3d, 0\hat{c}] \odot [4\bar{d}, 0\hat{c}] \odot \cdots,
\end{aligned}
\tag{13.37}
$$

with states $|m\bar{c}\rangle$ and $|m\bar{d}\rangle$ defined in (13.18). One can equally well put the split at an earlier time, by using histories

$$
\begin{aligned}
\bar{Z}^c &= [0a, 0\hat{c}] \odot [1c, 0\hat{c}] \odot [2c, 0\hat{c}] \odot [3c, 1\hat{c}] \odot [4\bar{c}, 1\hat{c}] \odot \cdots, \\
\bar{Z}^d &= [0a, 0\hat{c}] \odot [1d, 0\hat{c}] \odot [2d, 0\hat{c}] \odot [3d, 0\hat{c}] \odot [4\bar{d}, 0\hat{c}] \odot \cdots,
\end{aligned}
\tag{13.38}
$$

which resemble those in (13.17) in that the particle is in the c or in the d arm from the moment it leaves the first beam splitter.

One can also introduce a second split at the second beam splitter, to produce a family with support

$$
\begin{aligned}
\bar{Z}^{ce} &= [0a, 0\hat{c}] \odot [1c, 0\hat{c}] \odot [2c, 0\hat{c}] \odot [3c, 1\hat{c}] \odot [4e, 1\hat{c}] \odot [5e, 1\hat{c}] \odot \cdots, \\
\bar{Z}^{cf} &= [0a, 0\hat{c}] \odot [1c, 0\hat{c}] \odot [2c, 0\hat{c}] \odot [3c, 1\hat{c}] \odot [4f, 1\hat{c}] \odot [5f, 1\hat{c}] \odot \cdots, \\
\bar{Z}^{de} &= [0a, 0\hat{c}] \odot [1d, 0\hat{c}] \odot [2d, 0\hat{c}] \odot [3d, 0\hat{c}] \odot [4e, 0\hat{c}] \odot [5e, 0\hat{c}] \odot \cdots, \\
\bar{Z}^{df} &= [0a, 0\hat{c}] \odot [1d, 0\hat{c}] \odot [2d, 0\hat{c}] \odot [3d, 0\hat{c}] \odot [4f, 0\hat{c}] \odot [5f, 0\hat{c}] \odot \cdots.
\end{aligned}
\tag{13.39}
$$

This family is consistent, in contrast to (13.19), because the projectors of the different histories at some final time τ are mutually orthogonal: the orthogonal final states of the detector prevent the inconsistency which would arise, as in (13.20), if one only had particle states. In addition, one could place another detector in one of the output channels. However, when used with a family analogous to (13.39) this detector would simply confirm the arrival of the particle in the corresponding channel with the same probability as if the detector had been absent, so one would learn nothing new.

Inserting a detector into the c arm of the interferometer provides an instance of what is often called *decoherence*. The states $|m\bar{a}\rangle$ and $|m\bar{b}\rangle$ defined in (13.4) are *coherent* superpositions of the states $|mc\rangle$ and $|md\rangle$ in which the particle is localized in one or the other arm of the interferometer, and the relative phases in the superposition are of physical significance, since in the absence of a detector one of these superpositions will result in the particle emerging in the f channel, and the other in its emerging in e. However, when something like a cosmic ray interacts with the particle in a sufficiently different way in the c and the d arm, it destroys the coherence (the influence of the relative phase), and thus produces decoherence.

The scattering of light in Feynman's version of the double-slit experiment is an example of decoherence in this sense, and it results in interference effects being washed out. However, decoherence is usually not an "all or nothing" affair. The weakly-coupled detectors discussed in Sec. 13.5 provide an example of *partial decoherence*. As well as washing out interference effects, decoherence can expand the range of possibilities for constructing consistent families. Thus the family based on (13.19) in which the particle is in a definite arm inside the interferometer and emerges from the interferometer in a definite channel is inconsistent, whereas its counterpart in (13.39), with decoherence taking place inside the interferometer, is consistent. Some additional discussion of decoherence will be found in Ch. 26.

13.5 Weak detectors in internal arms

As noted in Sec. 13.1, Feynman in his discussion of double-slit interference tells us that as the intensity of the light behind the double slits is reduced, one will find that those electrons which do not scatter a photon will, when they arrive in the diffraction zone, exhibit the same interference pattern as when the light is off. Let us try to understand this effect by placing *weakly-coupled* or *weak* detectors in the c and d arms of the toy Mach–Zehnder interferometer.

A simple toy weak detector has two orthogonal states, $|0\hat{c}\rangle$ "ready" and $|1\hat{c}\rangle$ "triggered", and the weak coupling is arranged by replacing the unitary transformation R_c in (13.32) with R'_c, which is the identity except for

$$
\begin{aligned}
R'_c|2c, 0\hat{c}\rangle &= \alpha|2c, 0\hat{c}\rangle + \beta|2c, 1\hat{c}\rangle, \\
R'_c|2c, 1\hat{c}\rangle &= \gamma|2c, 0\hat{c}\rangle + \delta|2c, 1\hat{c}\rangle,
\end{aligned}
\tag{13.40}
$$

where α, β, γ, and δ are (in general complex) numbers forming a unitary 2×2 matrix

$$
\begin{pmatrix} \alpha & \beta \\ \gamma & \delta \end{pmatrix}.
\tag{13.41}
$$

The "strongly-coupled" or "strong" detector used previously is a special case in which $\beta = 1 = \gamma$, $\alpha = \delta = 0$. Making $|\beta|$ small results in a weak coupling, since the probability that the detector will be triggered by the presence of a particle at site

$2c$ is $|\beta|^2$. (One can also modify the time-elapse detector of Sec. 12.3 to make it a weakly-coupled detector, by modifying (12.40) in a manner analogous to (13.40), but we will not need it for the present discussion.) It is convenient for purposes of exposition to assume a symmetrical arrangement in which there is a second detector, with ready and triggered states $|0\hat{d}\rangle$ and $|1\hat{d}\rangle$, in the d arm of the interferometer, with its coupling to the particle governed by a unitary transformation R'_d equal to the identity except for

$$R'_d|2d, 0\hat{d}\rangle = \alpha|2d, 0\hat{d}\rangle + \beta|2d, 1\hat{d}\rangle,$$
$$R'_d|2d, 1\hat{d}\rangle = \gamma|2d, 0\hat{d}\rangle + \delta|2d, 1\hat{d}\rangle, \tag{13.42}$$

where the numerical coefficients α, β, γ, and δ are the same as in (13.40).

The overall unitary time development of the entire system $\mathcal{M} \otimes \mathcal{C} \otimes \mathcal{D}$ consisting of the particle and the two detectors is determined by the operator

$$T = S_i R'_c R'_d = S_i R'_d R'_c, \tag{13.43}$$

where S_i (rather than S'_i) means the phase shifts ϕ_c and ϕ_d are 0. The unitary time evolution,

$$|\Omega_t\rangle = T^t|\Omega_0\rangle, \quad |\Omega_0\rangle = |0a, 0\hat{c}, 0\hat{d}\rangle, \tag{13.44}$$

of an initial state $|\Omega_0\rangle$ in which the particle is at $[0a]$ and both detectors are in their ready states results in

$$|\Omega_4\rangle = \alpha\,|4f, 0\hat{c}, 0\hat{d}\rangle$$
$$+ \tfrac{1}{2}\beta\big(|4e, 1\hat{c}, 0\hat{d}\rangle + |4f, 1\hat{c}, 0\hat{d}\rangle - |4e, 0\hat{c}, 1\hat{d}\rangle + |4f, 0\hat{c}, 1\hat{d}\rangle\big) \tag{13.45}$$

at $t = 4$; for any later time $|\Omega_t\rangle$ is given by the same expression with 4 replaced by t.

Consider a family of two-time histories with initial state $|\Omega_0\rangle$ at $t = 0$, and at $t = 4$ a decomposition of the identity in which each detector is in a pointer state (ready or triggered) and the particle emerges in either the e or the f channel. Consistency follows from the fact that there are only two times, and the probabilities can be computed using (13.45) as a pre-probability. There is a finite probability $|\alpha|^2$ that at $t = 4$ neither detector has detected the particle, and in this case it always emerges in the f channel. On the other hand, if the particle has been detected by the c detector, it will emerge with equal probability in either the e or the f channel, and the same is true if it has been detected by the d detector.

All of this agrees with Feynman's discussion of electrons passing through a double slit and illuminated by a weak light source. Emerging in the f channel rather than the e channel is what happens when no detectors are present inside the interferometer, and represents an interference effect. By contrast, detection of the particle in either arm washes out the interference effect, and the particle emerges with equal probability in either the e or the f channel. Note that the probability is zero that *both* detectors will detect the particle. This is what one would expect, since the particle cannot be both *in* the c arm and *in* the d arm of the interferometer; quantum particles are never in two different places at the same time.

Additional complications arise when there is a weakly-coupled detector in only one arm, or when the numerical coefficients in (13.42) are different from those in (13.40). Sorting them out is best done using $T = S_i' R_c' R_d'$ or $T = S_i' R_c'$ in place of (13.43), and thinking about what happens when the phase shifts ϕ_c and ϕ_d are allowed to vary. Exploring this is left to the reader.

When weakly-coupled detectors are present, what can we say about the particle *while it is inside the interferometer*? Again assume, for simplicity, that ϕ_c and ϕ_d are zero. There are many possible frameworks, and we shall only consider one example, a consistent family whose support consists of the five histories

$$[\Omega_0] \odot [1c, 0\hat{c}, 0\hat{d}] \odot [2c, 0\hat{c}, 0\hat{d}] \odot [3c, 1\hat{c}, 0\hat{d}] \odot [4e, 1\hat{c}, 0\hat{d}],$$

$$[\Omega_0] \odot [1c, 0\hat{c}, 0\hat{d}] \odot [2c, 0\hat{c}, 0\hat{d}] \odot [3c, 1\hat{c}, 0\hat{d}] \odot [4f, 1\hat{c}, 0\hat{d}],$$

$$[\Omega_0] \odot [1d, 0\hat{c}, 0\hat{d}] \odot [2d, 0\hat{c}, 0\hat{d}] \odot [3d, 0\hat{c}, 1\hat{d}] \odot [4e, 0\hat{c}, 1\hat{d}], \qquad (13.46)$$

$$[\Omega_0] \odot [1d, 0\hat{c}, 0\hat{d}] \odot [2d, 0\hat{c}, 0\hat{d}] \odot [3d, 0\hat{c}, 1\hat{d}] \odot [4f, 0\hat{c}, 1\hat{d}],$$

$$[\Omega_0] \odot [1\bar{a}, 0\hat{c}, 0\hat{d}] \odot [2\bar{a}, 0\hat{c}, 0\hat{d}] \odot [3\bar{a}, 0\hat{c}, 0\hat{d}] \odot [4f, 0\hat{c}, 0\hat{d}].$$

(Consistency follows from the orthogonality of the final projectors, Sec. 11.3.) Using this family one can conclude that if at $t = 4$ the $\hat{c}$ detector has been triggered, the particle was earlier ($t = 1, 2$, or 3) in the c arm; if the $\hat{d}$ detector has been triggered, the particle was earlier in the d arm; and if neither detector has been triggered, the particle was earlier in a superposition state $|\bar{a}\rangle$. The corresponding statements for Feynman's double slit with a weak light source would be that if a photon scatters off an electron which has just passed through the slit system, then the electron previously passed through the slit indicated by the scattered photon, whereas if no photon scatters off the electron, it passed through the slit system in a coherent superposition.

While these results are not unreasonable, there is nonetheless something a bit odd going on. The projector $[1\bar{a}, 0\hat{c}, 0\hat{d}]$ at time $t = 1$ in the last history in (13.46) does not commute with the projectors at $t = 1$ in the other histories, even though the projectors for the histories themselves (on the history space $\breve{\mathcal{H}}$) do commute with each other, since their products are 0. This means that the Boolean algebra associated with (13.46) does not contain the projector $[1\bar{a}]_1$ for the particle to be in a coherent superposition state at the time $t = 1$, nor does it contain $[1c]_1$ or $[1d]_1$. Thus the events at $t = 1$, and also at $t = 2$ and $t = 3$, in these histories are *dependent* or *contextual* in the sense employed in Sec. 6.6 when discussing

(6.55). Within the framework represented by (13.46), they only make sense when discussed together with certain later events; they depend on the later outcomes of the weak measurements in a sense which will be discussed in Ch. 14.

14

Dependent (contextual) events

14.1 An example

Consider two spin-half particles a and b, and suppose that the corresponding Boolean algebra $\mathcal{L}$ of properties on the tensor product space $\mathcal{A} \otimes \mathcal{B}$ is generated by a sample space of four projectors,

$$[z_a^+] \otimes [z_b^+], \quad [z_a^+] \otimes [z_b^-], \quad [z_a^-] \otimes [x_b^+], \quad [z_a^-] \otimes [x_b^-], \tag{14.1}$$

which sum to the identity operator $I \otimes I$. Let $A = [z_a^+]$ be the property that $S_{az} = +1/2$ for particle a, and its negation $\tilde{A} = I - A = [z_a^-]$ the property that $S_{az} = -1/2$. Likewise, let $B = [z_b^+]$ and $\tilde{B} = I - B = [z_b^-]$ be the properties $S_{bz} = +1/2$ and $S_{bz} = -1/2$ for particle b. Together with the projectors AB and $A\tilde{B}$, the first two items in (14.1), the Boolean algebra $\mathcal{L}$ also contains their sum

$$A = AB + A\tilde{B} \tag{14.2}$$

and its negation $\tilde{A}$. On the other hand $\mathcal{L}$ does *not* contain the projector B or its negation $\tilde{B}$, as is obvious from the fact that these operators do not commute with the last two projectors in (14.1). Thus when using the framework $\mathcal{L}$ one can discuss whether S_{az} is $+1/2$ or $-1/2$ without making any reference to the spin of particle b. But it only makes sense to discuss whether S_{bz} is $+1/2$ or $-1/2$ when one knows that $S_{az} = +1/2$. That is, one cannot ascribe a value to S_{bz} in an *absolute* sense without making any reference to the spin of particle a.

If it makes sense to talk about a property B when a system possesses the property A but not otherwise, we shall say that B is a *contextual* property: it is meaningful only within a certain context. Also we shall say that B *depends on* A, and that A is the *base* of B. (One might also call A the *support* of B.) A slightly more restrictive definition is given in Sec. 14.3, and generalized to contextual events which do not have a base. It is important to notice that contextuality and the corresponding dependence is very much a function of the Boolean algebra $\mathcal{L}$ employed for constructing a quantum description. For example, the Boolean algebra $\mathcal{L}'$ generated by

$$[z_a^+] \otimes [z_b^+], \quad [z_a^+] \otimes [z_b^-], \quad [z_a^-] \otimes [z_b^+], \quad [z_a^-] \otimes [z_b^-] \qquad (14.3)$$

contains both $A = [z_a^+]$ and $B = [z_b^+]$, and thus in this algebra B does not depend upon A. And in the algebra $\mathcal{L}''$ generated by

$$[z_a^+] \otimes [z_b^+], \quad [z_a^-] \otimes [z_b^+], \quad [x_a^+] \otimes [z_b^-], \quad [x_a^-] \otimes [z_b^-], \qquad (14.4)$$

the property A is contextual and depends on B.

Since quantum theory does not prescribe a single "correct" Boolean algebra of properties to use in describing a quantum system, whether or not some property is contextual or dependent on another property is a consequence of the physicist's choice to describe a quantum system in a particular way and not in some other way. In particular, when B *depends on* A in the sense we are discussing, one should not think of B as being *caused by* A, as if the two properties were linked by a physical cause. The dependence is logical, not physical, and has to do with what other properties are or are not allowed as part of the description based upon a particular Boolean algebra.

14.2 Classical analogy

It is possible to construct an analogy for quantum contextual properties based on purely classical ideas. The analogy is somewhat artificial, but even its artificial character will help us understand better why dependency is to be expected in quantum theory, when it normally does not show up in classical physics. Let x and y be real numbers which can take on any values between 0 and 1, so that pairs (x, y) are points in the unit square, Fig. 14.1. In classical statistical mechanics one sometimes divides up the phase space into nonoverlapping cells (Sec. 5.1), and in a similar way we shall divide up the unit square into cells of finite area, and regard each cell as an element of the sample space of a probabilistic theory. The sample space corresponding to the cells in Fig. 14.1(a) consists of four mutually-exclusive properties:

$$\{0 \leq x < 1/2, \, 0 \leq y < 1/2\}, \quad \{0 \leq x < 1/2, \, 1/2 \leq y \leq 1\},$$
$$\{1/2 \leq x \leq 1, \, 0 \leq y < 1/2\}, \quad \{1/2 \leq x \leq 1, \, 1/2 \leq y \leq 1\}. \qquad (14.5)$$

Let A be the property $0 \leq x < 1/2$, so its complement $\tilde{A}$ is $1/2 \leq x \leq 1$, and let B be the property $0 \leq y < 1/2$, so $\tilde{B}$ is $1/2 \leq y \leq 1$. Then the four sets in (14.5) correspond to the properties $A \wedge B$, $A \wedge \tilde{B}$, $\tilde{A} \wedge B$, $\tilde{A} \wedge \tilde{B}$. It is then obvious that the Boolean algebra of properties generated by (14.5) contains both A and B, so (14.5) is analogous in this respect to the quantum sample space (14.3).

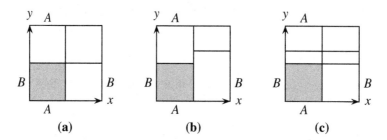

Fig. 14.1. Unit square in the x, y plane: (a) shows the set of cells in (14.5), (b) the set of cells in (14.6), and (c) the cells in a common refinement (see text). Property A is represented by the vertical rectangular cell on the left, and B by the horizontal rectangular cell (not present in (b)) on the bottom. The gray region represents $A \wedge B$.

An alternative choice for cells is shown in Fig. 14.1(b), where the four mutually-exclusive properties are

$$\{0 \leq x < 1/2, \ 0 \leq y < 1/2\}, \quad \{0 \leq x < 1/2, \ 1/2 \leq y \leq 1\},$$
$$\{1/2 \leq x \leq 1, \ 0 \leq y < 2/3\}, \quad \{1/2 \leq x \leq 1, \ 2/3 \leq y \leq 1\}. \tag{14.6}$$

If A and B are defined in the same way as before, the new algebra of properties generated by (14.6) contains A and $A \wedge B$, but does *not* contain B. In this respect it is analogous to (14.1) in the quantum case, and B is a contextual or dependent property: it only makes sense to ask whether the system has or does not have the property B when the property A is true, that is, when x is between 0 and $1/2$, but the same question does not make sense when x is between $1/2$ and 1, that is, when A is false.

Isn't this just some sort of formal nitpicking? Why not simply refine the sample space of Fig. 14.1(b) by using the larger collection of cells shown in Fig. 14.1(c)? The corresponding Boolean algebra of properties includes all those in (14.6), so we have not lost the ability to describe whatever we would like to describe, and now B as well as A is part of the algebra of properties, so dependency is no longer of any concern. Such a refinement of the sample space can always be employed in classical statistical mechanics. However, a similar type of refinement may or may not be possible in quantum mechanics. There is no way to refine the sample space in (14.1), for the four projectors in that list already project onto one-dimensional subspaces, which is as far as a quantum refinement can go. The move from (b) to (c) in Fig. 14.1, which conveniently gets rid of contextual properties in a classical context, will not work in the case of (14.1); the latter is an example of an *irreducible* contextuality.

To be more specific, the refinement in Fig. 14.1(c) is obtained by forming the products of the indicators for B, $\tilde{B}$, B', and $\tilde{B}'$ with one another and with A and $\tilde{A}$, where B' is the property $0 \leq y < 2/3$. The analogous process for (14.1) would require taking products of projectors such as $[z_b^+]$ and $[x_b^+]$, but since they do not commute with each other, their product is not a projector. That noncommutativity of the projectors is at the heart of the contextuality associated with (14.1) can also be seen by considering two *classical* spinning objects a and b with angular momenta $\mathbf{L}_a$ and $\mathbf{L}_b$, and interpreting $[z_a^+]$ and $[z_a^-]$ in (14.1) as $L_{az} \geq 0$ and $L_{az} < 0$, etc. In the classical case there is no difficulty refining the sample space of (14.1) to get rid of dependency, for $[z_b^+][x_b^+]$ is the property $L_{bz} \geq 0 \wedge L_{bx} \geq 0$, which makes perfectly good (classical) sense. But its quantum counterpart for a spin-half particle has no physical meaning.

14.3 Contextual properties and conditional probabilities

If A and B are elements of a Boolean algebra $\mathcal{L}$ for which a probability distribution is defined, then

$$\Pr(B \mid A) = \Pr(AB)/\Pr(A) \tag{14.7}$$

is defined provided $\Pr(A)$ is greater than 0. If, however, B is not an element of $\mathcal{L}$, then $\Pr(B)$ is not defined and, as a consequence, $\Pr(A \mid B)$ is also not defined. In view of these remarks it makes sense to define B as a contextual property which depends upon A, A is the base of B, provided $\Pr(B \mid A)$ is positive (which implies $\Pr(AB) > 0$), whereas $\Pr(B)$ is undefined. This definition is stricter than the one in Sec. 14.1, but the cases it eliminates — those with $\Pr(B \mid A) = 0$ — are in practice rather uninteresting. In addition, one is usually interested in situations where the dependence is irreducible, that is, it cannot be eliminated by appropriately refining the sample space, unlike the classical example in Sec. 14.2.

One can extend this definition to events which depend on other contextual events. For example, let A, B, and C be commuting projectors, and suppose A, AB, and ABC belong to the Boolean algebra, but B and C do not. Then as long as

$$\Pr(C \mid AB) = \Pr(ABC)/\Pr(AB) \tag{14.8}$$

is positive, we shall say that C depends on B (or on AB), and B depends on A. Note that if (14.8) is positive, so is $\Pr(AB)$, and thus $\Pr(B \mid A)$, (14.7), is also positive.

There are situations in which the properties A and B, represented by commuting projectors, are contextual even though neither can be said to depend upon or be the base of the other. That is, AB belongs to the Boolean algebra $\mathcal{L}$ and has positive probability, but neither A nor B belongs to $\mathcal{L}$. In this case neither $\Pr(A \mid B)$ nor $\Pr(B \mid A)$ is defined, so one cannot say that B depends on A or A on B, though one might refer to them as "codependent". As an example, let $\mathcal{A}$ and $\mathcal{B}$ be two Hilbert spaces of dimension 2 and 3, respectively, with orthonormal bases $\{|0a\rangle, |1a\rangle\}$ and $\{|0b\rangle, |1b\rangle, |2b\rangle\}$. In addition, define

$$|+b\rangle = \big(|0b\rangle + |1b\rangle\big)/\sqrt{2}, \quad |-b\rangle = \big(|0b\rangle - |1b\rangle\big)/\sqrt{2}, \tag{14.9}$$

and $|+a\rangle$ and $|-a\rangle$ in a similar way. Then the six kets

$$\begin{aligned} &|0a\rangle \otimes |0b\rangle, \quad |1a\rangle \otimes |+b\rangle, \quad |+a\rangle \otimes |2b\rangle, \\ &|0a\rangle \otimes |1b\rangle, \quad |1a\rangle \otimes |-b\rangle, \quad |-a\rangle \otimes |2b\rangle, \end{aligned} \tag{14.10}$$

form an orthonormal basis for $\mathcal{A} \otimes \mathcal{B}$, and the corresponding projectors generate a Boolean algebra $\mathcal{L}$. If $A = [0a] \otimes I$ and $B = I \otimes [0b]$, then $\mathcal{L}$ contains AB, corresponding to the first ket in (14.10), but neither A nor B belongs to $\mathcal{L}$, since $[0a]$ does not commute with $[+a]$, and $[0b]$ does not commute with $[+b]$. More complicated cases of "codependency" are also possible, as when $\mathcal{L}$ contains the product ABC of three commuting projectors, but none of the six projectors A, B, C, AB, BC, and AC belong to $\mathcal{L}$.

14.4 Dependent events in histories

In precisely the same way that quantum properties can be dependent upon other quantum properties of a system at a single time, a quantum event — a property of a quantum system at a particular time — can be dependent upon a quantum event at some different time. That is, in the family of consistent histories used to describe the time development of a quantum system, it may be the case that the projector B for an event at a particular time does not occur by itself in the Boolean algebra $\mathcal{L}$ of histories, but is only present if some other event A at some different time is present in the same history. Then B depends on A, or A is the base of B, using the terminology introduced earlier. And there are situations in which a third event C at still another time depends on B, so that it only makes sense to discuss C as part of a history in which both A and B occur. Sometimes this contextuality can be removed by refining the history sample space, but in other cases it is irreducible, either because a refinement is prevented by noncommuting projectors, or because it would result in a violation of consistency conditions.

   Families of histories often contain contextual events that depend upon a base that occurs at an *earlier* time. Such a family is said to show "branch dependence". A particular case is a family of histories with a single initial state Ψ_0. If one uses the Boolean algebra suggested for that case in Sec. 11.5, then *all* the later events in all the histories of interest are (ultimately) dependent upon the initial event Ψ_0. This is because the only history in which the negation $\tilde{\Psi}_0 = I - \Psi_0$ of the initial event occurs is the history Z in (11.14), and in that history only the identity occurs

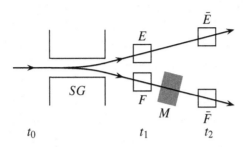

Fig. 14.2. Upper and lower beams emerging from a Stern–Gerlach magnet SG. An atom in the lower beam passes through an additional region of uniform magnetic field M. The square boxes indicate regions in space, and the time when the atom will pass through a given region is indicated at the bottom of the figure.

at later times. It may or may not be possible to refine such a family in order to remove some or all of the dependence upon Ψ_0.

An example of branch dependence involving something other than the initial state is shown in Fig. 14.2. A spin-half particle passes through a Stern–Gerlach magnet (Sec. 17.2) and emerges moving at an upwards angle if $S_z = +1/2$, or a downwards angle if $S_z = -1/2$. Let E and F be projectors on two regions in space which include the upward- and downward-moving wave packets at time t_1, assuming a state $|\Psi_0\rangle$ (space-and-spin wave function of the particle) at time t_0. In the interval between t_1 and t_2 the downward-moving wave packet passes through a region M of uniform magnetic field which causes the spin state to rotate by $90°$ from $S_z = -1/2$ to $S_x = +1/2$. This situation can be described using a consistent family whose support is the two histories

$$\Psi_0 \odot E \odot [z^+], \qquad \Psi_0 \odot F \odot [x^+],$$ (14.11)

which can also be written in the form

$$\Psi_0 \odot \begin{cases} E \odot [z^+], \\ F \odot [x^+], \end{cases}$$ (14.12)

where the initial element common to both histories is indicated only once. Consistency follows from the fact that the spatial wave functions at the final time t_2 have negligible overlap, even though they are not explicitly referred to in (14.12). Whatever may be the zero-weight histories, it is at once evident that neither of the two histories

$$\Psi_0 \odot I \odot [z^+], \qquad \Psi_0 \odot I \odot [x^+]$$ (14.13)

can occur in the Boolean algebra, since the projector for the first history in (14.13) does not commute with that for the second history in (14.12), and the second history in (14.13) is incompatible with the first history in (14.12). Consequently, in the consistent family (14.12) $[z^+]$ at t_2 depends upon E at t_1, and $[x^+]$ at t_2 depends upon F at t_1. Furthermore, as the necessity for this dependency can be traced to noncommuting projectors, the dependency is irreducible: one cannot get rid of it by refining the consistent family.

An alternative way of thinking about the same gedanken experiment is to note that at t_2 the wave packets do not overlap, so we can find mutually orthogonal projectors $\bar{E}$ and $\bar{F}$ on nonoverlapping regions of space, Fig. 14.2, which include the upward- and downward-moving parts of the wave packet at this time. Consider the consistent family whose support is the two histories

$$\Psi_0 \odot I \odot \{[z^+]\bar{E}, [x^+]\bar{F}\}, \tag{14.14}$$

where the notation is a variant of that in (14.12): the two events inside the curly brackets are both at the time t_2, so one history ends with the projector $[z^+]\bar{E}$, the other with the projector $[x^+]\bar{F}$. Once again, the final spin states $[z^+]$ and $[x^+]$ are dependent events, but now $[z^+]$ depends upon $\bar{E}$ and $[x^+]$ upon $\bar{F}$, so the bases occur at the same time as the contextual events which depend on them. This is a situation which resembles (14.1), with $\bar{E}$ and $\bar{F}$ playing the roles of $[z_a^+]$ and $[z_a^-]$, respectively, while the spin projectors in (14.14) correspond to those of the b particle in (14.1). One could also move the regions $\bar{E}$ and $\bar{F}$ further to the right in Fig. 14.2, and obtain a family of histories

$$\Psi_0 \odot I \odot \begin{cases} [z^+] \odot \bar{E}, \\ [x^+] \odot \bar{F}, \end{cases} \tag{14.15}$$

for the times $t_0 < t_1 < t_2 < t_3$, in which $[z^+]$ and $[x^+]$ are dependent on the *later* events $\bar{E}$ and $\bar{F}$.

Dependence on later events also arises, for certain families of histories, in the next example we shall consider, which is a variant of the toy model discussed in Sec. 13.5. Figure 14.3 shows a device which is like a Mach–Zehnder interferometer, but the second beam splitter has been replaced by a weakly-coupled measuring device M, with initial ("ready") state $|M\rangle$. The relevant unitary transformations are

$$|\Psi_0\rangle = |0a\rangle \otimes |M\rangle \mapsto \big(|1c\rangle + |1d\rangle\big)/\sqrt{2} \otimes |M\rangle \tag{14.16}$$

for the time interval t_0 to t_1, and

$$\begin{aligned} |1c\rangle \otimes |M\rangle &\mapsto |2f\rangle \otimes \big(|M\rangle + |M^c\rangle\big)/\sqrt{2}, \\ |1d\rangle \otimes |M\rangle &\mapsto |2e\rangle \otimes \big(|M\rangle + |M^d\rangle\big)/\sqrt{2} \end{aligned} \tag{14.17}$$

for t_1 to t_2. Here $|0a\rangle$ is a wave packet approaching the beam splitter in channel a at

t_0, $|1c\rangle$ is a wave packet in the c arm at time t_1, and so forth. The time t_1 is chosen so that the particle is inside the device, somewhere between the initial beam splitter and the detector M, whereas at t_2 it has emerged in e or f. The states $|M\rangle$, $|M^c\rangle$, and $|M^d\rangle$ of the detector are mutually orthogonal and normalized. Combining (14.16) and (14.17) yields a unitary time development

$$|\Psi_0\rangle \mapsto \left(|2e\rangle \otimes |M^d\rangle + |2f\rangle \otimes |M^c\rangle + \sqrt{2}|2s\rangle|M\rangle\right)/2 \qquad (14.18)$$

from t_0 to t_2, where

$$|2s\rangle = \left(|2e\rangle + |2f\rangle\right)/\sqrt{2} \qquad (14.19)$$

is a superposition state of the final particle wave packets.

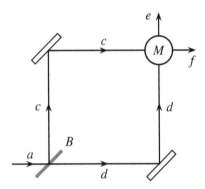

Fig. 14.3. Mach–Zehnder interferometer with the second beam splitter replaced by a measuring device M.

Consider the consistent family for $t_0 < t_1 < t_2$ whose support is the three histories

$$\Psi_0 \odot I \odot \left\{[2e] \otimes M^d,\ [2f] \otimes M^c,\ [2s] \otimes M\right\}. \qquad (14.20)$$

Since the projector $[2s]$ does not commute with the projectors $[2e]$ and $[2f]$, it is clear that $[2e]$, $[2f]$, and $[2s]$ are dependent upon the detector states M^d, M^c, and M at the (same) time t_2, and one has conditional probabilities

$$\Pr(2e \mid M_2^d) = \Pr(2f \mid M_2^c) = \Pr(2s \mid M_2) = 1. \qquad (14.21)$$

On the other hand, $\Pr(M_2^d \mid 2e)$, $\Pr(M_2^c \mid 2f)$, and $\Pr(M_2 \mid 2s)$ are not defined. (Following our usual practice, Ψ_0 is not shown explicitly as one of the conditions.) One could also say that $2e$ and $2f$ are both dependent upon the state M^t with projector $M^c + M^d$, corresponding to the fact that the detector has detected something.

Some understanding of the physical significance of this dependency can be obtained by supposing that later experiments are carried out to confirm (14.21). One can check that the particle emerging from M is in the e channel if the detector state is M^d, or in f if the detector is in M^c, by placing detectors in the e and f channels. One could also verify that the particle emerges in the superposition state s in a case in which it is *not* detected (the detector is still in state M at t_2) by the strategy of adding two more mirrors to bring the e and f channels back together again at a beam splitter which is followed by detectors. Of course, this last measurement cannot be carried out if there are already detectors in the e and f channels, reflecting the fact that the property $2s$ is incompatible with $2e$ and $2f$. (A similar pair of incompatible measurements is discussed in Sec. 18.4, see Fig. 18.3.)

An alternative consistent family for $t_0 < t_1 < t_2$ has support

$$\Psi_0 \odot \begin{cases} [1c] \odot M^c, \\ [1d] \odot M^d, \\ [1r] \odot M, \end{cases} \tag{14.22}$$

where

$$|1r\rangle = \big(|1c\rangle + |1d\rangle\big)/\sqrt{2} \tag{14.23}$$

is a superposition state of the particle before it reaches M. From the fact that $[1r]$ does not commute with $[1c]$ or $[1d]$, it is obvious that the particle states at the intermediate time t_1 in (14.22) must depend upon the later detector states: $[1c]$ upon M^c, $[1d]$ upon M^d, and $[1r]$ upon M. Indeed,

$$\Pr(1c \mid M_2^c) = \Pr(1d \mid M_2^d) = \Pr(1r \mid M_2) = 1, \tag{14.24}$$

whereas $\Pr(M_2^c \mid 1c)$, $\Pr(M_2^d \mid 1d)$ and $\Pr(M_2 \mid 1r)$, the conditional probabilities with their arguments in reverse order, are not defined. A very similar dependence upon later events occurs in the family (13.46) associated with weak measurements in the arms of a Mach–Zehnder interferometer, Sec. 13.5.

It may seem odd that earlier contextual events can depend on later events. Does this mean that the future is somehow influencing the past? As already noted in Sec. 14.1, it is important not to confuse the term *depends on*, used to characterize the logical relationship among events in a consistent family, with a notion of *physical influence* or *causality*. The following analogy may be helpful. Think of a historian writing a history of the French revolution. He will not limit himself to the events of the revolution itself, but will try and show that these events were preceded by others which, while their significance may not have been evident at the time, can in retrospect be seen as useful for understanding what happened later. In selecting the type of prior events which enter his account, the historian will use his knowledge of what happened later. It is not a question of later events somehow "causing" the earlier events, at least as causality is ordinarily understood. Instead, those earlier events are introduced into the account which are useful for understanding the later events. While classical histories cannot provide a perfect analogy with quantum histories, this example may help in understanding how the earlier particle states in (14.22) can be said to "depend on" the later states of M without being "caused by" them.

To be sure, one often encounters quantum descriptions in which earlier events, such as the initial state, are the bases of later dependent events, and it is rather natural in such cases to think of (at least some of) the later events as actually caused by the earlier events. This may be why later contextual events that depend on earlier events somehow seem more intuitively reasonable than the reverse. Nonetheless, the ideas of causation and contextuality are quite distinct, and confusing the two can lead to paradoxes.

15

Density matrices

15.1 Introduction

Density matrices are employed in quantum mechanics to give a *partial descrip-*
tion of a quantum system, one from which certain details have been omitted. For
example, in the case of a composite quantum system consisting of two or more
subsystems, one may find it useful to construct a quantum description of just one
of these subsystems, either at a single time or as a function of time, while ignoring the other subsystem(s). Or it may be the case that the exact initial state of a
quantum system is not known, and one wants to use a probability distribution or
pre-probability as an initial state.

Probability distributions are used in classical statistical mechanics in order to
construct partial descriptions, and density matrices play a somewhat similar role in
quantum statistical mechanics, a subject which lies outside the scope of this book.
In this chapter we shall mention a few of the ways in which density matrices are
used in quantum theory, and discuss their physical significance.

Positive operators and density matrices were defined in Sec. 3.9. To recapitulate,
a positive operator is a Hermitian operator whose eigenvalues are nonnegative, and
a density matrix ρ is a positive operator whose trace (the sum of its eigenvalues) is
1. If R is a positive operator but not the zero operator, its trace is greater than 0,
and one can define a corresponding density matrix by means of the formula

$$\rho = R/\operatorname{Tr}(R). \tag{15.1}$$

The eigenvalues of a density matrix ρ must lie between 0 and 1. If one of the eigenvalues is 1, the rest must be 0, and $\rho = \rho^2$ is a projector onto a one-dimensional
subspace of the Hilbert space. Such a density matrix is called a *pure state*. Otherwise there must be at least two nonzero eigenvalues, and the density matrix is called
a *mixed state*.

202

Density matrices very often function as pre-probabilities which can be used to generate probability distributions in different bases, and averages of different observables. This is discussed in Sec. 15.2. Density matrices arise rather naturally when one is trying to describe a subsystem $\mathcal{A}$ of a larger system $\mathcal{A} \otimes \mathcal{B}$, and Secs. 15.3–15.5 are devoted to this topic. The use of a density matrix to describe an isolated system is considered in Sec. 15.6. Section 15.7 on conditional density matrices discusses a more advanced topic related to correlations between subsystems.

15.2 Density matrix as a pre-probability

Recall that in some circumstances a quantum wave function or ket $|\psi\rangle$ need not denote an actual physical property $[\psi]$ of the quantum system; instead it can serve as a *pre-probability*, a mathematical device which allows one to calculate various probabilities. See the discussion in Sec. 9.4, and various examples in Sec. 12.1 and Ch. 13. In most cases (see the latter part of Sec. 15.6 for one of the exceptions) a density matrix is best thought of as a pre-probability. Thus while it provides useful information about a quantum system, one should not think of it as corresponding to an actual physical property; it does not represent "quantum reality". For this reason, referring to a density matrix as the "state" of a quantum system can be misleading. However, in classical statistical mechanics it is customary to refer to probability distributions as "states", even though a probability distribution is obviously not a physical property, and hence it is not unreasonable to use the same term for a density matrix functioning as a quantum pre-probability.

A density matrix which is a pre-probability can be used to generate a probability distribution in the following way. Given a sample space corresponding to a decomposition of the identity

$$ I = \sum_j P^j \tag{15.2} $$

into orthogonal projectors, the probability of the property P^j is

$$ p_j = \mathrm{Tr}(P^j \rho P^j) = \mathrm{Tr}(\rho P^j), \tag{15.3} $$

where the traces are equal because of cyclic permutation, Sec. 3.8. The operator $P^j \rho P^j$ is positive — use the criterion (3.86) — and therefore its trace, the sum of its eigenvalues, cannot be negative. Thus (15.3) defines a set of probabilities: nonnegative real numbers whose sum, in view of (15.2), is equal to 1, the trace of ρ. In particular, if for each j the projector $P^j = [j]$ is onto a state belonging to an orthonormal basis $\{|j\rangle\}$, then

$$ p_j = \mathrm{Tr}\big(\rho |j\rangle\langle j|\big) = \langle j|\rho|j\rangle \tag{15.4} $$

is the jth diagonal element of ρ in this basis. Hence the diagonal elements of ρ in an orthonormal basis form a probability distribution when this basis is used as the quantum sample space. As a special case, the probabilities given by the Born rule, Secs. 9.3 and 9.4, are of the form (15.4) when $\rho = |\psi_1\rangle\langle\psi_1|$ and $|j\rangle = |\phi_1^j\rangle$ in the notation used in (9.35).

From (15.3) it is evident that the average $\langle V\rangle$, see (5.42), of an observable

$$V = V^\dagger = \sum_j v_j P^j \tag{15.5}$$

can be written in a very compact form using the density matrix:

$$\langle V\rangle = \sum_j p_j v_j = \mathrm{Tr}(\rho V). \tag{15.6}$$

If ρ is a pure state $|\psi_1\rangle\langle\psi_1|$, then $\langle V\rangle$ is $\langle\psi_1|V|\psi_1\rangle$, as in (9.38). It is worth emphasizing that while the trace in (15.6) can be carried out using any basis, interpreting $\langle V\rangle$ as the average of a physical variable requires at least an implicit reference to a basis (or decomposition of the identity) in which V is diagonal. Thus if two observables V and W do not commute with each other, the two averages $\langle V\rangle$ and $\langle W\rangle$ cannot be thought of as pertaining to a single (stochastic) description of a quantum system, for they necessarily involve incompatible quantum sample spaces, and thus different probability distributions. The comments made about averages in Ch. 9 while discussing the Born rule, towards the end of Sec. 9.3 and in connection with (9.38), also apply to averages calculated using density matrices.

15.3 Reduced density matrix for subsystem

Suppose we are interested in a composite system (Ch. 6) with a Hilbert space $\mathcal{A}\otimes\mathcal{B}$. For example, $\mathcal{A}$ might be the Hilbert space of a particle, and $\mathcal{B}$ that of some system (possibly another particle) with which it interacts. At t_0 let $|\Psi_0\rangle$ be a normalized state of the combined system which evolves, by Schrödinger's equation, to a state $|\Psi_1\rangle$ at time t_1. Assume that we are interested in histories for two times, t_0 and t_1, of the form $\Psi_0 \odot (A^j \otimes I)$, where Ψ_0 stands for the projector $[\Psi_0] = |\Psi_0\rangle\langle\Psi_0|$ and the A^j form a decomposition of the identity of the subsystem $\mathcal{A}$:

$$I_\mathcal{A} = \sum_j A^j. \tag{15.7}$$

The probability that system $\mathcal{A}$ will have the property A^j at t_1 can be calculated using the generalization of the Born rule found in (10.34):

$$\mathrm{Pr}(A^j) = \langle\Psi_1|A^j \otimes I|\Psi_1\rangle = \mathrm{Tr}\big[\Psi_1(A^j \otimes I)\big]. \tag{15.8}$$

The trace on the right side of (15.8) can be carried out in two steps, see Sec. 6.5:

first a partial trace over $\mathcal{B}$ to yield an operator on $\mathcal{A}$, followed by a trace over $\mathcal{A}$. In the first step the operator A^j, since it acts on $\mathcal{A}$ rather than $\mathcal{B}$, can be taken out of the trace, so that

$$\text{Tr}_\mathcal{B}\big[\Psi_1(A^j \otimes I)\big] = \rho A^j, \qquad (15.9)$$

where

$$\rho = \text{Tr}_\mathcal{B}(\Psi_1) \qquad (15.10)$$

is called the *reduced density matrix*, because it is used to describe the subsystem $\mathcal{A}$ rather than the whole system $\mathcal{A} \otimes \mathcal{B}$. Since ρ is the partial trace of a positive operator, it is itself a positive operator: apply the test in (3.86). In addition, the trace of ρ is

$$\text{Tr}_\mathcal{A}(\rho) = \text{Tr}(\Psi_1) = \langle \Psi_1 | \Psi_1 \rangle = 1, \qquad (15.11)$$

so ρ is a density matrix. Upon taking the trace of both sides of (15.9) over $\mathcal{A}$, one obtains, see (15.8), the expression

$$\text{Pr}(A^j) = \text{Tr}_\mathcal{A}\left(\rho A^j\right) \qquad (15.12)$$

for the probability of the property A^j, in agreement with (15.3). Note that $|\Psi_1\rangle$, the counterpart of $|\psi_1\rangle$ in the discussion of the Born rule in Sec. 9.4, functions as a pre-probability, not as a physical property, and its partial trace ρ also functions as a pre-probability, which can be used to calculate probabilities for any sample space of the form (15.7). In the same way one can define the reduced density matrix

$$\rho' = \text{Tr}_\mathcal{A}(\Psi_1) \qquad (15.13)$$

for system $\mathcal{B}$ and use it to calculate probabilities of various properties of system $\mathcal{B}$.

Let us consider a simple example. Let $\mathcal{A}$ and $\mathcal{B}$ be the spin spaces for two spin-half particles a and b, and let

$$|\Psi_1\rangle = \alpha|z_a^+\rangle \otimes |z_b^-\rangle + \beta|z_a^-\rangle \otimes |z_b^+\rangle, \qquad (15.14)$$

where the subscripts identify the particles, and the coefficients satisfy

$$|\alpha|^2 + |\beta|^2 = 1, \qquad (15.15)$$

so that $|\Psi_1\rangle$ is normalized. The corresponding projector is

$$\begin{aligned}
\Psi_1 = |\Psi_1\rangle\langle\Psi_1| &= |\alpha|^2|z_a^+\rangle\langle z_a^+| \otimes |z_b^-\rangle\langle z_b^-| + |\beta|^2|z_a^-\rangle\langle z_a^-| \otimes |z_b^+\rangle\langle z_b^+| \\
&\quad + \alpha\beta^*|z_a^+\rangle\langle z_a^-| \otimes |z_b^-\rangle\langle z_b^+| + \alpha^*\beta|z_a^-\rangle\langle z_a^+| \otimes |z_b^+\rangle\langle z_b^-|.
\end{aligned} \qquad (15.16)$$

The partial trace in (15.10) is easily evaluated by noting that

$$\text{Tr}_\mathcal{B}\left(|z_b^-\rangle\langle z_b^+|\right) = \langle z_b^+|z_b^-\rangle = 0, \qquad (15.17)$$

etc.; thus

$$\rho = |\alpha|^2[z_a^+] + |\beta|^2[z_a^-]. \tag{15.18}$$

This is a positive operator, since its eigenvalues are $|\alpha|^2$ and $|\beta|^2$, and its trace is equal to 1, (15.15). If both α and β are nonzero, ρ is a mixed state.

Employing either (15.8) or (15.12), one can show that if the decomposition $[z_a^+]$, $[z_a^-]$, the S_{az} framework, is used as a sample space, the corresponding probabilities are $|\alpha|^2$ and $|\beta|^2$, whereas if one uses $[x_a^+]$, $[x_a^-]$, the S_{ax} framework, the probability of each is $1/2$. Of course it makes no sense to suppose that these two sets of probabilities refer simultaneously to the same particle, as the two sample spaces are incompatible. Using either the S_{ax} or the S_{az} framework precludes treating Ψ_1 at t_1 as a physical property when α and β are both nonzero, since as a projector it does not commute with $[w_a^+]$ for any direction w. Thus Ψ_1 and its partial trace ρ should be thought of as pre-probabilities.

Except when $|\alpha|^2 = |\beta|^2$ there is a unique basis, $|z_a^+\rangle$, $|z_a^-\rangle$, in which ρ is diagonal. However, ρ can be used to assign a probability distribution for any basis, and thus there is nothing special about the basis in which it is diagonal. In this respect ρ differs from operators that represent physical variables, such as the Hamiltonian, for which the eigenfunctions do have a particular physical significance.

The expression on the right side of (15.14) is an example of the Schmidt form

$$|\Psi_1\rangle = \sum_j \lambda_j |\hat{a}_j\rangle \otimes |\hat{b}_j\rangle \tag{15.19}$$

introduced in (6.18), where $\{|\hat{a}_j\rangle\}$ and $\{|\hat{b}_k\rangle\}$ are special choices of orthonormal bases for $\mathcal{A}$ and $\mathcal{B}$. The reduced density matrices ρ and ρ' for $\mathcal{A}$ and $\mathcal{B}$ are easily calculated from the Schmidt form using (15.10) and (15.13), and one finds:

$$\rho = \sum_j |\lambda_j|^2[\hat{a}_j], \quad \rho' = \sum_j |\lambda_j|^2[\hat{b}_j]. \tag{15.20}$$

One can check that ρ in (15.18) is, indeed, given by this expression.

Relative to the physical state of the subsystem $\mathcal{A}$ at time t_1, ρ contains the same amount of information as Ψ_1. However, relative to the total system $\mathcal{A} \otimes \mathcal{B}$, ρ is much less informative. Suppose that

$$I_{\mathcal{B}} = \sum_k B^k \tag{15.21}$$

is some decomposition of the identity for subsystem $\mathcal{B}$, and we are interested in histories of the form $\Psi_0 \odot (A^j \otimes B^k)$. Then the joint probability distribution

$$\Pr(A^j \wedge B^k) = \mathrm{Tr}\big[\Psi_1(A^j \otimes B^k)\big] \tag{15.22}$$

can be calculated using Ψ_1, whereas from ρ we can obtain only the marginal distribution

$$\Pr(A^j) = \sum_k \Pr(A^j \wedge B^k). \tag{15.23}$$

The other marginal distribution, $\Pr(B^k)$, can be obtained using the reduced density matrix ρ' for subsystem $\mathcal{B}$. However, from a knowledge of both ρ and ρ', one still cannot calculate the correlations between the two subsystems. For instance, in the two-spin example of (15.14), if we use a framework in which S_{az} and S_{bz} are both defined at t_1, Ψ_1 implies that $S_{az} = -S_{bz}$, a result which is not contained in ρ or ρ'. This illustrates the fact, pointed out in the introduction, that density matrices typically provide partial descriptions of quantum systems, descriptions from which certain features are omitted.

Rather than a projector on a one-dimensional subspace, Ψ_1 could itself be a density matrix on $\mathcal{A} \otimes \mathcal{B}$. For example, if the total quantum system with Hilbert space $\mathcal{A} \otimes \mathcal{B} \otimes \mathcal{C}$ consists of three subsystems $\mathcal{A}$, $\mathcal{B}$, and $\mathcal{C}$, and unitary time evolution beginning with a normalized initial state $|\Phi_0\rangle$ at t_0 results in a state $|\Phi_1\rangle$ with projector Φ_1 at t_1, then

$$\Psi_1 = \text{Tr}_{\mathcal{C}}(\Phi_1) \tag{15.24}$$

is a density matrix. The partial traces of Ψ_1, (15.10) and (15.13), again define density matrices ρ and ρ' appropriate for calculating probabilities of properties of $\mathcal{A}$ or $\mathcal{B}$, since, for example,

$$\rho = \text{Tr}_{\mathcal{B}}(\Psi_1) = \text{Tr}_{\mathcal{B}\mathcal{C}}(\Phi_1) \tag{15.25}$$

can be obtained from Ψ_1 or directly from Φ_1. Even when $\mathcal{A} \otimes \mathcal{B}$ is not part of a larger system it can be described by means of a density matrix as discussed in Sec. 15.6.

15.4 Time dependence of reduced density matrix

There is, of course, nothing very special about the time t_1 used in the discussion in Sec. 15.3. If $|\Psi_t\rangle$ is a solution to the Schrödinger equation as a function of time t for the composite system $\mathcal{A} \otimes \mathcal{B}$, and Ψ_t the corresponding projector, then one can define a density matrix

$$\rho_t = \text{Tr}_{\mathcal{B}}(\Psi_t) \tag{15.26}$$

for subsystem $\mathcal{A}$ at any time t, and use it to calculate the probability of a history of the form $\Psi_0 \odot A^j$ based on the two times 0 and t, where A^j is a projector on $\mathcal{A}$. One should not think of ρ_t as some sort of physical property which develops in time. Instead, it is somewhat analogous to the classical single-time probability distribution $\rho_t(s)$ at time t for a particle undergoing a random walk, or $\rho_t(\mathbf{r})$ for a Brownian particle, discussed in Sec. 9.2. In particular, ρ_t provides no information about correlations of quantum properties at successive times. To discuss such correlations requires the use of quantum histories, see Sec. 15.5.

In general, ρ_t as a function of time does not satisfy a simple differential equation. An exception is the case in which $\mathcal{A}$ is itself an isolated subsystem, so that the time development operator for $\mathcal{A} \otimes \mathcal{B}$ factors,

$$T(t', t) = T_A(t', t) \otimes T_B(t', t), \tag{15.27}$$

or, equivalently, the Hamiltonian is of the form

$$H = H_A \otimes I + I \otimes H_B \tag{15.28}$$

during the times which are of interest. This would, for example, be the case if $\mathcal{A}$ and $\mathcal{B}$ were particles (or larger systems) flying away from each other after a collision. Using the fact that

$$\Psi_t = |\Psi_t\rangle\langle\Psi_t| = T(t, 0)\Psi_0 T(0, t), \tag{15.29}$$

one can show (e.g., by writing Ψ_0 as a sum of product operators of the form $P \otimes Q$) that when $T(t, 0)$ factors, (15.27),

$$\rho_t = T_A(t, 0)\rho_0 T_A(0, t). \tag{15.30}$$

Upon differentiating this equation one obtains

$$i\hbar \frac{d\rho_t}{dt} = [H_A, \rho_t], \tag{15.31}$$

since for an isolated system $T_A(t, 0)$ satisfies (7.45) and (7.46) with H_A in place of H. Note that (15.31) is also valid when H_A depends on time. If H_A is independent of time and diagonal in the orthonormal basis $\{|e_n\rangle\}$,

$$H_A = \sum_n E_n |e_n\rangle\langle e_n|, \tag{15.32}$$

one can use (7.48) to rewrite (15.30) in the form

$$\rho_t = e^{-itH_A/\hbar} \rho_0 e^{itH_A/\hbar}, \tag{15.33}$$

or the equivalent in terms of matrix elements:

$$\langle e_m|\rho_t|e_n\rangle = \langle e_m|\rho_0|e_n\rangle \exp[-i(E_m - E_n)t/\hbar]. \tag{15.34}$$

There are situations in which (15.28) is only true in a first approximation, and there is an additional weak interaction between $\mathcal{A}$ and $\mathcal{B}$, so that $\mathcal{A}$ is not truly isolated. Under such circumstances it may still be possible, given a suitable system

$\mathcal{B}$, to write an approximate differential equation for ρ_t in which additional terms appear on the right side. A discussion of open systems of this type lies outside the scope of this book.

15.5 Reduced density matrix as initial condition

Let Ψ_0 be a projector representing an initial pure state at time t_0 for the composite system $\mathcal{A} \otimes \mathcal{B}$, and assume that for $t > t_0$ the subsystem $\mathcal{A}$ is isolated from $\mathcal{B}$, so that the time development operator factors, (15.27). We shall be interested in histories of the form

$$Z^{\alpha} = \Psi_0 \odot Y^{\alpha}, \tag{15.35}$$

where

$$Y^{\alpha} = A_1^{\alpha_1} \odot A_2^{\alpha_2} \odot \cdots A_f^{\alpha_f} \tag{15.36}$$

is a history of $\mathcal{A}$ at the times $t_1 < t_2 < \cdots t_f$, with $t_1 > t_0$, and each of the projectors $A_j^{\alpha_j}$ at time t_j comes from a decomposition of the identity

$$I_{\mathcal{A}} = \sum_{\alpha_j} A_j^{\alpha_j} \tag{15.37}$$

of subsystem $\mathcal{A}$. A history of the form Z^{α} says nothing at all about what is going on in $\mathcal{B}$ after the initial time t_0, even though there might be nontrivial correlations between $\mathcal{A}$ and $\mathcal{B}$.

The Heisenberg chain operator for Z^{α}, Sec. 11.4, using a reference time $t_r = t_0$, can be written in the form

$$\hat{K}(Z^{\alpha}) = \left[\hat{K}_{\mathcal{A}}(Y^{\alpha}) \otimes I \right] \Psi_0, \tag{15.38}$$

where

$$\hat{K}_{\mathcal{A}}(Y^{\alpha}) = \hat{A}_f^{\alpha_f} \cdots \hat{A}_2^{\alpha_2} \hat{A}_1^{\alpha_1} \tag{15.39}$$

is the Heisenberg chain operator for Y^{α}, considered as a history of $\mathcal{A}$, with

$$\hat{A}_j^{\alpha_j} = T_{\mathcal{A}}(t_0, t_j) A_j^{\alpha_j} T_{\mathcal{A}}(t_j, t_0) \tag{15.40}$$

the Heisenberg counterpart of the Schrödinger operator $A_j^{\alpha_j}$, see (11.7).

By first taking a partial trace over $\mathcal{B}$, one can write the operator inner products needed to check consistency and calculate weights for the histories in (15.35) in the form

$$\langle \hat{K}(Z^{\alpha}), \hat{K}(Z^{\tilde{\alpha}}) \rangle = \text{Tr} \left[\Psi_0 \hat{K}_{\mathcal{A}}^{\dagger}(Y^{\alpha}) \hat{K}_{\mathcal{A}}(Y^{\tilde{\alpha}}) \right]$$

$$= \text{Tr}_{\mathcal{A}} \left[\rho \hat{K}_{\mathcal{A}}^{\dagger}(Y^{\alpha}) \hat{K}_{\mathcal{A}}(Y^{\tilde{\alpha}}) \right] = \langle \hat{K}_{\mathcal{A}}(Y^{\alpha}), \hat{K}_{\mathcal{A}}(Y^{\tilde{\alpha}}) \rangle_{\rho}, \tag{15.41}$$

where the operator inner product $\langle , \rangle_\rho$ is defined for any pair of operators A and $\bar{A}$ on $\mathcal{A}$ by

$$\langle A, \bar{A} \rangle_\rho := \text{Tr}_{\mathcal{A}}(\rho A^\dagger \bar{A}), \tag{15.42}$$

using the reduced density matrix

$$\rho = \text{Tr}_{\mathcal{B}}(\Psi_0). \tag{15.43}$$

The definition (15.42) yields an inner product with all of the usual properties, including $\langle A, A \rangle_\rho \geq 0$, except that it might be possible (depending on ρ) for $\langle A, A \rangle_\rho$ to vanish when A is not zero.

The consistency conditions for the histories in (15.35) take the form

$$\langle \hat{K}_{\mathcal{A}}(Y^\alpha), \hat{K}_{\mathcal{A}}(Y^{\bar{\alpha}}) \rangle_\rho = 0 \text{ for } \alpha \neq \bar{\alpha}, \tag{15.44}$$

and the probability of occurrence of Z^α or, equivalently, Y^α is given by

$$\text{Pr}(Z^\alpha) = \text{Pr}(Y^\alpha) = \langle \hat{K}_{\mathcal{A}}(Y^\alpha), \hat{K}_{\mathcal{A}}(Y^\alpha) \rangle_\rho. \tag{15.45}$$

Thus as long as we are only interested in histories of the form (15.35) that make no reference at all to $\mathcal{B}$ (aside from the initial state Ψ_0), the consistency conditions and weights can be evaluated with formulas which only involve $\mathcal{A}$ and make no reference to $\mathcal{B}$. They are of the same form employed in Ch. 10, except for replacing the operator inner product $\langle , \rangle$ defined in (10.12) by $\langle , \rangle_\rho$ defined in (15.42). It is also possible to write (15.44) and (15.45) using the Schrödinger chain operators $K(Y^\alpha)$ in place of the Heisenberg operators $\hat{K}(Y^\alpha)$, and this alternative form is employed in (15.48) and (15.50).

If $\mathcal{A}$ is a small system and $\mathcal{B}$ is large, the second trace in (15.41) will be much easier to evaluate than the first. Thus using a density matrix can simplify what might otherwise be a rather complicated problem. To be sure, calculating ρ from Ψ_0 using (15.43) may be a nontrivial task. However, it is often the case that Ψ_0 is not known, so what one does is to assume that ρ has some form involving adjustable parameters, which might, for example, be chosen on the basis of experiment. Thus even if one does not know its precise form, the very fact that ρ exists can assist in analyzing a problem.

In the special case $f = 1$ in which the histories Y^α involve only a single time t, and the consistency conditions (15.44) are automatically satisfied, the probability (15.45) can be written in the form (15.3),

$$\text{Pr}(A^j, t) = \text{Tr}_{\mathcal{A}}(\rho_t A^j), \tag{15.46}$$

where ρ_t is a solution of (15.31), or given by (15.33) in the case in which $H_{\mathcal{A}}$ is independent of time. In this equation ρ_t is functioning as a time-dependent preprobability; see the comments at the beginning of Sec. 15.4.

15.6 Density matrix for isolated system

It is also possible to use a density matrix ρ, thought of as a pre-probability, as the initial state of an isolated system which is not regarded as part of a larger, composite system. In such a case ρ embodies whatever information is available about the system, and this information does not have to be in the form of a particular property represented by a projector, or a probability distribution associated with some decomposition of the identity. As an example, the canonical density matrix

$$\rho = e^{-H/k\theta}/\mathrm{Tr}(e^{-H/k\theta}), \qquad (15.47)$$

where k is Boltzmann's constant and H the time-independent Hamiltonian, is used in quantum statistical mechanics to describe a system in thermal equilibrium at an absolute temperature θ. While one often pictures such a system as being in contact with a thermal reservoir, and thus part of a larger, composite system, the density matrix (15.47) makes perfectly good sense for an isolated system, and a system of macroscopic size can constitute its own thermal reservoir.

The formulas employed in Sec. 15.5 can be used, with some obvious modifications, to check consistency and assign probabilities to histories of an isolated system for which ρ is the initial pre-probability at the time t_0. Thus for a family of histories of the form (15.36) at the times $t_1 < t_2 < \cdots t_f$, with $t_1 \geq t_0$, the consistency condition takes the form

$$\langle K(Y^\alpha), K(Y^{\tilde{\alpha}})\rangle_\rho = \mathrm{Tr}\big[\rho K^\dagger(Y^\alpha)K(Y^{\tilde{\alpha}})\big] = 0 \text{ for } \alpha \neq \tilde{\alpha}, \qquad (15.48)$$

where the (Schrödinger) chain operator is defined by

$$K(Y^\alpha) = A_f^{\alpha_f} T(t_f, t_{f-1}) \cdots A_2^{\alpha_2} T(t_2, t_1) A_1^{\alpha_1} T(t_1, t_0), \qquad (15.49)$$

and the inner product $\langle , \rangle_\rho$ is the same as in (15.42), except for omitting the subscript on Tr. If the consistency conditions are satisfied, the probability of occurrence of a history Y^α is equal to its weight:

$$W(Y^\alpha) = \langle K(Y^\alpha), K(Y^\alpha)\rangle_\rho = \mathrm{Tr}\big[\rho K^\dagger(Y^\alpha)K(Y^\alpha)\big]. \qquad (15.50)$$

One could equally well use Heisenberg chain operators $\hat{K}$ in (15.48) and (15.50), as in the analogous formulas (15.44) and (15.45) in Sec. 15.5. Note that (15.48) and (15.50) are essentially the same as the corresponding formulas (10.20) and (10.11) in Ch. 10, aside from the presence of the density matrix ρ inside the trace defining the operator inner product $\langle , \rangle_\rho$.

In the special case of histories involving only a *single* time $t > t_0$ and a decomposition of the identity $I = \sum A^j$ at this time, consistency is automatic, and the corresponding probabilities take the form

$$\Pr(A^j, t) = \mathrm{Tr}(\rho_t A^j), \qquad (15.51)$$

or $\langle j|\rho_t|j\rangle$ when $A^j = |j\rangle\langle j|$ is a projector on a pure state, where ρ_t is a solution to the Schrödinger equation (15.31) with the subscript $\mathcal{A}$ omitted from H, or of the form (15.33) when the Hamiltonian H is independent of time. One should, however, not make the mistake of thinking that ρ_t as a function of time represents anything like a complete description of the time development of a quantum system; see the remarks at the beginning of Sec. 15.4. In order to discuss correlations it is necessary to employ histories with two or more times following t_0. For these the consistency conditions (15.48) are not automatic, and probabilities must be worked out using (15.50). Both of these formulas require more information about time development than is contained in ρ_t.

There are also situations in which information about the initial state of an isolated system is given in the form of a probability distribution on a set of initial states, and an initial density matrix is generated from this probability distribution. The basic idea can be understood by considering a family of histories

$$[\psi_0^j] \odot [\phi_1^k] \tag{15.52}$$

involving two times t_0 and t_1, where $\{|\psi_0^j\rangle\}$ and $\{|\phi_1^k\rangle\}$ are orthonormal bases, and the initial condition is that $[\psi_0^j]$ occurs with probability p_j. The probability that $[\phi_1^k]$ occurs at time t_1 is given by

$$\Pr(\phi_1^k) = \sum_j \Pr(\phi_1^k \mid \psi_0^j)\, p_j, \tag{15.53}$$

where the conditional probabilities come from the Born formula

$$\Pr(\phi_1^k \mid \psi_0^j) = |\langle\phi_1^k|T(t_1,t_0)|\psi_0^j\rangle|^2. \tag{15.54}$$

An alternative method for calculating $\Pr(\phi_1^k)$ is to define a density matrix

$$\rho_0 = \sum_j p_j[\psi_0^j] \tag{15.55}$$

at t_0 using the initial probability distribution. Since each summand is a positive operator, the sum is positive, Sec.3.9, and the trace of ρ_0 is $\sum_j p_j = 1$. Unlike the situations discussed previously, the eigenvalues of ρ_0 are of direct physical significance, since they are the probabilities of the initial distribution, and the eigenvectors are the physical properties of the system at t_0 for this family of histories. Next, let

$$\rho_1 = T(t_1,t_0)\rho_0 T(t_0,t_1) \tag{15.56}$$

be the result of integrating Schrödinger's equation, (15.31) with H in place of $H_\mathcal{A}$, from t_0 to t_1. Then the probabilities (15.53) can be written as

$$\Pr(\phi_1^k) = \mathrm{Tr}\big(\rho_1[\phi_1^k]\big). \tag{15.57}$$

In this expression the density matrix ρ_1, in contrast to ρ_0, functions as a pre-probability, and its eigenvalues and eigenvectors have no particular physical significance.

The expression (15.57) is more compact than (15.53), as it does not involve the collection of conditional probabilities in (15.54). On the other hand, the description of the quantum system provided by ρ_1 is also less detailed. For example, one cannot use it to calculate correlations between the various initial and final states, or conditional probabilities such as

$$\Pr(\psi_0^j \mid \phi_1^k). \tag{15.58}$$

To be sure, a less detailed description is often more useful than one that is more detailed, especially when one is not interested in the details. The point is that a density matrix provides a partial description, and it is in principle possible to construct a more detailed description if one is interested in doing so.

15.7 Conditional density matrices

Suppose that at time t_0 a particle $\mathcal{A}$ has interacted with a device $\mathcal{B}$ and is moving away from it, so that the two no longer interact, and assume that the projectors $\{B^k\}$ in the decomposition of the identity (15.21) for $\mathcal{B}$ represent some states of physical significance. Given that $\mathcal{B}$ is in the state B^k at time t_0, what can one say about the future behavior of $\mathcal{A}$? For example, $\mathcal{B}$ might be a device which emits a spin-half particle with a spin polarization $S_v = +1/2$, where the direction v depends on some setting of the device indicated by the index k of B^k.

The question of interest to us can be addressed using a family of histories of the form

$$Z^{k\alpha} = B^k \odot Y^{k\alpha}, \tag{15.59}$$

defined for the times $t_0 < t_1 < \cdots$, where the $Y^{k\alpha}$ are histories of $\mathcal{A}$ of the sort defined in (15.36), except that they are labeled with k as well as with α to allow for the possibility that the decomposition of the identity in (15.7) could depend upon k. (One could also employ a set of times $t_1 < t_2 < \cdots$ that depend on k.)

Assume that the combined system $\mathcal{A} \otimes \mathcal{B}$ is described at time t_0 by an initial density matrix Ψ_0, which functions as a pre-probability. For example, Ψ_0 could result from unitary time evolution of an initial state defined at a still earlier time. Let

$$p_k = \mathrm{Tr}(\Psi_0 B^k) \tag{15.60}$$

be the probability of the event B^k. If p_k is greater than 0, the kth *conditional density matrix* is an operator on $\mathcal{A}$ defined by the partial trace

$$\rho^k = (1/p_k)\operatorname{Tr}_\mathcal{B}(\Psi_0 B^k). \tag{15.61}$$

Each conditional density matrix gives rise to an inner product

$$\langle A, \bar{A}\rangle_k := \operatorname{Tr}_\mathcal{A}(\rho^k A^\dagger \bar{A}) \tag{15.62}$$

of the form (15.42).

Using the same sort of analysis as in Sec. 15.5, one can show that the family of histories (15.59) is consistent provided

$$\langle \hat{K}(Y^{k\alpha}), \hat{K}(Y^{k\bar{\alpha}})\rangle_k = 0 \text{ for } \alpha \neq \bar{\alpha} \tag{15.63}$$

is satisfied for every k with $p_k > 0$, where the Heisenberg chain operators $\hat{K}(Y^{k\alpha})$ are defined as in (15.39), but with the addition of a superscript k for each projector on the right side. Schrödinger chain operators could also be used, as in Sec. 15.6. Note that one does not have to check "cross terms" involving chain operators of histories with different values of k. If the consistency conditions are satisfied, the behavior of $\mathcal{A}$ given that $\mathcal{B}$ is in the state B^k at t_0 is described by the conditional probabilities

$$\Pr(Y^{k\alpha} \mid B^k) = \langle \hat{K}(Y^{k\alpha}), \hat{K}(Y^{k\alpha})\rangle_k. \tag{15.64}$$

The physical interpretation of the conditional density matrix is essentially the same as that of the simple density matrix ρ discussed in Sec. 15.5. Indeed, the latter can be thought of as a special case in which the decomposition of the identity of $\mathcal{B}$ in (15.21) consists of nothing but the identity itself. Note in particular that the eigenvalues and eigenvectors of ρ^k play no (direct) role in its physical interpretation, since ρ^k functions as a pre-probability.

Time-dependent conditional density matrices can be defined in the obvious way,

$$\rho_t^k = T_\mathcal{A}(t, t_0)\rho^k T_\mathcal{A}(t_0, t), \tag{15.65}$$

as solutions of the Schrödinger equation (15.31). One can use ρ_t^k to calculate the probability of an event A in $\mathcal{A}$ at time t conditional upon B_k, but not correlations between events in $\mathcal{A}$ at several different times. The comments about ρ_t at the beginning of Sec. 15.4 also apply to ρ_t^k.

The simple or "unconditional" density matrix of $\mathcal{A}$ at time t_0,

$$\rho = \operatorname{Tr}_\mathcal{B}(\Psi_0), \tag{15.66}$$

is an average of the conditional density matrices:

$$\rho = \sum_k p_k \rho^k. \tag{15.67}$$

While ρ can be used to check consistency and calculate probabilities of histories in $\mathcal{A}$ which make no reference to $\mathcal{B}$, for these purposes there is no need to introduce the refined family (15.59) in place of the coarser (15.35). To put it somewhat differently, the context in which the average (15.67) might be of interest is one in which ρ is not the appropriate mathematical tool for addressing the questions one is likely to be interested in.

Let us consider the particular case in which $\Psi_0 = |\Psi_0\rangle\langle\Psi_0|$ and the projectors

$$B^k = |b^k\rangle\langle b^k| \tag{15.68}$$

are pure states. Then one can expand $|\Psi_0\rangle$ in terms of the $|b^k\rangle$ in the form

$$|\Psi_0\rangle = \sum_k \sqrt{p_k}|\alpha^k\rangle \otimes |b^k\rangle, \tag{15.69}$$

where p_k was defined in (15.60). Inserting the coefficient $\sqrt{p_k}$ in (15.69) means that the $\{|\alpha^k\rangle\}$ are normalized, $\langle\alpha^k|\alpha^k\rangle = 1$, but there is no reason to expect $|\alpha^k\rangle$ and $|\alpha^l\rangle$ to be orthogonal for $k \neq l$. The conditional density matrices are now pure states represented by the dyads

$$\rho^k = |\alpha^k\rangle\langle\alpha^k|, \tag{15.70}$$

and (15.67) takes the form

$$\rho = \sum_k p_k|\alpha^k\rangle\langle\alpha^k| = \sum_k p_k[\alpha^k]. \tag{15.71}$$

The expression (15.71) is sometimes interpreted to mean that the system $\mathcal{A}$ is in the state $|\alpha^k\rangle$ with probability p_k at time t_0. However, this is a bit misleading, because in general the $|\alpha^k\rangle$ are not mutually orthogonal, and if two quantum states are not orthogonal to each other, it does not make sense to ask whether a system is in one or the other, as they do not represent mutually-exclusive possibilities; see Sec. 4.6. Instead, one should assign a probability p_k at time t_0 to the state $|\alpha^k\rangle \otimes |b^k\rangle$ of the combined system $\mathcal{A} \otimes \mathcal{B}$. Such states are mutually orthogonal because the $|b^k\rangle$ are mutually orthogonal. In general, $|\alpha^k\rangle$ is an event *dependent on* $|b^k\rangle$ in the sense discussed in Ch. 14, so it does not make sense to speak of $[\alpha^k]$ as a property of $\mathcal{A}$ by itself without making at least implicit reference to the state $|b^k\rangle$ of $\mathcal{B}$. If one wants to ascribe a probability to $|\alpha^k\rangle \otimes |b^k\rangle$, this ket or the corresponding projector must be an element of an appropriate sample space. The projector does not appear in (15.59), but one can insert it by replacing $B^k = [b^k]$ with $[\alpha^k] \otimes [b^k]$. The resulting collection of histories then forms the support of what is, at least technically, a different consistent family of histories. However, the consistency conditions and the probabilities in the new family are the same as those in the original family (15.59), so the distinction is of no great importance.

16

Quantum reasoning

16.1 Some general principles

There are some important differences between quantum and classical reasoning which reflect the different mathematical structure of the two theories. The most precise classical description of a mechanical system is provided by a point in the classical phase space, while the most precise quantum description is a ray or one-dimensional subspace of the Hilbert space. This in itself is not an important difference. What is more significant is the fact that two distinct points in a classical phase space represent mutually exclusive properties of the physical system: if one is a true description of the sytem, the other must be false. In quantum theory, on the other hand, properties are mutually exclusive in this sense only if the corresponding projectors are mutually orthogonal. Distinct rays in the Hilbert space need not be orthogonal to each other, and when they are not orthogonal, they do not correspond to mutually exclusive properties. As explained in Sec. 4.6, if the projectors corresponding to the two properties do not commute with one another, and are thus not orthogonal, the properties are (mutually) incompatible. The relationship of incompatibility means that the properties cannot be logically compared, a situation which does not arise in classical physics. The existence of this nonclassical relationship of incompatibility is a direct consequence of assuming (following von Neumann) that the negation of a property corresponds to the orthogonal complement of the corresponding subspace of the Hilbert space; see the discussion in Sec. 4.6.

Quantum reasoning is (at least formally) identical to classical reasoning when using a single quantum *framework*, and for this reason it is important to be aware of the framework which is being used to construct a quantum description or carry out quantum reasoning. A framework is a Boolean algebra of commuting projectors based upon a suitable sample space, Sec. 5.2. The sample space is a collection of mutually orthogonal projectors which sum to the identity, and thus form a decomposition of the identity. A sample space of histories must also satisfy the consistency conditions discussed in Ch. 10.

In quantum theory there are always many possible frameworks which can be used to describe a given quantum system. While this situation can also arise in classical physics, as when one considers alternative coarse grainings of the phase space, it does not occur very often, and in any case classical frameworks are always mutually compatible, in the sense that they possess a common refinement. For reasons discussed in Sec. 16.4, compatible frameworks do not give rise to conceptual difficulties. By contrast, different quantum frameworks are generally incompatible, which means that the corresponding descriptions cannot be combined. As a consequence, when constructing a quantum description of a physical system it is necessary to restrict oneself to a single framework, or at least not mix results from incompatible frameworks. This *single-framework rule* or *single-family rule* has no counterpart in classical physics. Alternatively, one can say that in classical physics the single-framework rule is always satisfied, for reasons indicated in Sec. 26.6, so one never needs to worry about it.

Quantum dynamics differs from classical Hamiltonian dynamics in that the latter is deterministic: given a point in phase space at some time, there is a unique trajectory in phase space representing the states of the system at earlier or later times. In the quantum case, the dynamics is stochastic: even given a precise state of the system at one time, various alternatives can occur at other times, and the theory only provides probabilities for these alternatives. (Only in the exceptional case of unitary histories, see Secs. 8.7 and 10.3, is there a unique (probability 1) possibility at each time, and thus a deterministic dynamics.) Stochastic dynamics requires both the specification of an appropriate sample space or family of histories, as discussed in Ch. 8, and also a rule for assigning probabilities to histories. The latter, see Chs. 9 and 10, involves calculating weights for the histories using the unitary time development operators $T(t', t)$, equivalent to solving Schrödinger's equation, and then combining these with contingent data, typically an initial condition. Consequently, the reasoning process involved in applying the laws of quantum dynamics is somewhat different from that used for a deterministic classical system.

Probabilities can be consistently assigned to a family of histories of an isolated quantum system using the laws of quantum dynamics only if the family is represented by a Boolean algebra of projectors satisfying the *consistency conditions* discussed in Ch. 10. A family which satisfies these conditions is known as a *consistent family* or *framework*. Each framework has its own sample space, and the single-framework rule says that the probabilities which apply to one framework cannot be used for a different framework, even for events or histories which are represented in both frameworks. It is, however, often possible to assign probabilities to elements of two or more distinct frameworks using the same initial data, as discussed below.

The laws of logic allow one to draw correct conclusions from some initial propositions, or "data", *assuming the latter are correct*. (Following the rules does not by itself always lead to the right answer; the principle of "garbage in, garbage out" was known to ancient logicians, though no doubt they worded it differently.) This is the sort of quantum reasoning with which we are concerned in this chapter. Given some facts or features of a quantum system, the "initial data", what else can we say about it? What conclusions can we draw by applying the principles of quantum theory? For example, an atom is in its ground state and a fast muon passes by 1 nm away: Will the atom be ionized? The "initial data" may simply be the initial state of the quantum system, but could also include information about what happens later, as in the specific example discussed in Sec. 16.2. Thus "initial" refers to what is given at the beginning of the logical argument, not necessarily some property of the quantum system which occurred before something else that one is interested in.

The first step in drawing conclusions from initial data consists in expressing the latter in proper quantum mechanical terms. In a typical situation the data are embedded in a sample space of mutually-exclusive possibilities by assigning probabilities to the elements of this space. This includes the case in which the initial data identify a unique element of the sample space that is assigned a probability of 1, while all other elements have probability 0. If the initial data include information about the system at different times, the Hilbert space must, of course, be the Hilbert space of histories, and the sample space will consist of histories. See the example in Sec. 16.2. Initial data can also be expressed using a density matrix thought of as a pre-probability, see Sec. 15.6. Initial data which cannot be expressed in appropriate quantum terms cannot be used to initiate a quantum reasoning process, even if they make good classical sense.

Once the initial data have been embedded in a sample space, and probabilities have been assigned in accordance with quantum laws, *the reasoning process follows the usual rules of probability theory*. This means that, in general, the conclusion of the reasoning process will be a set of probabilities, rather than a definite result. However, if a consequence can be inferred with probability 1, we call it "true", while if some event or history has probability 0, it is "false", always assuming that the initial data are "true".

It is worth emphasizing once again that the peculiarities of quantum theory do not manifest themselves as long as one is using a *single* sample space and the corresponding event algebra. Instead, they come about because there are *many different* sample spaces in which one can embed the initial data. Hence the conclusions one can draw from those data depend upon which sample space is being used. This multiplicity of sample spaces poses some special problems for quantum reasoning, and these will be discussed in Secs. 16.3 and 16.4, after considering a specific example in the next section.

There are many other sorts of reasoning which go on when quantum theory is applied to a particular problem; e.g., the correct choice of boundary conditions for solving a differential equation, the appropriate approximation to be employed for calculating the time development, the use of symmetries, etc. These are not included in the present discussion because they are the same as in classical physics.

16.2 Example: Toy beam splitter

Consider the toy beam splitter with a detector in the c output channel shown in Fig. 12.2 on page 166 and discussed in Sec. 12.2. Suppose that the initial state at $t = 0$ is $|0a, 0\hat{c}\rangle$: the particle in the a entrance channel to the beam splitter, and the detector in its $0\hat{c}$ "ready" state. Also suppose that at $t = 3$ the detector is in its $1\hat{c}$ state indicating that the particle has been detected. These pieces of information about the system at $t = 0$ and $t = 3$ constitute the initial data as that term was defined in Sec. 16.1. We shall also make use of a certain amount of "background" information: the structure of the toy model and its unitary time transformation, as found in Sec. 12.2.

In order to draw conclusions from the initial data, they must be embedded in an appropriate sample space. A useful approach is to begin with a relatively coarse sample space, and then refine it in different ways depending upon the sorts of questions one is interested in. One choice for the initial, coarse sample space is the set of histories

$$
\begin{aligned}
X^* &= [0a, 0\hat{c}] \odot I \odot I \odot [1\hat{c}], \\
X^\circ &= [0a, 0\hat{c}] \odot I \odot I \odot [0\hat{c}], \\
X^z &= \quad R \quad \odot I \odot I \odot \quad I
\end{aligned}
\tag{16.1}
$$

for the times $t = 0, 1, 2, 3$, where $R = I - [0a, 0\hat{c}]$. Here the superscript $*$ stands for the triggered and $\circ$ for the ready state of the detector at $t = 3$. The sum of these projectors is the history identity $\breve{I}$, and it is easy to see that the consistency conditions are satisfied in view of the orthogonality of the initial and final states, Sec. 11.3. Since X^* is the only member of (16.1) consistent with the initial data, it is assigned probability 1, and the others are assigned probability 0.

Where was the particle at $t = 1$? The histories in (16.1) tell us nothing about any property of the system at $t = 1$, since the identity I is uninformative. Thus in order to answer this question we need to refine the sample space. This can be done by replacing X^* with the three history projectors

$$X^{*c} = [0a, 0\hat{c}] \odot [1c] \odot I \odot [1\hat{c}],$$

$$X^{*d} = [0a, 0\hat{c}] \odot [1d] \odot I \odot [1\hat{c}], \qquad (16.2)$$

$$X^{*p} = [0a, 0\hat{c}] \odot \ P \ \odot I \odot [1\hat{c}],$$

whose sum is X^*, where

$$P = I - [1c] - [1d] \qquad (16.3)$$

is the projector for the particle to be someplace other than sites $1c$ or $1d$. The weights of X^{*d} and X^{*p} are 0, given the dynamics as specified in Sec. 12.2. The history X° can be refined in a similar way, and the weights of $X^{\circ c}$ and $X^{\circ p}$ are 0. (We shall not bother to refine X^z, though this could also be done if one wanted to.) Consistency is easily checked.

When one refines a sample space, the probability associated with each of the elements of the original space is divided up among their replacements in proportion to their weights, as explained in Sec. 9.1. Consequently, in the refined sample space, X^{*c} has probability 1, and all the other histories have probability 0. Note that while X^{*d} and X^{*p} are consistent with the initial data, the fact that they have zero weight (are dynamically impossible) means that they have zero probability. From this we conclude that the initial data imply that the particle has the property $[1c]$, meaning that it is at the site $1c$, at $t = 1$. That is, $[1c]$ at $t = 1$ is true if one assumes the initial data are true.

Given the same initial data, one can ask a different question: At $t = 1$, was the particle in one or the other of the two states

$$|1\bar{a}\rangle = \big(|1c\rangle + |1d\rangle\big)/\sqrt{2}, \quad |1\bar{b}\rangle = \big(-|1c\rangle + |1d\rangle\big)/\sqrt{2} \qquad (16.4)$$

resulting from the unitary evolution of $|0a\rangle$ and $|0b\rangle$ (see (12.2))? To answer this question, we use an alternative refinement of the sample space (16.1), in which X^* is replaced with the three histories

$$X^{*a} = [0a, 0\hat{c}] \odot [1\bar{a}] \odot I \odot [1\hat{c}],$$

$$X^{*b} = [0a, 0\hat{c}] \odot [1\bar{b}] \odot I \odot [1\hat{c}], \qquad (16.5)$$

$$X^{*p} = [0a, 0\hat{c}] \odot \ P \ \odot I \odot [1\hat{c}],$$

with P again given by (16.3). (Note that $[1\bar{a}] + [1\bar{b}]$ is the same as $[1c] + [1d]$.) A similar refinement can be carried out for X°. Both X^{*b} and X^{*p} have zero weight, so the initial data imply that the history X^{*a} has probability 1. Consequently, we can conclude that the particle is in the superposition state $[1\bar{a}]$ with probability 1 at $t = 1$. That is, $[1\bar{a}]$ at $t = 1$ is true if one assumes the initial data are true.

However, the family which includes (16.5) is incompatible with the one which includes (16.2), as is obvious from the fact that $[1c]$ and $[1\bar{a}]$ do not commute with each other. Hence the probability 1 (true) conclusion obtained using one family cannot be combined with the probability 1 conclusion obtained using the other family. We cannot deduce from the initial data that at $t = 1$ the particle was in the state $[1c]$ and also in the state $[1\bar{a}]$, for this is quantum nonsense. Putting together results from two incompatible frameworks in this way violates the single-framework rule. So which is the *correct* family to use in order to work out the *real* state of the particle at $t = 1$: should one employ (16.2) or (16.5)? This is not a meaningful question in the context of quantum theory, for reasons which will be discussed in Sec. 16.4.

Now let us ask a third question based on the same initial data used previously. Where was the particle at $t = 2$: was it at $2c$ or at $2d$? The answer is obvious. All we need to do is to replace (16.2) with a different refinement

$$
\begin{aligned}
X^{*c'} &= [0a, 0\hat{c}] \odot I \odot [2c] \odot [1\hat{c}], \\
X^{*d'} &= [0a, 0\hat{c}] \odot I \odot [2d] \odot [1\hat{c}], \\
X^{*p'} &= [0a, 0\hat{c}] \odot I \odot P' \odot [1\hat{c}],
\end{aligned}
\tag{16.6}
$$

with $P' = I - [2c] - [2d]$. Since $X^{*c'}$ has probability 1, it is certain, given the initial data, that the particle was at $2c$ at $t = 2$.

The same answer can be obtained starting with the sample space which includes the histories in (16.2), and refining it to include the history

$$
X^{*cc} = [0a, 0\hat{c}] \odot [1c] \odot [2c] \odot [1\hat{c}],
\tag{16.7}
$$

which has probability 1, along with additional histories with probability 0. In the same way, one could start with the sample space which includes the histories in (16.5), and refine it so that it contains

$$
X^{*ac} = [0a, 0\hat{c}] \odot [1\bar{a}] \odot [2c] \odot [1\hat{c}],
\tag{16.8}
$$

whose probability (conditional upon the initial data) is 1, plus others whose probability is 0. It is obvious that the sample space containing (16.7) is incompatible with that containing (16.8), since these two history projectors do not commute with each other. Nonetheless, either family can be used to answer the question "Where is the particle at $t = 2$?", and both give precisely the same answer: the initial data imply that it is at $2c$, and not someplace else.

16.3 Internal consistency of quantum reasoning

The example in Sec. 16.2 illustrates the principles of quantum reasoning introduced in Sec. 16.1. It also exhibits some important ways in which reasoning about
quantum systems differs from what one is accustomed to in classical physics. In
deterministic classical mechanics one is used to starting from some initial state and
integrating the equations of motion to produce a trajectory in which at each time
the system is described by a single point in its phase space. Given this trajectory
one can answer any question of physical interest such as, for example, the time
dependence of the kinetic energy.

In quantum theory one typically (unitary histories are an exception) uses a rather
different strategy. Instead of starting with a single well-defined temporal development which can answer all questions, one has to start with the physical questions
themselves and use these questions to generate an appropriate framework in which
they make sense. Once this framework is specified, the principles of stochastic
quantum dynamics can be brought to bear in order to supply answers, usually in
the form of probabilities, to the questions one is interested in.

One cannot use a single framework to answer all possible questions about a
quantum system, because answering one question will require the use of a framework that is incompatible with another framework needed to address some other
question. But even a particular question can often be answered using more than one
framework, as illustrated by the third (last) question in Sec. 16.2. This multiplicity
of frameworks, along with the rule which requires that a quantum description, or
the reasoning from initial data to a conclusion, use only a *single framework*, raises
two somewhat different issues. The first issue is that of internal consistency: if
many frameworks are available, will one get the same answer to the same question
if one works it out in different frameworks? We shall show that this is, indeed, the
case. The second issue, discussed in the next section, is the intuitive significance
of the fact that alternative incompatible frameworks can be employed for one and
the same quantum system.

The internal consistency of quantum reasoning can be shown in the following
way. Assume that $\mathcal{F}_1, \mathcal{F}_2, \ldots \mathcal{F}_n$ are different consistent families of histories,
which may be incompatible with one another, each of which contains the initial
data and the other events, or histories, that are needed to answer a particular physical question. Each framework is a set of projectors which forms a Boolean algebra,
and one can define $\mathcal{F}$ to be their set-theoretic intersection:

$$\mathcal{F} = \mathcal{F}_1 \cap \mathcal{F}_2 \cap \cdots \mathcal{F}_n. \tag{16.9}$$

That is, a projector Y is in $\mathcal{F}$ if and only if it is also in each $\mathcal{F}_j$, for $1 \leq j \leq n$.
It is straightforward to show that $\mathcal{F}$ is a Boolean algebra of commuting history

projectors: It contains the history identity $\breve{I}$; if it contains a projector Y, then it also contains its negation $I - Y$; and if it contains Y and Y', then it also contains $YY' = Y'Y$. These assertions follow at once from the fact that they are true of each of the $\mathcal{F}_j$. Furthermore, the fact that each $\mathcal{F}_j$ is a consistent family means that $\mathcal{F}$ is consistent; one can use the criterion in (10.21).

Since each $\mathcal{F}_j$ contains the projectors needed to represent the initial data, along with those needed to express the conclusions one is interested in, the same is true of $\mathcal{F}$. Consequently, the task of assigning probabilities using the initial data together with the dynamical weights of the histories, and then using probabilistic arguments to reach certain conclusions, can be carried out in $\mathcal{F}$. But since it can be done in $\mathcal{F}$, it can also be done in an identical fashion in any of the $\mathcal{F}_j$, as the latter contains all the projectors of $\mathcal{F}$. Furthermore, any history in $\mathcal{F}$ will be assigned the same weight in $\mathcal{F}$ and in any $\mathcal{F}_j$, since the weight $W(Y)$ is defined directly in terms of the history projector Y using a formula, (10.11), that makes no reference to the family which contains the projector. Consequently, the conclusions one draws from initial data about physical properties or histories will be identical in all frameworks which contain the appropriate projectors.

This internal consistency is illustrated by the discussion of the third (last) question in Sec. 16.2: $\mathcal{F}$ is the family based on the sample space containing (16.6), and $\mathcal{F}_1$ and $\mathcal{F}_2$ are two mutually incompatible refinements containing the histories in (16.7) and (16.8), respectively. One can use either $\mathcal{F}_1$ or $\mathcal{F}_2$ to answer the question "Where is the particle at $t = 2$?", and the answer is the same.

As well as providing a proof of consistency, the preceding remarks suggest a certain strategy for carrying out quantum reasoning of the type we are concerned with: Use the smallest, or coarsest framework which contains both the initial data and the additional properties of interest in order to analyze the problem. Any other framework which can be used for the same purpose will be a refinement of the coarsest one, and will give the same answers, so there is no point in going to extra effort. If one has some specific initial data in mind, but wants to consider a variety of possible conclusions, some of which are incompatible with others, then start off with the coarsest framework $\mathcal{E}$ which contains all the initial data, and refine it in the different ways needed to draw different conclusions.

This was the strategy employed in Sec. 16.2, except that the coarsest sample space that contains the initial data X^* consists of the two projectors X^* and $\breve{I} - X^*$, whereas we used a sample space (16.1) containing three histories rather than just two. One reason for using X° and X^z in this case is that each has a straightforward physical interpretation, unlike their sum $\breve{I} - X^*$. The argument for consistency given above shows that there is no harm in using a more refined sample space as a starting point for further refinements, as long as it allows one to answer the

questions one is interested in, for in the end one will always get precisely the same
answer to any particular question.

16.4 Interpretation of multiple frameworks

The example of Sec. 16.2 illustrates a situation which arises rather often in reasoning about quantum systems. The initial data $\mathcal{D}$ can be used in various different
frameworks $\mathcal{F}_1, \mathcal{F}_2, \ldots$, to yield different conclusions $\mathcal{C}_1, \mathcal{C}_2, \ldots$. The question
then arises as to the relationship among these different conclusions. In particular,
can one say that they all apply simultaneously to the same physical system? Generally the conclusions are expressed in terms of probabilities that are greater than
0 and less than 1, and thus involve some uncertainty. But sometimes, and we de-
liberately focused on this situation in the example in Sec. 16.2, one concludes that
an event (or history) has probability 1, in which case it is natural to interpret this
as meaning that the event actually occurs, or is a "true" consequence of the initial
data. Similarly, probability 0 can be interpreted to mean that the event does not
occur, or is "false".

If two or more frameworks are *compatible*, there is nothing problematical in
supposing that the corresponding conclusions apply simultaneously to the same
physical system. The reason is that compatibility implies the existence of a common refinement, a framework $\mathcal{G}$ which contains the projectors necessary to describe
the initial data and all of the conclusions. The consistency of quantum reasoning,
Sec. 16.3, means that the conclusions $\mathcal{C}_j$ will be identical in $\mathcal{F}_j$ and in $\mathcal{G}$. Consequently one can think of $\mathcal{F}_1, \mathcal{F}_2, \ldots$ as representing alternative "views" or "per-
spectives" of the same physical system, much as one can view an object, such as a
teacup, from various different angles. Certain details are visible from one perspective and others from a different perspective, but there is no problem in supposing
that they all form part of a single correct description, or that they are all simultaneously true, for the object in question.

In the example considered in Sec. 16.2, $\mathcal{F}_1$ could be the framework based on
(16.2), which allows one to describe the position of the particle at $t = 1$, but not
for any other $t > 0$, and $\mathcal{F}_2$ the one based on (16.6), which provides a description
of the position of the particle at $t = 2$, but not at $t = 1$. Their common refinement
provides a description of the position of the particle at $t = 1$ and $t = 2$, and $\mathcal{F}_1$ and
$\mathcal{F}_2$ can be thought of as supplying complementary parts of this description.

Conceptual difficulties arise, however, when two or more frameworks are *incompatible*. Again with reference to the example in Sec. 16.2, let $\mathcal{F}_3$ be the framework
based on (16.5). It is incompatible with $\mathcal{F}_1$, because X^{*c} in (16.2) and X^{*a} in
(16.5) do not commute with each other, since the projectors $[1c]$ and $[1\bar{a}]$ at $t = 1$
do not commute. From the initial data one can conclude using $\mathcal{F}_1$ that the particle

possesses the property [1c] at $t = 1$ with probability 1. Using $\mathcal{F}_2$ and the same initial data, one concludes that the particle has the property [1$\bar{a}$] at $t = 1$, again with probability 1. But even though [1c] and [1$\bar{a}$] are both "true" (probability 1) consequences of the initial data, one cannot think of them as representing properties of the particle which are simultaneously true in the same sense one is accustomed to when thinking about classical systems, for there is no property corresponding to [1c] AND [1$\bar{a}$], just as there is no property corresponding to $S_z = +1/2$ AND $S_x = +1/2$ for a spin-half particle.

The conceptual difficulty goes away if one supposes that the two incompatible frameworks are being used to describe two distinct physical systems that are described by the same initial data, or the same system during two different runs of an experiment. In the case of two separate but identical systems, each with Hilbert space $\mathcal{H}$, the combination is described by a tensor product $\mathcal{H} \otimes \mathcal{H}$, and employing $\mathcal{F}_1$ for the first and $\mathcal{F}_3$ for the second is formally the same as a single consistent family for the combination. This is analogous to the fact that while $S_z = +1/2$ AND $S_x = +1/2$ for a spin-half particle is quantum nonsense, there is no problem with the statement that $S_z = +1/2$ for one particle and $S_x = +1/2$ for a different particle. In the same way, different experimental runs for a single system must occur during different intervals of time, and the tensor product $\breve{\mathcal{H}} \odot \breve{\mathcal{H}}$ of two history Hilbert spaces plays the same role as $\mathcal{H} \otimes \mathcal{H}$ for two distinct systems.

Incompatible frameworks do give rise to conceptual problems when one tries to apply them to the *same* system during the *same* time interval. To be sure, there is never any harm in constructing as many alternative descriptions of a quantum system as one wants to, and writing them down on the same sheet of paper. The difficulty comes about when one wants to think of the results obtained using incompatible frameworks as all referring simultaneously to the same physical system, or tries to combine the results of reasoning based upon incompatible frameworks. It is this which is forbidden by the single-framework rule of quantum reasoning.

Note, by the way, that in view of the internal consistency of quantum reasoning discussed in Sec. 16.3, it is never possible, even using incompatible frameworks, to derive *contradictory* results starting from the same initial data. Thus for the example in Sec. 16.2, the fact that there is a framework in which one can conclude with certainty that the particle is at the site 1c at $t = 1$ means there cannot be another framework in which one can conclude that the particle is someplace else at $t = 1$, or that it can be at site 1c with some probability less than 1. Any framework which contains both the initial data and the possibility of discussing whether the particle is or is not at the site 1c at $t = 1$ will lead to precisely the same conclusion as $\mathcal{F}_1$. This does not contradict the fact that in $\mathcal{F}_3$ the particle is predicted to be in a state [1$\bar{a}$] at $t = 1$: $\mathcal{F}_3$ does not contain [1c], and thus in this framework one cannot address the question of whether the particle is at the site 1c at $t = 1$.

Even though the single-framework rule tells us that the result [1*c*] from framework $\mathcal{F}_1$ and the result [1*ā*] from $\mathcal{F}_3$ cannot be combined or compared, this state of affairs is intuitively rather troubling, for the following reason. In classical physics whenever one can draw the conclusion through one line of reasoning that a system has a property *P*, and through a different line of reasoning that it has the property *Q*, then it is correct to conclude that the system possesses both properties simultaneously. Thus if *P* is true (assuming the truth of some initial data) and *Q* is also true (using the same data), then it is always the case that *P* AND *Q* is true. By contrast, in the case we have been discussing, [1*c*] is true (a correct conclusion from the data) in $\mathcal{F}_1$, [1*ā*] is true if we use $\mathcal{F}_3$, while the combination [1*c*] AND [1*ā*] is not even meaningful as a quantum property, much less true!

When viewed from the perspective of quantum theory, see Ch. 26, classical physics is an approximation to quantum theory in certain circumstances in which the corresponding quantum description requires only a single framework (or, which amounts to the same thing, a collection of *compatible* frameworks). Thus the problem of developing rules for correct reasoning when one is confronted with a multiplicity of *incompatible* frameworks never arises in classical physics, or in our everyday "macroscopic" experience which classical physics describes so well. But this is precisely why the rules of reasoning which are perfectly adequate and quite successful in classical physics cannot be depended upon to provide reliable conceptual tools for thinking about the quantum domain. However deep-seated may be our intuitions about the meaning of "true" and "false" in the classical realm, these cannot be uncritically extended into quantum theory.

As probabilities can only be defined once a sample space has been specified, probabilistic reasoning in quantum theory necessarily depends upon the sample space and its associated framework. As a consequence, if "true" is to be identified with "probability 1", then the notion of "truth" in quantum theory, in the sense of deriving true conclusions from initial data that are assumed to be true, must necessarily depend upon the framework which one employs. This feature of quantum reasoning is sometimes regarded as unacceptable because it is hard to reconcile with an intuition based upon classical physics and ordinary everyday experience. But classical physics cannot be the arbiter for the rules of quantum reasoning. Instead, these rules must conform to the mathematical structure upon which quantum theory is based, and as has been pointed out repeatedly in previous chapters, this structure is significantly different from that of a classical phase space. To acquire a good "quantum intuition", one needs to work through various quantum examples in which a system can be studied using different incompatible frameworks. Several examples have been considered in previous chapters, and there are some more in later chapters. I myself have found the example of a beam splitter insider a box, Fig. 18.3 on page 253, particularly helpful. For additional comments on multiple incompatible frameworks, see Secs. 18.4 and 27.3.

17

Measurements I

17.1 Introduction

I place a tape measure with one end on the floor next to a table, read the height of the table from the tape, and record the result in a notebook. What are the essential features of this *measurement process*? The key point is the establishment of a *correlation* between a *physical property* (the height) of a *measured system* (the table) and a suitable *record* (in the notebook), which is itself a physical property of some other system. It will be convenient in what follows to think of this record as part of the measuring apparatus, which consists of everything essential to the measuring process apart from the measured system. Human beings are not essential to the measuring process. The height of a table could be measured by a robot. In the modern laboratory, measurements are often carried out by automated equipment, and the results stored in a computer memory or on magnetic tape, etc. While scientific progress requires that human beings pay attention to the resulting data, this may occur a long time after the measurements are completed.

In this and the next chapter we consider measurements as physical processes in which a *property* of some quantum system, which we shall usually think of as some sort of "particle", becomes correlated with the *outcome* of the measurement, itself a property of another quantum system, the "apparatus". Both the measured system and the apparatus which carries out the measurement are to be thought of as parts of a single closed quantum mechanical system. This makes it possible to apply the principles of quantum theory developed in earlier chapters. There are no special principles which apply to measurements in contrast to other quantum processes. We need an appropriate Hilbert space for the measured system plus apparatus, some sort of initial state, unitary time development operators, and a suitable framework or consistent family of histories. There are, as always, many possible frameworks. A correct quantum description of the measuring process must employ a *single framework*; mixing results from incompatible frameworks will only cause confusion.

In practice it is necessary to make a number of idealizations and approximations in order to discuss measurements as quantum mechanical processes. This should not be surprising, for the same is true of classical physics. For example, the motion of the planets in the solar system can be described to quite high precision by treating them as point masses subject to gravitational forces, but of course this is not an exact description. The usual procedure followed by a physicist is to first work out an approximate description of some situation in order to get an idea of the various magnitudes involved, and then see how this first approximation can be improved, if greater precision is needed, by including effects which have been ignored. We shall follow this approach in this and the following chapter, sometimes pointing out how a particular approximation can be improved upon, at least in principle. The aim is physical insight, not a precise formalism which will cover all cases.

Quantum measurements can be divided into two broad categories: nondestructive and destructive. In *nondestructive measurements*, also called nondemolition measurements, the measured property is preserved, so the particle has the same, or almost the same property after the measurement is completed as it had before the measurement. While it is easy to make nondestructive measurements on macroscopic objects, such as tables, nondestructive measurements of microscopic quantum systems are much more difficult. Even when a quantum measurement is nondestructive for a particular property, it will be destructive for many other properties, so that the term nondestructive can only be defined relative to some property or properties, and does not refer to all conceivable properties of the quantum system.

In *destructive measurements* the property of interest is altered during the measurement process, often in an uncontrolled fashion, so that after the measurement the particle no longer has this property. For example, the kinetic energy of an energetic particle can be measured by bringing it to rest in a scintillator and finding the amount of light produced. This tells one what the energy of the particle was before it entered the scintillator, whereas at the end of the measurement process the kinetic energy of the particle is zero. In this and other examples of destructive measurements it is clear that the correlation of interest is between a property the particle had *before* the measurement took place, and the state of the apparatus *after* the measurement, and thus involves properties at two *different* times. The absence of a systematic way of treating correlations involving different times, except in very special cases, is the basic reason why the theory of measurement developed by von Neumann, Sec. 18.2, is not very satisfactory.

17.2 Microscopic measurement

The measurement of the spin of a spin-half particle illustrates many of the principles of the quantum theory of measurement, so we begin with this simple case, using a certain number of approximations to keep the discussion from becoming too complicated. Consider a neutral spin-half particle, e.g., a silver atom in its ground state, moving through the inhomogeneous magnetic field of a Stern–Gerlach apparatus, shown schematically in Fig. 17.1. We shall assume the magnetic field is such that if the z-component S_z of the spin is $+1/2$, there is an upwards force on the particle, and it emerges from the magnet moving upwards, whereas if $S_z = -1/2$, the force is in the opposite direction, and the particle moves downward as it leaves the magnet.

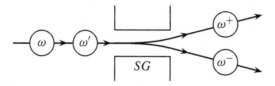

Fig. 17.1. Spin-half particle passing through a Stern–Gerlach magnet.

This can be described in quantum mechanical terms as follows. The spin states of the particle corresponding to $S_z = \pm 1/2$ are $|z^+\rangle$ and $|z^-\rangle$ in the notation of Sec. 4.2. Let t_0 and t_1 be two successive times preceding the moment at which the particle enters the magnetic field, see Fig. 17.1, and t_2 a later time after it has emerged from the magnetic field. Assume that the unitary time development from t_0 to t_1 to t_2 is given by

$$
\begin{aligned}
|z^+\rangle|\omega\rangle &\mapsto |z^+\rangle|\omega'\rangle \mapsto |z^+\rangle|\omega^+\rangle, \\
|z^-\rangle|\omega\rangle &\mapsto |z^-\rangle|\omega'\rangle \mapsto |z^-\rangle|\omega^-\rangle,
\end{aligned}
\tag{17.1}
$$

where $|\omega\rangle, |\omega'\rangle, |\omega^+\rangle, |\omega^-\rangle$ are wave packets for the particle's center of mass, at the locations indicated in Fig. 17.1. (One could also write these as $\omega(\mathbf{r})$, etc.)

One can think of the center of mass of the particle as the "apparatus". The two possible outcomes of the measurement are that the particle emerges from the magnet in one of the two spatial wave packets $|\omega^+\rangle$ or $|\omega^-\rangle$. It is important that the outcome wave packets be orthogonal,

$$
\langle \omega^+ | \omega^- \rangle = 0,
\tag{17.2}
$$

as otherwise we cannot speak of them as mutually-exclusive possibilities. This condition will be fulfilled if the wave packets have negligible overlap, as suggested by the sketch in Fig. 17.1.

In calculating the unitary time development in (17.1) we assume that the Hamiltonian for the particle includes an interaction with the magnetic field, and this field is assumed to be "classical"; that is, it provides a potential for the particle's motion, but does not itself need to be described using an appropriate field-theoretical Hilbert space. Similarly, we have omitted from our quantum description the atoms of the magnet which actually produce this magnetic field. These "inert" parts of the apparatus could, in principle, be included in the sort of quantum description discussed in Sec. 17.3, but this is an unnecessary complication, since their essential role is included in the unitary time development in (17.1).

The process shown in Fig. 17.1 can be thought of as a measurement because the value of S_z before the measurement, the property being measured, is correlated with the spatial wave packet of the particle after the measurement, which forms the output of the measurement. It is also the case that S_z before the measurement is correlated with its value after the measurement, and this means the measurement is nondestructive for the properties $S_z = \pm 1/2$. One can easily imagine a destructive version of the same measurement by supposing that the wave packets emerging from the field gradient of the main magnet pass through some regions of uniform magnetic field, which do not affect the center of mass motion, but do cause a precession of the spin. Consequently, at the end the process the location of the wave packet for the center of mass will still serve to indicate the value of S_z before the measurement began, even though the final value of S_z need not be the same as the initial value.

Suppose that the initial spin state is not one of the possibilities $S_z = \pm 1/2$, but instead

$$|x^+\rangle = \left(|z^+\rangle + |z^-\rangle\right)/\sqrt{2} \tag{17.3}$$

corresponding to $S_x = +1/2$. What happens during the measuring process? The unitary time development of the initial state

$$|\psi_0\rangle = |x^+\rangle|\omega\rangle = \left(|z^+\rangle|\omega\rangle + |z^-\rangle|\omega\rangle\right)/\sqrt{2} \tag{17.4}$$

is obtained by taking a linear combination of the two cases in (17.1):

$$|\psi_0\rangle \mapsto |x^+\rangle|\omega'\rangle \mapsto \left(|z^+\rangle|\omega^+\rangle + |z^-\rangle|\omega^-\rangle\right)/\sqrt{2}. \tag{17.5}$$

The unitary history in (17.5) cannot be used to describe the measuring process, because the measurement outcomes, $|\omega^+\rangle$ and $|\omega^-\rangle$, are clearly incompatible with the final state in (17.5). A quantum mechanical description of a measurement with particular outcomes must, obviously, employ a framework in which these outcomes are represented by appropriate projectors, as in the consistent family whose support consists of the two histories

$$[\psi_0] \odot [x^+][\omega'] \odot \begin{cases} [z^+][\omega^+], \\ [z^-][\omega^-]. \end{cases} \tag{17.6}$$

The notation, see (14.12), indicates that the two histories are identical at the times t_0 and t_1, but contain different events at t_2. While this family contains the measurement outcomes $[\omega^+]$ and $[\omega^-]$, it is still not satisfactory for discussing the process in Fig. 17.1 as a *measurement*, because it does not allow us to relate these outcomes to the spin states $[z^+]$ and $[z^-]$ of the particle *before* the measurement took place. Since the properties $S_z = \pm 1/2$ are incompatible with a spin state $[x^+]$ at t_1, (17.6) does not allow us to say anything about S_z before the particle enters the magnetic field gradient. It is true that S_z at t_2 is correlated with the measurement outcome if we use (17.6). But this would also be true if the apparatus had somehow produced a particle in a certain spin state without any reference to its previous properties, and calling that a "measurement" would be rather odd.

A more satisfactory description of the process in Fig. 17.1 as a measurement is obtained by using an alternative consistent family whose support is the two histories

$$[\psi_0] \odot \begin{cases} [z^+][\omega'] \odot [z^+][\omega^+], \\ [z^-][\omega'] \odot [z^-][\omega^-]. \end{cases} \tag{17.7}$$

As both histories have positive weights, one sees that

$$\begin{aligned} \Pr(z_1^+ \,|\, \omega_2^+) = 1 = \Pr(z_1^- \,|\, \omega_2^-), \\ \Pr(\omega_2^+ \,|\, z_1^+) = 1 = \Pr(\omega_2^- \,|\, z_1^-), \end{aligned} \tag{17.8}$$

where we follow our usual convention that square brackets can be omitted and subscripts refer to times: e.g., z_1^+ is the same as $[z^+]_1$ and means $S_z = +1/2$ at t_1. (In addition, the initial state ψ_0 could be included among the conditions, but, as usual, we omit it.) These conditional probabilities tell us that if the measurement outcome is ω^+ at t_2, we can be certain that the particle had $S_z = +1/2$ at t_1, and vice versa; likewise, ω^- at t_2 implies $S_z = -1/2$ at t_1. (For an initial spin state $|z^+\rangle$ the conditional probabilities involving z^- and ω^- are undefined, and those involving z^+ and ω^+ are undefined for an initial $|z^-\rangle$.)

It is (17.8) which tells us that what we have been referring to as a *measurement process* actually deserves that name, for it shows that the result of this process is a correlation between specific outcomes and appropriate properties of the measured system before the measurement took place. If these probabilities were slightly less than 1, it would still be possible to speak of an *approximate measurement*, and in practice all measurements are to some degree approximate.

In conclusion it is worth emphasizing that in order to describe a quantum process as a measurement it is necessary to employ a framework which includes *both* the measurement outcomes (pointer positions) *and* the properties of the measured system before the measurement took place, by means of suitable projectors. These requirements are satisfied by (17.7), whereas (17.6), even though it is an improvement over a unitary family, cannot be used to derive the correlations (17.8) that are characteristic of a measurement.

17.3 Macroscopic measurement, first version

If the results are to be of use to scientists, measurements of the properties of microscopic quantum systems must eventually produce macroscopic results visible to the eye or at least accessible to the computer. This requires devices that amplify microscopic signals and produce some sort of macroscopic record. These processes are thermodynamically irreversible, and this irreversibility contributes to the permanence of the resulting records. Thus even though the production of certain correlations, which is the central feature of the measuring process, can occur on a microscopic scale, as discussed in the previous section, macroscopic systems must be taken into account when quantum theory is used to describe practical measurements. A full and detailed quantum mechanical description of the processes going on in a macroscopic piece of apparatus containing 10^{23} particles is obviously not possible. Nonetheless, by making a certain number of plausible assumptions it is possible to explore what such a description might contain, and this is what we shall do in this and the next section, for a macroscopic version of the measurement of the spin of a spin-half particle.

Once again, assume that the particle passes through a magnetic field gradient, Fig. 17.1, which splits the center of mass wave packet into two pieces which are eventually separated by a macroscopic distance. The macroscopic measurement is then completed by adding particle detectors to determine whether the particle is in the upper or lower beam as it leaves the magnetic field. One could, for example, suppose that light from a laser ionizes a silver atom as it travels along one of the paths emerging from the apparatus, and the resulting electron is accelerated in an electric field and made to produce a macroscopic current by a cascade process. Detection of single atoms in this fashion is technically feasible, though it is not easy. Of course, one must expect that in such a measurement process the spin direction of the atom will not be preserved; indeed, the atom itself is broken up by the ionization process. Hence such a measurement is destructive.

Let us assume once again that three times t_0, t_1, and t_2 are used in a quantum description of the measurement process. The times t_0 and t_1 precede the entry of the particle into the magnetic field, Fig. 17.1, whereas t_2 is long enough after the particle has emerged from the magnetic field to allow its detection, and the result indicating the channel in which it emerged to be recorded in some macroscopic device, say a pointer easily visible to the naked eye. Assume that before the measurement takes place the pointer points in a horizontal direction, and at the completion of the measurement it either points upwards, indicating that the particle emerged in the upper channel corresponding to $S_z = +1/2$, or downwards, indicating that the particle emerged in the $S_z = -1/2$ channel. Of course, no one would build an apparatus in this fashion nowadays, but when discussing conceptual questions there is an advantage in using something easily visualized, rather than the direction of magnetization in some region on a magnetic tape or disk. The principles are in any case the same.

As a first attempt at a quantum description of such a macroscopic measurement, assume that at t_0 the apparatus plus the center of mass of the particle whose spin is to be measured is in a quantum state $|\Omega\rangle$. Then we might expect that the unitary time development of the apparatus plus particle would be similar to (17.1), that is, of the form

$$
\begin{aligned}
|z^+\rangle|\Omega\rangle \mapsto |z^+\rangle|\Omega'\rangle \mapsto |\Omega^+\rangle, \\
|z^-\rangle|\Omega\rangle \mapsto |z^-\rangle|\Omega'\rangle \mapsto |\Omega^-\rangle,
\end{aligned}
\tag{17.9}
$$

where $|\Omega^+\rangle$ is some state of the apparatus in which the pointer points upwards, and $|\Omega^-\rangle$ a state in which the pointer points downwards. The difference between $|\Omega\rangle$ and $|\Omega'\rangle$ reflects both the fact that the position of the center of mass of the particle changes between t_0 and t_1, and that the apparatus itself is evolving in time. The only assumption we have made is that this time evolution is not influenced by the direction of the spin of the particle, which seems plausible. In contrast to (17.1), the particle spin does not appear at time t_2 in (17.9). This is because we are dealing with a destructive measurement, and the value of the particle's spin at t_2 is irrelevant. Indeed, the concept may not even be well defined. Thus $|\Omega^+\rangle$ and $|\Omega^-\rangle$ are defined on a slightly different Hilbert space than $|\Omega\rangle$ and $|\Omega'\rangle$.

The counterpart of (17.2) is

$$
\langle\Omega^+|\Omega^-\rangle = 0,
\tag{17.10}
$$

a consequence of unitary time development: since the two states in (17.9) at time t_0 are orthogonal to each other, those at t_2 must also be orthogonal. But (17.10) is also what one would expect on physical grounds for quantum states corresponding to distinct macroscopic situations, in this case different orientations of the pointer. The orthogonality in (17.2) was justified by assuming that the two emerging wave packets in Fig. 17.1 have negligible overlap. Two distinct pointer positions will mean that there are an enormous number of atoms whose wave packets have negligible overlap, and thus (17.10) will be satisfied to an excellent approximation.

It follows from (17.9) and our assumption about the way in which $|\Omega^+\rangle$ and $|\Omega^-\rangle$ are related to the pointer position that if the particle starts off with $S_z = +1/2$ at t_0, the pointer will be pointing upwards at t_2, while if the particle starts off with $S_z = -1/2$, the pointer will later point downwards. But what will happen if the initial spin state is not an eigenstate of S_z? Let us assume a spin state $|x^+\rangle$, (17.3), at t_0 corresponding to $S_x = +1/2$. The unitary time development of the initial state

$$|\Psi_0\rangle = |x^+\rangle|\Omega\rangle = \big(|z^+\rangle|\Omega\rangle + |z^-\rangle|\Omega\rangle\big)/\sqrt{2}, \qquad (17.11)$$

the macroscopic counterpart of (17.4), is given by

$$|\Psi_0\rangle \mapsto |x^+\rangle|\Omega'\rangle \mapsto |\bar{\Omega}\rangle = \big(|\Omega^+\rangle + |\Omega^-\rangle\big)/\sqrt{2}. \qquad (17.12)$$

The state $|\bar{\Omega}\rangle$ on the right side is a macroscopic quantum superposition (MQS) of states representing distinct macroscopic situations: a pointer pointing up and a pointer pointing down. It is incompatible with the measurement outcomes Ω^+ and Ω^- in the same way as the right side of (17.5) is incompatible with ω^+ and ω^-, so it cannot be used for describing the possible outcomes of the measurement. See the discussion in Sec. 9.6.

The measurement outcomes can be discussed using a family resembling (17.6) with support

$$[\Psi_0] \odot [x^+][\Omega'] \odot \begin{cases} [\Omega^+], \\ [\Omega^-]. \end{cases} \qquad (17.13)$$

However, as pointed out in connection with (17.6), the presence of $[x^+]$ in the histories in this family at times preceding the measurement makes it impossible to discuss S_z. Thus one cannot employ (17.13) to obtain a correlation between the measurement outcomes and the value of S_z before the measurement took place.

Hence we are led to consider yet another family, the counterpart of (17.7), whose support is the two histories:

$$[\Psi_0] \odot \begin{cases} [z^+][\Omega'] \odot [\Omega^+], \\ [z^-][\Omega'] \odot [\Omega^-]. \end{cases} \qquad (17.14)$$

From it we can deduce the conditional probabilities

$$\begin{aligned} \Pr(z_1^+ \mid \Omega_2^+) &= 1 = \Pr(z_1^- \mid \Omega_2^-), \\ \Pr(\Omega_2^+ \mid z_1^+) &= 1 = \Pr(\Omega_2^- \mid z_1^-), \end{aligned} \qquad (17.15)$$

which are the analogs of (17.8). The initial state Ψ_0 can be thought of as one of the conditions, though it is not shown explicitly.

However, (17.15), while technically correct, does not really provide the sort of result one wants from a macroscopic theory of measurement. What one would like to say is: "Given the initial state and the fact that the pointer points up at the time t_2, S_z must have had the value $+1/2$ at t_1." While the state $|\Omega^+\rangle$ is, indeed, a state of the apparatus for which the pointer is up, it does not mean the same thing as "the pointer points up". There are an enormous number of quantum states of the apparatus consistent with "the pointer points up", and $|\Omega^+\rangle$ is just one of these, so it contains a lot of information in addition to the direction of the pointer. It provides a very precise description of the state of the apparatus, whereas what we would like to have is a conditional probability whose condition involves only a relatively coarse "macroscopic" description of the apparatus. One can also fault the use of the family (17.14) on the grounds that $|\Psi_0\rangle$ is itself a very precise description of the initial state of the apparatus. In practice it is impossible to set up an apparatus in such a way that one can be sure it is in such a precise initial state.

What we need are conditional probabilities which lead to the same conclusions as (17.15), but with conditions which involve a much less detailed description of the apparatus at t_0 and t_2. Such coarse-grained descriptions in classical physics are provided by statistical mechanics. While quantum statistical mechanics lies outside the scope of this book, the histories formalism developed earlier provides tools which are adequate for the task at hand, and we shall use them in the next section to provide an improved version of macroscopic measurements.

17.4 Macroscopic measurement, second version

Physical properties in quantum theory are associated with subspaces of the Hilbert space, or the corresponding projectors. Often these are projectors on relatively small subspaces. However, it is also possible to consider projectors which correspond to macroscopic properties of a piece of apparatus, such as "the pointer points upwards". We shall call such projectors "macro projectors", since they single out regions of the Hilbert space corresponding to macroscopic properties.

Let Z be a macro projector onto the initial state of the apparatus ready to carry out a measurement of the spin of the particle. It projects onto an enormous subspace $\mathcal{Z}$ of the Hilbert space, one with a dimension, $\mathrm{Tr}[Z]$, which is of the order of $e^{S/k}$, where S is the (absolute) thermodynamic entropy of the apparatus, and k is Boltzmann's constant. Thus $\mathrm{Tr}[Z]$ could be 10 raised to the power 10^{23}. Such a macro projector is not uniquely defined, but the ambiguity is not important for the argument which follows. It is convenient to include in $\mathcal{Z}$ the information about the center of mass of the particle at t_0, but not its spin. Similarly, the apparatus after the measurement can be described by the macro projectors Z^+, projecting on a subspace $\mathcal{Z}^+$ for which the pointer points up, and Z^-, projecting on a subspace $\mathcal{Z}^-$ for which the pointer points down. For reasons indicated in Sec. 17.3, any state in which the pointer is directed upwards will surely be orthogonal to any state in which it is directed downwards, and thus

$$Z^+ Z^- = 0. \tag{17.16}$$

Let $\{|\Omega_j\rangle\}$, $j = 1, 2, \ldots$ be an orthonormal basis for $\mathcal{Z}$. We assume that the unitary time evolution from t_0 to t_1 to t_2 takes the form

$$\begin{aligned}
|z^+\rangle|\Omega_j\rangle &\mapsto |z^+\rangle|\Omega'_j\rangle \mapsto |\Omega_j^+\rangle, \\
|z^-\rangle|\Omega_j\rangle &\mapsto |z^-\rangle|\Omega'_j\rangle \mapsto |\Omega_j^-\rangle,
\end{aligned} \tag{17.17}$$

for $j = 1, 2, \ldots$, and that for every j,

$$Z^+|\Omega_j^+\rangle = |\Omega_j^+\rangle, \quad Z^-|\Omega_j^-\rangle = |\Omega_j^-\rangle. \tag{17.18}$$

That is to say, whatever may be the precise initial state of the apparatus at t_0, if $S_z = +1/2$ at this time, then at t_2 the apparatus pointer will be directed upwards, whereas if $S_z = -1/2$ at t_0, the pointer will later be pointing downwards. Note that combining (17.16) with (17.18) tells us that for every j

$$Z^+|\Omega_j^-\rangle = 0 = Z^-|\Omega_j^+\rangle. \tag{17.19}$$

Since the $\{|\Omega_j^+\rangle\}$ are mutually orthogonal — (17.17) represents a unitary time development — they span a subspace of the Hilbert space having the same dimension, $\text{Tr}[Z]$, as $\mathcal{Z}$. Hence (17.18) can only be true if the subspace $\mathcal{Z}^+$ onto which Z^+ projects has a dimension $\text{Tr}[Z^+]$ at least as large as $\text{Tr}[Z]$, and the same comment applies to $\mathcal{Z}^-$. We expect the process which results in moving the pointer to a particular position to be irreversible in the thermodynamic sense: the entropy of the apparatus will increase during this process. Since, as noted earlier, the trace of a macro projector is on the order of $e^{S/k}$, where S is the thermodynamic entropy, even a modest (macroscopic) increase in entropy is enough to make $\text{Tr}[Z^+]$ (and likewise $\text{Tr}[Z^-]$) enormously larger than the already very large $\text{Tr}[Z]$: the ratio $\text{Tr}[Z^+]/\text{Tr}[Z]$ will be 10 raised to a large power. There is thus no difficulty in supposing that (17.18) is satisfied, as there is plenty of room in $\mathcal{Z}^+$ and $\mathcal{Z}^-$ to hold all the states which evolve unitarily from $\mathcal{Z}$, and in this respect the unitary time development assumed in (17.17) is physically plausible.

Now let us consider various families of histories based upon an initial state represented by the projector

$$\Phi_0 = [x^+] \otimes Z, \tag{17.20}$$

which in physical terms means that the particle has $S_x = +1/2$ and the apparatus is ready to carry out the measurement. Note that Φ_0, in contrast to the pure state Ψ_0 used in Sec. 17.3, is a projector on a very large subspace, and thus a relatively imprecise description of the initial state of the apparatus.

Consider first the case of unitary time evolution starting with Φ_0 at t_0 and leading to a state

$$\Phi_2 = T(t_2, t_0)\Phi_0 T(t_0, t_2) = \sum_j |\bar{\Omega}_j\rangle\langle\bar{\Omega}_j|, \qquad (17.21)$$

at t_2, where

$$|\bar{\Omega}_j\rangle = \left(|\Omega_j^+\rangle + |\Omega_j^-\rangle\right)/\sqrt{2} \qquad (17.22)$$

is an MQS state. None of the terms in the sum in (17.21) commutes with Z^+ and Z^-, and it is easy to show that the same is true of the sum itself (that is, there are no cancellations). Since Φ_2 does not commute with Z^+ and Z^-, a history using unitary time evolution precludes any discussion of measurement outcomes. Another way of stating this is that whatever the initial apparatus state, unitary time evolution will inevitably lead to an MQS state in which the pointer positions have no meaning. Hence it is essential to employ a nonunitary history in order to discuss the measurement process; using macro projectors does not change our earlier conclusion in this respect.

Likewise the counterpart of the family (17.13), in which one can discuss measurement outcomes but not the value of S_z at t_1, is unsatisfactory as a description of a measurement process for the same reason indicated earlier. Thus we are led to consider a family analogous to that in (17.14), whose support consists of the two histories

$$\Phi_0 \odot \begin{cases} [z^+] \odot Z^+, \\ [z^-] \odot Z^- \end{cases} \qquad (17.23)$$

involving events at the times t_0, t_1, t_2. Note that (in contrast to (17.14)) no mention is made of an apparatus state at t_1, and of course no mention is made of a spin state at t_2. The histories $\Phi_0 \odot [z^-] \odot Z^+$ and $\Phi_0 \odot [z^+] \odot Z^-$ have zero weight in view of (17.19).

As both histories in (17.23) have positive weight, it is clear that

$$\begin{aligned}
\Pr(z_1^+ \mid Z_2^+) &= 1 = \Pr(z_1^- \mid Z_2^-), \\
\Pr(Z_2^+ \mid z_1^+) &= 1 = \Pr(Z_2^- \mid z_1^-),
\end{aligned} \qquad (17.24)$$

where the initial state Φ_0 can be thought of as one of the conditions, even though it is not shown explicitly. Thus if the pointer is directed upwards at t_2, then S_z had the value $+1/2$ at t_1, while a pointer directed downwards at t_2 means that S_z was $-1/2$ at t_1. These results are formally the same as those in (17.15), but (17.24) is more satisfactory from a physical point of view in that the conditions (including the implicit Φ_0) only involve "macroscopic" information about the measuring apparatus. Note that (17.14) is not misleading, even though its physical interpretation is less satisfactory than (17.24), and the former is somewhat easier to derive. It is often the case that one can model a macroscopic measurement process in somewhat simplistic terms, and nonetheless obtain a plausible answer. Of course, if there are any doubts about this procedure, it is a good idea to check it using macro projectors.

An alternative to the preceding discussion is an approach based upon statistical mechanics, which in its simplest form consists in choosing an appropriate basis $\{|\Omega_j\rangle\}$ for the subspace corresponding to the initial state of the apparatus (the space on which Z projects), and assigning a probability p_j to the state $|\Omega_j\rangle$. Assuming the correctness of (17.17) and (17.18), one can use the consistent family supported by the (enormous) collection of histories of the form

$$
\begin{aligned}
[x^+] \otimes [\Omega_j] \odot [z^+] \odot Z^+, \\
[x^+] \otimes [\Omega_j] \odot [z^-] \odot Z^-,
\end{aligned}
\tag{17.25}
$$

with $j = 1, 2, \ldots$, to obtain (17.24). Note that consistency is ensured by $Z^+ Z^- = 0$ along with the fact that the initial states for histories ending in Z^+ are mutually orthogonal, and likewise those ending in Z^-.

Yet another approach to the same problem is to describe the measuring apparatus at t_0 by means of a density matrix ρ thought of as a pre-probability, as discussed in Sec. 15.6. Since ρ describes an apparatus in an initial ready state, the probability, computed from ρ, that the apparatus will *not* be in this state must be zero:

$$
\text{Tr}[\rho(I - Z)] = 0.
\tag{17.26}
$$

Since both ρ and $I - Z$ are positive operators, (17.26) implies, see (3.92), that $\rho(I - Z) = 0$, or

$$
Z\rho = \rho,
\tag{17.27}
$$

which means that the support of ρ (Sec. 3.9) falls in the subspace $\mathcal{Z}$ on which Z projects. Consequently, ρ may be written in the diagonal form

$$
\rho = \sum_j p_j |\Omega_j\rangle\langle\Omega_j|,
\tag{17.28}
$$

where $\{|\Omega_j\rangle\}$ is an orthonormal basis of $\mathcal{Z}$. To be sure, this could be a different basis of $\mathcal{Z}$ from the one introduced earlier, but since the vectors in any basis can be expressed as linear combinations of vectors in the other, it follows that (17.17) and (17.18) will still be true.

Given ρ in the form (17.28), the measurement process can be analyzed using the procedures of Sec. 15.6, including (15.48) for consistency conditions and (15.50) for weights. Using these one can show that the two histories

$$[x^+] \odot \begin{cases} [z^+] \odot Z^+, \\ [z^-] \odot Z^- \end{cases} \tag{17.29}$$

form the support of a consistent family. This family resembles (17.23), except that the initial state $[x^+]$ at t_0 contains no reference to the apparatus, since the initial state of the apparatus is represented by a density matrix. The weights of the histories in (17.29) are the same as for their counterparts in (17.23), and once again lead to the conditional probabilities in (17.24).

17.5 General destructive measurements

The preceding discussion of the measurement of S_z for a spin-half particle can be easily extended to a schematic description of an idealized measuring process for a more complicated system S which interacts with a measuring apparatus $\mathcal{M}$. The measured properties will correspond to some orthonormal basis $\{|s^k\rangle\}$, $k = 1, 2, \ldots n$ of S, and we shall assume that the measurement process corresponds to a unitary time development from t_0 to t_1 to t_2 of the form

$$|s^k\rangle \otimes |M_0\rangle \mapsto |s^k\rangle \otimes |M_1\rangle \mapsto |N^k\rangle, \tag{17.30}$$

where $|M_0\rangle$ and $|M_1\rangle$ are states of the apparatus at t_0 and t_1 before it interacts with S, and the $\{|N^k\rangle\}$ are orthonormal states on $S \otimes \mathcal{M}$ for which a measurement pointer indicates a definite outcome of the measurement. (The $\{|N^k\rangle\}$ are a *pointer basis* in the notation introduced at the end of Sec. 9.5.)

Assume that at t_0 the initial state of $S \otimes \mathcal{M}$ is

$$|\Psi_0\rangle = |s_0\rangle \otimes |M_0\rangle, \tag{17.31}$$

where

$$|s_0\rangle = \sum_k c_k |s^k\rangle, \tag{17.32}$$

with $\sum_k |c_k|^2 = 1$, is an arbitrary superposition of the basis states of S. Unitary time evolution will then result in a state

$$|\Psi_2\rangle = T(t_2, t_0)|\Psi_0\rangle = \sum_k c_k |N^k\rangle \tag{17.33}$$

at t_2. Using the two-time family

$$\Psi_0 \odot I \odot \{N^1, N^2, \ldots\}, \tag{17.34}$$

where square brackets have been omitted from $[N^k]$, and regarding $|\Psi_2\rangle$ as a pre-probability, one finds

$$\Pr(N_2^k) = |c_k|^2 \tag{17.35}$$

for the probability of the kth outcome of the measurement at the time t_2.

One can refine (17.34) to a consistent family with support

$$\Psi_0 \odot \begin{cases} [s^1] \odot N^1, \\ [s^2] \odot N^2, \\ \dots \\ [s^n] \odot N^n, \end{cases} \tag{17.36}$$

and from it derive the conditional probabilities

$$\Pr(s_1^j \mid N_2^k) = \delta_{jk} = \Pr(N_2^k \mid s_1^j), \tag{17.37}$$

assuming $\Pr(N_2^k) > 0$. That is, given the measurement outcome N^k at t_2, S was in the state $|s^k\rangle$ before the measurement took place. Thus the measurement interaction results in an appropriate correlation between the later apparatus output and the earlier state of the measured system.

The preceding analysis can be generalized to a measurement of properties which are not necessarily pure states, but form a decomposition of the identity

$$I_S = \sum_k S^k \tag{17.38}$$

for system S, where some of the projectors are onto subspaces of dimension greater than 1. This might arise if one were interested in the measurement of a physical variable of the form

$$V = \sum_k v_k' S^k, \tag{17.39}$$

see (5.24), some of whose eigenvalues are degenerate.

Let us assume that the subspace onto which S^k projects is spanned by an orthonormal collection $\{|s^{kl}\rangle, l = 1, 2, \dots\}$, so that

$$S^k = \sum_l |s^{kl}\rangle\langle s^{kl}|. \tag{17.40}$$

Assume that the counterpart of (17.30) is a unitary time development

$$|s^{kl}\rangle \otimes |M_0\rangle \mapsto |s^{kl}\rangle \otimes |M_1\rangle \mapsto |N^{kl}\rangle, \tag{17.41}$$

where $\{|N^{kl}\rangle\}$ is an orthonormal collection of states on $S \otimes M$ labeled by both k

and l, and

$$N^k = \sum_l |N^{kl}\rangle\langle N^{kl}| \qquad (17.42)$$

represents a property of $\mathcal{M}$ corresponding to the kth measurement outcome.

The counterpart of (17.36) is the consistent family

$$\Psi_0 \odot \begin{cases} S^1 \odot N^1, \\ S^2 \odot N^2, \\ \dots \\ S^n \odot N^n, \end{cases} \qquad (17.43)$$

where $|\Psi_0\rangle$ is given by (17.31), with

$$|s_0\rangle = \sum_{kl} c_{kl}|s^{kl}\rangle \qquad (17.44)$$

the obvious counterpart of (17.32). Corresponding to (17.37) one has

$$\Pr(S_1^j \mid N_2^k) = \delta_{jk} = \Pr(N_2^k \mid S_1^j), \qquad (17.45)$$

with the physical interpretation that a measurement outcome N^k at t_2 implies that S had the property S^k at t_1, and vice versa.

The measurement schemes discussed in this section can be extended to a genuinely macroscopic description of the measuring apparatus in a straightforward manner using either of the approaches discussed in Sec. 17.4.

18

Measurements II

18.1 Beam splitter and successive measurements

Sometimes a quantum system, hereafter referred to as a "particle", is destroyed during a measurement process, but in other cases it continues to exist in an identifiable
form after interacting with the measuring apparatus, with some of its properties
unchanged or related in a nontrivial way to properties which it possessed before
this interaction. In such a case it is interesting to ask what will happen if a second
measurement is carried out on the particle: how will the outcome of the second
measurement be related to the outcome of the first measurement, and to properties
of the particle between the two measurements?

Let us consider a specific example in which a particle (photon or neutron) passes
through a beam splitter B and is then subjected to a measurement by nondestructive
detectors located in the c and d output channels as shown in Fig. 18.1. Assume
that the unitary time development of the particle in the absence of any measuring
devices is given by

$$|0a\rangle \mapsto \big(|1c\rangle + |1d\rangle\big)/\sqrt{2} \mapsto \big(|2c\rangle + |2d\rangle\big)/\sqrt{2} \mapsto \cdots \qquad (18.1)$$

as time progresses from t_0 to t_1 to $t_2 \ldots$. Here the kets denote wave packets whose
approximate locations are shown by the circles in Fig. 18.1, and the labels are
similar to those used for the toy model in Ch. 12.

The detectors are assumed to register the passage of the particle while having a
negligible influence on the time development of its wave packet. Toy detectors with
this property were introduced earlier, in Secs. 7.4 and 12.3. To actually construct
such a device in the laboratory is much more difficult, but, at least for some types
of particle, not out of the question. We assume that the interaction of the particle
with the detector C in Fig. 18.1 leads to a unitary time development during the

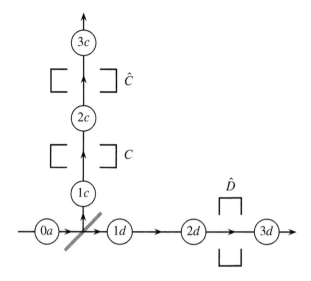

Fig. 18.1. Beam splitter followed by nondestructive measuring devices. The circles indicate the locations of wave packets corresponding to different kets.

interval from t_1 to t_2 of the form

$$
\begin{aligned}
|1c\rangle|C^\circ\rangle &\mapsto |2c\rangle|C^*\rangle, \\
|1d\rangle|C^\circ\rangle &\mapsto |2d\rangle|C^\circ\rangle,
\end{aligned}
\tag{18.2}
$$

where $|C^\circ\rangle$ denotes the "ready" or "untriggered" state of the detector, and $|C^*\rangle$ the "triggered" state orthogonal to $|C^\circ\rangle$. (The tensor product symbol, as in $|1c\rangle \otimes |C^\circ\rangle$, has been omitted.) The behavior of the other detectors $\hat{C}$ and $\hat{D}$ is similar, and thus an initial state

$$
|\Psi_0\rangle = |0a\rangle|C^\circ\rangle|\hat{C}^\circ\rangle|\hat{D}^\circ\rangle
\tag{18.3}
$$

develops unitarily to

$$
|\Psi_1\rangle = \left(|1c\rangle + |1d\rangle\right)|C^\circ\rangle|\hat{C}^\circ\rangle|\hat{D}^\circ\rangle/\sqrt{2},
\tag{18.4}
$$

$$
|\Psi_2\rangle = \left(|2c\rangle|C^*\rangle|\hat{C}^\circ\rangle|\hat{D}^\circ\rangle + |2d\rangle|C^\circ\rangle|\hat{C}^\circ\rangle|\hat{D}^\circ\rangle\right)/\sqrt{2},
\tag{18.5}
$$

$$
|\Psi_3\rangle = \left(|3c\rangle|C^*\rangle|\hat{C}^*\rangle|\hat{D}^\circ\rangle + |3d\rangle|C^\circ\rangle|\hat{C}^\circ\rangle|\hat{D}^*\rangle\right)/\sqrt{2}
\tag{18.6}
$$

at the times t_1, t_2, t_3.

We shall be interested in families of histories based on the initial state $|\Psi_0\rangle$. The simplest one to understand in physical terms is a family $\mathcal{F}$ in which at the times t_1, t_2, and t_3 every detector is either ready or triggered, and the particle is represented by a wave packet in one of the two output channels. The support of $\mathcal{F}$ consists of the two histories

$$Y^c = \Psi_0 \odot [1c]C^\circ \hat{C}^\circ \hat{D}^\circ \odot [2c]C^* \hat{C}^\circ \hat{D}^\circ \odot [3c]C^* \hat{C}^* \hat{D}^\circ,$$
$$Y^d = \Psi_0 \odot [1d]C^\circ \hat{C}^\circ \hat{D}^\circ \odot [2d]C^\circ \hat{C}^\circ \hat{D}^\circ \odot [3d]C^\circ \hat{C}^\circ \hat{D}^*, \tag{18.7}$$

where square brackets have been omitted from $[\Psi_0]$, $[C^\circ]$, etc., so that the formula remains valid if one employs macro projectors, as in Sec. 17.4. In Y^c the particle moves out along channel c and triggers the detectors C and $\hat{C}$ as it passes through them, while in Y^d the particle moves along channel d and triggers $\hat{D}$.

The situation described by these histories is essentially the same as it would be if a *classical* particle were scattered at random by the beam splitter into either the c or the d channel, and then traveled out along the channel triggering the corresponding detector(s). Thus if C is triggered at time t_2 the particle is surely in the c channel, and will later trigger $\hat{C}$, whereas if C is still in its ready state at t_2, this means the particle is in the d channel, and will later trigger $\hat{D}$. That these assertions are indeed correct for a quantum particle can be seen by working out various conditional probabilities, e.g.

$$\Pr([1c]_1 \mid C_2^*) = 1 = \Pr([1d]_1 \mid C_2^\circ), \tag{18.8}$$
$$\Pr([2c]_2 \mid C_2^*) = 1 = \Pr([2d]_2 \mid C_2^\circ), \tag{18.9}$$
$$\Pr([3c]_3 \mid C_2^*) = 1 = \Pr([3d]_3 \mid C_2^\circ), \tag{18.10}$$

where the subscripts indicate the times, t_1 or t_2 or t_3, at which the events occur. Thus the location of the particle either before or after t_2 can be inferred from whether it has or has not been detected by C at t_2. There are, in addition, correlations between the outcomes of the different measurements:

$$\Pr(\hat{C}_3^* \mid C_2^*) = 1, \quad \Pr(\hat{D}_3^* \mid C_2^*) = 0, \tag{18.11}$$
$$\Pr(\hat{C}_3^\circ \mid C_2^\circ) = 1, \quad \Pr(\hat{D}_3^* \mid C_2^\circ) = 1. \tag{18.12}$$

Thus whether $\hat{C}$ or $\hat{D}$ will later detect the particle is determined by whether it was or was not detected earlier by C.

The conditional probabilities in (18.8)–(18.12) are straightforward consequences of the fact that all histories in $\mathcal{F}$ except for the two in (18.7) have zero probability. Since these conditional probabilities, with the exception of (18.9), involve more than two times — note that the initial Ψ_0 is implicit in the condition — they cannot be obtained by using the Born rule, and are therefore inaccessible to older approaches to quantum theory which lack the formalism of Ch. 10. These older approaches employ a notion of "wave function collapse" in order to get results comparable to (18.9)–(18.12), and this is the subject of the next section.

18.2 Wave function collapse

Quantum measurements have often been analyzed using the following idea, which goes back to von Neumann. Consider an isolated system $\mathcal{S}$, and suppose that its wave function evolves unitarily, so that it is $|s_1\rangle$ at a time t_1. At this time, or very shortly thereafter, $\mathcal{S}$ interacts with a measuring apparatus $\mathcal{M}$ designed to determine whether $\mathcal{S}$ is in one of the states of a collection $\{|s^k\rangle\}$ forming an orthonormal basis of the Hilbert space of $\mathcal{S}$. The measurement will have an outcome k with probability $|\langle s_1|s^k\rangle|^2$, and if the outcome is k the effect of the measurement will be to "collapse" or "reduce" $|s_1\rangle$ to $|s^k\rangle$.

This collapse picture of a measurement proceeds in the following way when $\mathcal{S}$ is the particle and $\mathcal{M}$ the detector C in Fig. 18.1. The particle undergoes unitary time evolution until it encounters the measuring apparatus, and thus at t_1 it is in a state

$$|1a\rangle = (|1c\rangle + |1d\rangle)/\sqrt{2}. \tag{18.13}$$

The detector at time t_2 is either still in its ready state $|C^\circ\rangle$, or else in its triggered state $|C^*\rangle$ indicating that it has detected the particle. If the particle has been detected, its wave function will have collapsed from its earlier delocalized state $|1a\rangle$ into the $|2c\rangle$ wave packet localized in the c channel and moving towards detector $\hat{C}$, which will later detect the particle. If, on the other hand, the particle has *not* been detected by C, its wave function will have collapsed into the packet $|2d\rangle$ localized in the d channel and moving towards the $\hat{D}$ detector, which will later register the passage of the particle. Consequently C^* at t_2 results in $\hat{C}^*$ at t_3, whereas C° at t_2 implies $\hat{D}^*$ at t_3, in agreement with the conditional probabilities in (18.11) and (18.12).

This "collapse" picture has long been regarded by many quantum physicists as rather unsatisfactory for a variety of reasons, among them the following. First, it seems somewhat arbitrary to abandon the state $|\Psi_2\rangle$ obtained by unitary time evolution, (18.5), without providing some better reason than the fact that a measurement occurs; after all, what is special about a quantum measurement? Any real measurement apparatus is constructed out of aggregates of particles to which the laws of quantum mechanics apply, so the apparatus ought to be described by those laws, and not used to provide excuse for their breakdown. Second, while it might seem plausible that an interaction sufficient to trigger a measuring apparatus could somehow localize a particle wave packet somewhere in the vicinity of the apparatus, it is much harder to understand how the same apparatus by *not* detecting the particle manages to localize it in some region which is very far away.

This second, nonlocal aspect of the collapse picture is particularly troublesome, and has given rise to an extensive discussion of "interaction-free measurements" in which some property of a quantum system can be inferred from the fact that it did *not* interact with a measuring device. (We shall return to this subject in Sec. 21.5.) Since one can imagine the gedanken experiment in Fig. 18.1 set up in outer space with the wave packets $|2c\rangle$ and $|2d\rangle$ an enormous distance apart, there is also the problem that if wave function collapse takes place instantaneously it will conflict with the principle of special relativity according to which no influence can travel faster than the speed of light.

By contrast, the analysis given in Sec. 18.1 based upon the family $\mathcal{F}$, (18.7), shows no signs of any nonlocal effects. If C has not detected the particle at time t_2, this is because the particle is moving out the d channel, not the c channel. In the case of a classical particle such an "interaction free measurement" of the channel in which it is moving gives rise to no conceptual difficulties or conflicts with relativity theory. As pointed out in Sec. 18.1, the family $\mathcal{F}$ provides a quantum description which resembles that of a classical particle, and thus by using this family one avoids the nonlocality difficulties of wave function collapse.

Another way to avoid these difficulties is to think of wave function collapse not as a *physical effect* produced by the measuring apparatus, but as a *mathematical procedure* for calculating statistical correlations of the type shown in (18.9)–(18.12). That is, "collapse" is something which takes place in the theorist's notebook, rather than the experimentalist's laboratory. In particular, if the wave function is thought of as a pre-probability (Sec. 9.4), then it is perfectly reasonable to collapse it to a different pre-probability in the middle of a calculation.

With reference to the arrangement in Fig. 18.1, the idea of wave function collapse corresponds fairly closely to a consistent family $\mathcal{V}$ with support

$$\Psi_0 \odot \Psi_1 \odot \begin{cases} [2c]C^*\hat{C}^\circ\hat{D}^\circ \odot [3c]C^*\hat{C}^*\hat{D}^\circ, \\ [2d]C^\circ\hat{C}^\circ\hat{D}^\circ \odot [3d]C^\circ\hat{C}^\circ\hat{D}^*. \end{cases} \tag{18.14}$$

These two histories represent unitary time evolution of the initial state, so they are identical up to the time t_1, before the particle interacts (or fails to interact) with C, but are thereafter distinct. As a consequence of the internal consistency of quantum reasoning, Sec. 16.3, this family gives the same results for the conditional probabilities in (18.9)–(18.12) as does $\mathcal{F}$. (Those in (18.8) are not defined in $\mathcal{V}$.) In particular, either family can be used to predict the outcomes of later $\hat{C}$ and $\hat{D}$ measurements based upon the outcome of the earlier C measurement.

One can imagine constructing the framework $\mathcal{V}$ in successive steps as follows. Use unitary time development up to t_2, but think of $|\Psi_2\rangle$ in (18.5) as a pre-probability (rather than as representing an MQS property) useful for assigning probabilities to the two histories

$$\Psi_0 \odot \Psi_1 \odot \{C^*, C^\circ\}, \tag{18.15}$$

which form the support of a consistent family whose projectors at t_2 represent the two possible measurement outcomes. This is the minimum modification of a unitary family which can exhibit these outcomes. Next refine this family by including the corresponding particle properties at t_2 along with the ready states of the other detectors:

$$\Psi_0 \odot \Psi_1 \odot \begin{cases} [2c]C^*\hat{C}^\circ\hat{D}^\circ, \\ [2d]C^\circ\hat{C}^\circ\hat{D}^\circ. \end{cases} \tag{18.16}$$

Finally, use unitary extensions of these histories, Sec. 11.7, to obtain the family $\mathcal{V}$ of (18.14). In a more general situation the step from (18.15) to (18.16) can be more complicated: one may need to use conditional density matrices rather than projectors onto particle properties, as discussed in Sec. 15.7. But the general idea is still the same: information from the outcome of a measurement is used to construct a new initial state of the particle, which is then employed for calculating results at still later times. Wave function collapse is, in essence, an algorithm for constructing this new initial state given the outcome of the measurement.

Wave function collapse is in certain respects analogous to the "collapse" of a classical probability distribution when it is conditioned on the basis of new information. Once again think of a classical particle randomly scattered by the beam splitter into the c or d channel. Before the particle (possibly) passes through C, it is delocalized in the sense that the probability is $1/2$ for it to be in either the c or the d channel. But when the probability for the location of the particle is conditioned on the measurement outcome it collapses in the sense that the particle is either in the c channel, given C^*, or the d channel, given C°. This collapse of the classical probability distribution is obviously not a physical effect, and only in some metaphorical sense can it be said to be "caused" by the measurement. This becomes particularly clear when one notes that conditioning on the measurement outcome collapses the probability distribution at a time t_1 *before* the measurement occurs in the same way that it collapses it at t_2 or t_3 *after* the measurement occurs. Thinking of the collapse as being caused by the measurement would lead to an odd situation in which an effect precedes its cause.

Precisely the same comment applies to the collapse of a quantum wave function. A quantum description conditioned on a particular outcome of a measurement will generally provide more detail, and thus appear to be "collapsed", in comparison with one constructed without this information. But since the outcome of a quantum measurement can also tell one something about the properties of the measured particle prior to the measurement process (assuming a framework in which these properties can be discussed) one should not think of the collapse as some sort of physical effect with a physical cause. To be sure, in the family (18.14) it is not possible to discuss the location (c or d) of the particle before the measurement, because in this particular framework the location does not make sense. The implicit use of this type of family for discussions of quantum measurements is probably one reason why wave function collapse has often been confused with a physical effect. The availability of other families, such as $\mathcal{F}$ in (18.7), helps one avoid this mistake.

In summary, when quantum mechanics is formulated in a consistent way, wave function collapse is not needed in order to describe the interaction between a particle (or some other quantum system) and a measuring device. One can use a notion of collapse as a method of constructing a particular type of consistent family, as indicated in the steps leading from (18.15) to (18.16) to (18.14), or else as a picturesque way of thinking about correlations that in the more sober language of ordinary probability theory are written as conditional probabilities, as in (18.9)–(18.12). However, for neither of these purposes is it actually essential; any result that can be obtained by collapsing a wave function can also be obtained in a straightforward way by adopting an appropriate family of histories. The approach using histories is more flexible, and allows one to describe the measurement process in a natural way as one in which the properties of the particle before as well as after the measurement are correlated to the measurement outcomes.

While its picturesque language may have some use for pedagogical purposes or for constructing mnemonics, the concept of wave function collapse has given rise to so much confusion and misunderstanding that it would, in my opinion, be better to abandon it altogether, and instead use conditional states, such as the conditional density matrices discussed in Sec. 15.7 and in Sec. 18.5, and conditional probabilities. These are quite adequate for constructing quantum descriptions, and are much less confusing.

18.3 Nondestructive Stern–Gerlach measurements

The Stern–Gerlach apparatus for measuring one component of spin angular momentum of a spin-half atom was described in Ch. 17. Here we shall consider a modified version which, although it would be extremely difficult to construct in the laboratory, does not violate any principles of quantum mechanics, and is useful for understanding why quantum measurements that are nondestructive for certain properties will be destructive for other properties. Figure 18.2 shows the modified apparatus, which consists of several parts. First, a magnet with an appropriate field gradient like the one in Fig. 17.1 separates the incoming beam into two diverging beams depending upon the value of S_z, with the $S_z = +1/2$ beam going upwards and the $S_z = -1/2$ beam going downwards. There are then two additional magnets, with field gradients in a direction opposite to the gradient in the first magnet, to bend the separated beams in such a way that they are traveling parallel to each other. These beams pass through detectors D_a and D_b of the nondestructive sort employed in Fig. 18.1. We assume not only that these detectors produce a negligible perturbation of the spatial wave packets in each beam, but also that they do not perturb the z component of spin. (A detector in one beam and not the other would actually be sufficient, but using two emphasizes the symmetry of the situation.) The detectors are followed by a series of magnets which reverse the process produced by the first set of magnets and bring the two beams back together again.

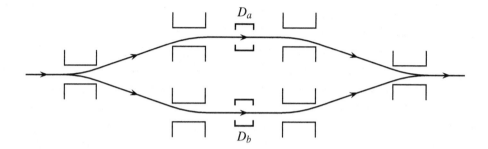

Fig. 18.2. Modified Stern–Gerlach apparatus for nondestructive measurements of S_z.

The net result is that an atom with either $S_z = +1/2$ or $S_z = -1/2$ will traverse the apparatus and emerge in the same beam at the other end. The only difference is in the detector which is triggered while the atom is inside the apparatus. The unitary time evolution corresponding to the measurement process is

$$|z^+\rangle|Z^\circ\rangle \mapsto |z^+\rangle|Z^+\rangle, \quad |z^-\rangle|Z^\circ\rangle \mapsto |z^-\rangle|Z^-\rangle, \tag{18.17}$$

where $|z^\pm\rangle$ are spin states corresponding to $S_z = \pm 1/2$, $|Z^\circ\rangle$ is the initial state of the apparatus, and $|Z^+\rangle$ and $|Z^-\rangle$ are mutually orthogonal apparatus states corresponding to detection by the upper or by the lower detector in Fig. 18.2. One could equally well use macro projectors for the apparatus states, as in Sec. 17.4, and for this reason we will employ Z° and $Z^\pm$ without square brackets as symbols for the corresponding projectors. In addition, the coordinate representing the center of mass of the atom is not shown in (18.17); omitting it will cause no confusion, and including it would merely clutter the notation. We shall assume that there are no magnetic fields outside the apparatus which could affect the atom's spin, and that the apparatus states $|Z^\circ\rangle$ and $|Z^\pm\rangle$ do not change with time except when interacting with the atom, (18.17). The latter assumption is convenient, but not essential.

It is obvious that the same type of apparatus can be used to measure other components of spin by using a different direction for the magnetic field gradient. For example, if the atom is thought of as moving along the y axis, then by simply rotating the apparatus about this axis it can be used to measure S_w for w any direction in the x, z plane. Alternatively, one could arrange for the atom to pass through regions of uniform magnetic field before entering and after leaving the apparatus sketched in Fig. 18.2, in order to cause a precession of an atom with $S_w = \pm 1/2$ into one with $S_z = \pm 1/2$, and then back again after the measurement is over.

We will consider various histories based upon an initial state

$$|\Psi_0\rangle = |u^+\rangle|Z^\circ\rangle, \tag{18.18}$$

at the time t_0, where the kets

$$\begin{aligned}|u^+\rangle &= +\cos(\vartheta/2)|z^+\rangle + \sin(\vartheta/2)|z^-\rangle, \\ |u^-\rangle &= -\sin(\vartheta/2)|z^+\rangle + \cos(\vartheta/2)|z^-\rangle, \end{aligned} \tag{18.19}$$

see (4.14), correspond to $S_u = +1/2$ and $-1/2$ for a direction u in the x, z plane at an angle ϑ to the $+z$ axis, so that S_u is equal to S_z when $\vartheta = 0$, and S_x when $\vartheta = \pi/2$.

Consider the consistent family with support

$$\Psi_0 \odot \begin{cases} [z^+] \odot Z^+ \odot [z^+], \\ [z^-] \odot Z^- \odot [z^-], \end{cases} \tag{18.20}$$

where the projectors refer to an initial time t_0, a time t_1 before the atom enters the apparatus, a time t_2 when it has left the apparatus, and a still later time t_3. The conditional probabilities

$$\Pr([z^+]_1 \mid Z_2^+) = 1 = \Pr([z^-]_1 \mid Z_2^-) \tag{18.21}$$

show that the properties $S_z = \pm 1/2$ before the measurement are correlated with the measurement outcomes, so that the apparatus does indeed carry out a measurement. In addition, the probabilities

$$\begin{aligned} \Pr([z^+]_3 \mid [z^+]_1) &= 1 = \Pr([z^-]_3 \mid [z^-]_1), \\ \Pr([z^-]_3 \mid [z^+]_1) &= 0 = \Pr([z^+]_3 \mid [z^-]_1) \end{aligned} \tag{18.22}$$

show that the measurement process carried out by this apparatus is nondestructive for the properties $[z^+]$ and $[z^-]$: they have the same values after the measurement as before.

Next consider a different family whose support consists of the four histories

$$\Psi_0 \odot [u^+] \odot \begin{cases} Z^+ \odot \{[u^+], [u^-]\}, \\ Z^- \odot \{[u^+], [u^-]\}. \end{cases} \tag{18.23}$$

Despite the fact that the four final projectors at t_3 are not all orthogonal to one another, the orthogonality of Z^+ and Z^- ensures consistency. It is straightforward to work out the weights associated with the different histories in (18.23) using the method of chain kets, Sec. 11.6. One result is

$$
\Pr([u^+]_3 \,|\, [u^+]_1) = |\langle u^+ | z^+ \rangle \langle z^+ | u^+ \rangle|^2 + |\langle u^+ | z^- \rangle \langle z^- | u^+ \rangle|^2
$$
$$
= (\cos(\vartheta/2))^4 + (\sin(\vartheta/2))^4 = (1 + \cos^2 \vartheta)/2. \qquad (18.24)
$$

Except for $\vartheta = 0$ or π, the probability of $[u^+]$ at t_3 is less than 1, meaning that the property $S_u = +1/2$ has a certain probability of being altered when the atom interacts with the apparatus designed to measure S_z. The disturbance is a maximum for $\vartheta = \pi/2$, which corresponds to $S_u = S_x$: indeed, the value of S_x after the atom has passed through the device is completely random, independent of its earlier value.

18.4 Measurements and incompatible families

As noted in Sec. 16.4, the relationship of incompatibility between quantum frameworks does not have a good classical analog, and thus it has to be understood in quantum mechanical terms and illustrated through quantum examples. Quantum measurements can provide useful examples, and in this section we consider two: one uses a beam splitter as in Sec. 18.1, the other employs nondestructive Stern–Gerlach devices of the type described in Sec. 18.3.

Think of a beam splitter, Fig. 18.3(a), similar to that in Fig. 18.1 except that there are no measuring devices in the output channels c and d. There is a consistent family whose support consists of the pair of histories

$$
[0a] \odot \{[1c], [1d]\} \qquad (18.25)
$$

at the times t_0 and t_1, where the notation is the same as in Sec. 18.1. The unitary time development in (18.1) implies that each history has a probability of $1/2$.

The closed box surrounding the apparatus in Fig. 18.3(a) means that we are thinking of it as an isolated quantum system. Because it is isolated, there is no direct way to check the probabilities associated with the family in (18.25). However, there is a strategy which can provide indirect evidence. Suppose that at some time later than t_1 and just before the particle would collide with one of the walls of the box, two holes are opened, as shown in Fig. 18.3(b), allowing the particle to escape and be detected by one of the two detectors C and D. If the particle is detected by C, it seems plausible that it was earlier traveling outwards through the c and not the d channel; similarly, detection by D indicates that it was earlier in the d channel. Data obtained by repeating the experiment a large number of times can be used to check that each history in (18.25) has a probability of $1/2$.

Could it be that opening the box along with the subsequent measurements perturbs the particle in such a way as to invalidate the preceding analysis? This is

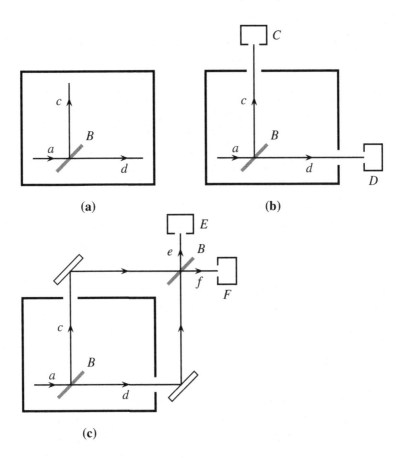

Fig. 18.3. Beam splitter inside closed box (a), with two possibilities (b) and (c) for a measurement if the particle is allowed to emerge through holes in the sides of the box.

a perfectly legitimate question, one which could also come up when one opens a "classical" box in order to determine what is going on inside it: think of a box containing unexposed photographic film, or a compressed gas. While there is no way of addressing the classical box-opening problem in a manner fully accept-able to sceptical philosophers, physicists will be content if they are able to achieve some reasonable understanding of what is likely to be going on during the open-ing process. This may require auxiliary experiments, mathematical modeling, and a certain amount of theoretical reasoning. On the basis of these physicists might be reasonably confident when inferring something about the state of affairs in-side a box before it is opened, using information from observations carried out afterwards.

Given the internal consistency of quantum reasoning, and the fact that quantum principles have been verified time and time again in innumerable experiments, it is not unreasonable to use quantum theory itself in order to examine what will happen if holes are opened in the box in Fig. 18.3, and whether the detection of the particle by C is a good reason to suppose that it was in the c channel at t_1. Carrying out such an analysis is not difficult if one assumes, as is plausible, that a timely opening of the holes has no effect upon the unitary time evolution of the particle's wave packet other than to allow it to propagate as it would have in the complete absence of any walls. The rest of the analysis is the same as in Sec. 18.1, and shows that the conditional probabilities (18.8) also apply to the present situation: if the particle is later detected by C, it was in the c channel inside the box at t_1.

An alternative consistent family has for its support the single unitary history

$$[0a] \odot [1\bar{a}], \tag{18.26}$$

where $|1\bar{a}\rangle$ is the superposition state defined in (18.13). This family is clearly incompatible with the one in (18.25) because $[1\bar{a}]$ does not commute with either $[1c]$ or $[1d]$. Nonetheless, (18.26) is just as good a quantum description of the particle moving inside the closed box as is the pair of possibilities in (18.25). An experimental test which will confirm that the history (18.26) does, indeed, occur is shown in Fig. 18.3(c), and is only slightly more complicated than the one used earlier. Once again, holes are opened in the walls of the box just before the arrival of the particle, but now there are mirrors outside the holes and a second beam splitter, so one has a Mach–Zehnder interferometer. Let the path lengths be such that a particle in the state $|1\bar{a}\rangle$ at time t_1 will emerge from the second beam splitter in the f channel and trigger the detector F, whereas a particle in the orthogonal state

$$|1\bar{b}\rangle = (-|1c\rangle + |1d\rangle)/\sqrt{2} \tag{18.27}$$

will emerge in channel e and trigger E. The experiment needs to be repeated many times in order to get a statistically significant result, and if in every, or almost every, run the particle is detected in F rather than E, one can infer that it was in the state $[1\bar{a}]$ at the earlier time t_1. That this is a plausible inference follows once again from the fact that quantum mechanics is a consistent theory abundantly confirmed by a variety of experimental tests.

It is obviously impossible to carry out the two types of measurements indicated in (b) and (c) of Fig. 18.3 on the same system during the same experimental run, and this is not surprising given the fact that while both (18.25) and (18.26) are valid quantum descriptions, they are mutually incompatible, so they cannot be applied to the same system at the same time. The "classical" macroscopic incompatibility of the two experimental arrangements, in the sense that setting up one of them prevents setting up the other, mirrors the quantum incompatibility of the microscopic events which are measured in the two cases. Thus an analysis using measurements can assist one in gaining an intuitive understanding of the incompatibility of quantum events and frameworks.

It has sometimes been suggested that certain conceptual difficulties associated with incompatible quantum frameworks could be resolved if there were a law of nature which specified the framework which had to be employed in any particular circumstance. That such an idea is not likely to work can be seen from the fact that either of the experiments indicated in Fig. 18.3 could in principle be carried out a large distance away and thus a long time after the particle emerges from the box, long enough to allow a choice to be made between the two experimental arrangements (see the discussion of delayed choice in Ch. 20). Thus were there such a law of nature, it would need to either determine the choice of the later experiment, or allow that later choice to influence the particle while it was still inside the box. Neither of these seems very satisfactory.

A second example in which measurements are useful for understanding quantum incompatibility is shown in Fig. 18.4(a), in which a spin-half atom moving parallel to the y axis passes successively through two nondestructive Stern–Gerlach devices, represented schematically by squares, of the form shown in Fig. 18.2. At the times t_0, t_1, and t_2 the atom is (approximately) at the positions indicated by the dots in the figure. The first device measures S_z, and its unitary time development during the interval from t_0 to t_1 is given by (18.17). The second device measures S_x, and its unitary time development from t_1 to t_2 is given by

$$|x^+\rangle|X^\circ\rangle \mapsto |x^+\rangle|X^+\rangle, \quad |x^-\rangle|X^\circ\rangle \mapsto |x^-\rangle|X^-\rangle, \tag{18.28}$$

where $|X^\circ\rangle$, $|X^+\rangle$ and $|X^-\rangle$ are the initial state of the device and the states representing possible outcomes of the measurement.

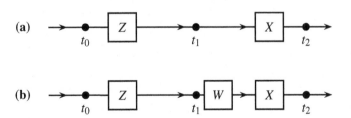

Fig. 18.4. Spin-half atom passing through successive nondestructive Stern–Gerlach devices.

Given the starting state

$$|\Psi_0\rangle = |x^+\rangle |Z^\circ\rangle |X^\circ\rangle \tag{18.29}$$

at t_0, and that at t_2 the detectors are in the states Z^+ and X^+, what can one say about the spin of the atom at the time t_1 when it is midway between the two devices? A relatively coarse family whose support is the four histories

$$\Psi_0 \odot I \odot \{Z^+X^+,\ Z^+X^-,\ Z^-X^+,\ Z^-X^-\} \tag{18.30}$$

is useful for representing the initial data (see Sec. 16.1) of Ψ_0 at t_0 and Z^+X^+ at t_2.

The consistent family (18.30) can be refined in various ways. One possibility is to include information about S_z at t_1:

$$\Psi_0 \odot \begin{cases} [z^+] \odot \{Z^+X^+,\ Z^+X^-\}, \\ [z^-] \odot \{Z^-X^+,\ Z^-X^-\}. \end{cases} \tag{18.31}$$

Using this family one sees that

$$\Pr([z^+]_1 \mid Z_2^+ X_2^+) = 1, \tag{18.32}$$

so that the initial data imply that $S_z = +1/2$ at t_1. A different refinement includes information about S_x at t_1:

$$\Psi_0 \odot \begin{cases} [x^+] \odot \{Z^+X^+,\ Z^-X^+\}, \\ [x^-] \odot \{Z^+X^-,\ Z^-X^-\}. \end{cases} \tag{18.33}$$

Using it one finds that

$$\Pr([x^+]_1 \mid Z_2^+ X_2^+) = 1, \tag{18.34}$$

so that in this framework the initial data imply that $S_x = +1/2$ at t_1. Since $[z^+]$ and $[x^+]$ do not commute, the frameworks (18.31) and (18.33) are incompatible, and the results (18.32) and (18.34) cannot be combined, even though each is correct in its own framework.

There is, of course, no experimental arrangement by means of which either (18.32) or (18.34) can be checked directly at the precise time t_1. The closest one can come is to insert another device W, as shown in Fig. 18.4(b), which carries out a nondestructive Stern–Gerlach measurement of S_w for some direction w at a time shortly after t_1. First consider the case $w = z$, so that the W apparatus repeats the measurement of the initial Z apparatus. One can show — the reader can easily work out the details — that with $w = z$, the Z and W devices have identical outcomes: Z^+W^+ or Z^-W^-. Thus if at t_2, when the atom has passed through all three devices, Z is in the state Z^+, W will be in the state W^+. This is precisely what one would have anticipated on the basis of (18.32): the property $S_z = +1/2$ at t_1 was confirmed by the W measurement a short time later. In this sense the W device with $w = z$ confirms the correctness of a conclusion reached on the basis of the consistent family in (18.31). On the other hand, if $w = x$, so that the W apparatus measures S_x, a similar analysis shows that the X and W devices must have identical outcomes. In particular, if at t_2 X is in the state X^+, W will be in the state W^+, and this confirms the correctness of (18.34). Since the device W must have its field gradient (the gradient in the first magnet in Fig. 18.2) in a particular direction, it is obvious that in a particular experimental run w is either in the x or in the z direction, and cannot be in both directions simultaneously. The situation is thus similar to what we found in the previous example: a classical macroscopic incompatibility of the two measurement possibilities reflects the quantum incompatibility of the two frameworks (18.31) and (18.33).

How can we know that at time t_1 the atom had the property revealed a bit later by the spin measurement carried out by W? The answer to this question is the same as for its analog in the previous example. Quantum theory itself provides a consistent description of the situation, including the relevant connection between a property of the atom before a measurement takes place and the outcome of the measurement. One must, of course, employ an appropriate framework for this connection to be evident. For example, in the case $w = x$ one should use a consistent family with $[x^+]$ and $[x^-]$ at time t_1, for a family with $[z^+]$ and $[z^-]$ at t_1 cannot, obviously, be used to discuss the value of S_x.

There is, however, another concern which did not arise in the previous example using the beam splitter. The device W in Fig. 18.4(b) is located where it might conceivably disturb the later S_x measurement carried out by X. Can we say that the outcome of the latter, X^+ or X^-, is the same as it would have been, for this particular experimental run, had the apparatus W been absent, as in Fig. 18.4(a)? This is a *counterfactual* question: given a situation in which W is in fact present, it asks what *would* have happened *if*, contrary to fact, W had been absent. Answering counterfactual questions requires a further development, found in Sec. 19.4, of the principles of quantum reasoning discussed in Ch. 16. By using it one can argue that both for the case $w = x$ and also for the case $w = z$, had W been absent the X measurement outcome would have been the same.

18.5 General nondestructive measurements

In Sec. 17.5 we discussed a fairly general scheme for measurements, in general destructive, of the properties of a quantum system S corresponding to an orthonormal basis $\{|s^k\rangle\}$, by a measuring apparatus $\mathcal{M}$ initially in the state $|M_0\rangle$. To construct a corresponding description of nondestructive measurements, suppose that the unitary time development from t_0 to t_1 to t_2 corresponding to (17.30) is of the form

$$|s^k\rangle \otimes |M_0\rangle \mapsto |s^k\rangle \otimes |M_1\rangle \mapsto |s^k\rangle \otimes |M^k\rangle, \tag{18.35}$$

where the interaction between $\mathcal{S}$ and $\mathcal{M}$ takes place during the time interval from t_1 to t_2, and $\{|M^k\rangle\}$ is an orthonormal collection of states of $\mathcal{M}$ corresponding to the different measurement outcomes.

Given some initial state $|s_0\rangle$ which is a linear combination of the $\{|s^k\rangle\}$, (17.32), one can set up a consistent family analogous to (17.36) with support

$$\Psi_0 \odot \begin{cases} [s^1] \odot [s^1] \otimes M^1, \\ [s^2] \odot [s^2] \otimes M^2, \\ \ldots \\ [s^n] \odot [s^n] \otimes M^n, \end{cases} \tag{18.36}$$

where $|\Psi_0\rangle$ is the state $|s_0\rangle|M_0\rangle$. Using this family one can show that M^k at t_2 implies s^k at t_1 — (17.37) is valid with N^k replaced by M^k — and, in addition,

$$\Pr(s_2^j \mid M_2^k) = \delta_{jk} = \Pr(s_2^j \mid s_1^k). \tag{18.37}$$

The first equality tells us that if at t_2 the measurement outcome is M^k, the system $\mathcal{S}$ at this time is in the state $|s^k\rangle$, whereas the second shows that this measurement is nondestructive for the properties $\{[s^j]\}$.

Provided $\mathcal{S}$ and $\mathcal{M}$ do not interact for $t > t_2$, the later time development of $\mathcal{S}$ (e.g., what will happen if it interacts with a second measuring apparatus $\mathcal{M}'$) can be discussed using the method of conditional density matrices described in Sec. 15.7, with appropriate changes in notation: t_0, $\mathcal{A}$, and $\mathcal{B}$ of Sec. 15.7 become t_2, $\mathcal{S}$, and $\mathcal{M}$. Given a measurement outcome M^k, the corresponding conditional density matrix, see (15.61), is

$$\rho^k = [s^k], \tag{18.38}$$

and this can be used (typically as a pre-probability) as an initial state for the further time development of system $\mathcal{S}$. (If there is a second measuring apparatus $\mathcal{M}'$, one must, of course, also specify its initial state.)

One can also formulate a nondestructive counterpart to the measurement of a general decomposition of the identity $I_S = \sum_k S^k$, (17.38), discussed in Sec. 17.5. Let the orthonormal basis $\{|s^{kl}\rangle\}$ be chosen so that $S^k = \sum_l [s^{kl}]$, (17.40), and assume a unitary time development

$$|s^{kl}\rangle \otimes |M_0\rangle \mapsto |s^{kl}\rangle \otimes |M_1\rangle \mapsto |s^{kl}\rangle \otimes |M^k\rangle, \tag{18.39}$$

where the apparatus state $|M^k\rangle$ corresponding to the kth outcome is assumed *not to depend upon l.* The counterpart of (17.43) is a consistent family with support

$$
\Psi_0 \odot \begin{cases} S^1 \odot S^1 \otimes M^1, \\ S^2 \odot S^2 \otimes M^2, \\ \cdots \\ S^n \odot S^n \otimes M^n, \end{cases} \tag{18.40}
$$

and it yields conditional probabilities

$$
\Pr(S_2^j \mid M_2^k) = \delta_{jk} = \Pr(S_2^j \mid S_1^k) \tag{18.41}
$$

that are the obvious counterpart of (18.37). In addition, the outcome M^k at t_2 implies the property S^k at t_1: (17.45) holds with N^k replaced by M^k.

It is possible to refine (18.40) to give a more precise description at t_2. Define

$$
|\sigma^k\rangle := S^k |s_0\rangle = \sum_l c_{kl} |s^{kl}\rangle, \tag{18.42}
$$

using the expression (17.44) for $|s_0\rangle$. Then the unitary time development in (18.39) implies that

$$
T(t_2, t_0)\big(|s_0\rangle \otimes |M_0\rangle\big) = \sum_k |\sigma^k\rangle \otimes |M^k\rangle. \tag{18.43}
$$

As a consequence, the histories $\Psi_0 \odot S^k \odot (I - [\sigma^k]) \otimes M^k$ have zero weight, and

$$
\Psi_0 \odot \begin{cases} S^1 \odot [\sigma^1] \otimes M^1, \\ S^2 \odot [\sigma^2] \otimes M^2, \\ \cdots \\ S^n \odot [\sigma^n] \otimes M^n \end{cases} \tag{18.44}
$$

is again the support of a consistent family. Indeed, one can produce an even finer family by replacing each S^k at t_1 with the corresponding $[\sigma^k]$.

In order to describe the later time development of $\mathcal{S}$, assuming no further interaction with $\mathcal{M}$ for $t > t_2$, one can again employ the method of conditional density matrices of Sec. 15.7, with

$$
\rho^k = [\sigma^k] \tag{18.45}
$$

at time t_2 corresponding to the measurement outcome M^k. If $\mathcal{S}$ is described by a density matrix ρ_0 at t_0, the corresponding result

$$
\rho^k = S^k \rho_0 S^k / \operatorname{Tr}(S^k \rho_0 S^k) \tag{18.46}
$$

is known as the *Lüders rule*. Note that the validity of both (18.41) and (18.42) depends on some fairly specific assumptions. If, for example, one were to suppose that

$$|s^{kl}\rangle \otimes |M_0\rangle \mapsto |s^{kl}\rangle \otimes |M_1\rangle \mapsto |s^{kl}\rangle \otimes |M^{kl}\rangle, \tag{18.47}$$

with the $\{|M^{kl}\rangle\}$ for different k and l an orthonormal collection, and define

$$M^k = \sum_l [M^{kl}] \tag{18.48}$$

as the projector corresponding to the kth measurement outcome, (18.41) would still be valid, but neither (18.45) nor (18.46) would (in general) be correct.

The results in this section, like those in Sec. 17.5, can be generalized to the case of a macroscopic measuring apparatus using the approaches discussed in Sec. 17.4.

19

Coins and counterfactuals

19.1 Quantum paradoxes

The next few chapters are devoted to resolving a number of quantum paradoxes in the sense of giving a reasonable explanation of a seemingly paradoxical result in terms of the principles of quantum theory discussed earlier in this book. None of these paradoxes indicates a defect in quantum theory. Instead, when they have been properly understood, they show us that the quantum world is rather different from the world of our everyday experience and of classical physics, in a way somewhat analogous to that in which relativity theory has shown us that the laws appropriate for describing the behavior of objects moving at high speed differ in significant ways from those of pre-relativistic physics.

An inadequate theory of quantum measurements is at the root of several quantum paradoxes. In particular, the notion that wave function collapse is a physical effect produced by a measurement, rather than a method of calculation, see Sec. 18.2, has given rise to a certain amount of confusion. Smuggling rules for classical reasoning into the quantum domain where they do not belong and where they give rise to logical inconsistencies is another common source of confusion. In particular, many paradoxes involve mixing the results from incompatible quantum frameworks.

Certain quantum paradoxes have given rise to the idea that the quantum world is permeated by mysterious influences that propagate faster than the speed of light, in conflict with the theory of relativity. They are mysterious in that they cannot be used to transmit signals, which means that they are, at least in any direct sense, experimentally unobservable. While relativistic quantum theory is outside the scope of this book, an analysis of nonrelativistic versions of some of the paradoxes which are supposed to show the presence of superluminal influences indicates that the real source of such ghostly effects is the need to correct logical errors arising from the assumption that the quantum world is behaving in some respects in a classical way. When the situation is studied using consistent quantum principles, the ghosts

261

disappear, and with them the corresponding difficulty in reconciling quantum mechanics with relativity theory. The reason why ghostly influences cannot be used to transmit signals faster than the speed of light is then obvious: there are no such influences.

Some quantum paradoxes are stated in a way that involves a free choice on the part of a human observer: e.g., whether to measure the x or the z component of spin angular momentum of some particle. Since the principles of quantum theory as treated in this book apply to a *closed system*, with all parts of it subject to quantum laws, a complete discussion of such paradoxes would require including the human observer as part of the quantum system, and using a quantum model of conscious human choice. This would be rather difficult to do given the current primitive state of scientific understanding of human consciousness. Fortunately, for most quantum paradoxes it seems possible to evade the issue of human consciousness by letting the outcome of a quantum coin toss "decide" what will be measured. As discussed in Sec. 19.2, the quantum coin is a purely physical device connected to a suitable servomechanism. By this means the stochastic nature of quantum mechanics can be used as a tool to model something which is indeterminate, which cannot be known in advance.

Certain quantum paradoxes are stated in terms of *counterfactuals*: what *would* have happened *if* some state of affairs had been different from what it actually was. Other paradoxes have both a counterfactual as well as in an "ordinary" form. In order to discuss counterfactual quantum paradoxes, one needs a quantum version of counterfactual reasoning. Unfortunately, philosophers and logicians have yet to reach agreement on what constitutes valid counterfactual reasoning in the classical domain. Our strategy will be to avoid the difficult problems which perplex the philosophers, such as "Would a kangaroo topple if it had no tail?", and focus on a rather select group of counterfactual questions which arise in a probabilistic context. These are of the general form: "What would have happened if the coin flip had resulted in heads rather than tails?" They are considered first from a classical (or everyday world) perspective in Sec. 19.3, and then translated into quantum terms in Sec. 19.4.

19.2 Quantum coins

In a world governed by classical determinism there are no truly random events. But quantum mechanics allows for events which are irreducibly probabilistic. For example, a photon is sent into a beam splitter and detected by one of two detectors situated on the two output channels. Quantum theory allows us to assign a probability that one detector or the other will detect the photon, but provides no deterministic prediction of which detector will do so in any particular realization of the experiment. This system generates a random output in the same way as tossing a coin, which is why it is reasonable to call it a quantum coin. One can arrange things so that the probabilities for the two outcomes are not the same, or so that there are three or even more random outcomes, with equal or unequal probabilities. We shall use the term "quantum coin" to refer to any such device, and "quantum coin toss" to refer to the corresponding stochastic process. There is no reason in principle why various experiments involving statistical sampling (such as drug trials) should not be carried out using the "genuine randomness" of quantum coins.

To illustrate the sort of thing we have in mind, consider the gedanken experiment in Fig. 19.1, in which a particle, initially in a wave packet $|0a\rangle$, is approaching a point P where a beam splitter B may or may not be located depending upon the outcome of tossing a quantum coin Q shortly before the particle arrives at P. If the outcome of the toss is Q', the beam splitter is left in place at B', whereas if it is Q'', a servomechanism rapidly moves the beam splitter to B'' out of the path of the particle, which continues in a straight line.

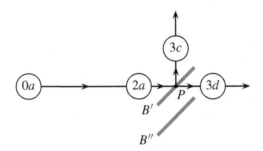

Fig. 19.1. Particle paths approaching and leaving a beam splitter which is either left in place, B', or moved out of the way, B'', before the arrival of the particle.

Let us describe this in quantum terms in the following way. Suppose that $|Q\rangle$ is the initial state of the quantum coin and the attached servomechanism at time t_0, and that between t_0 and t_1 there is a unitary time evolution

$$|Q\rangle \mapsto (|Q'\rangle + |Q''\rangle)/\sqrt{2}. \qquad (19.1)$$

Next, let $|B'\rangle$ and $|B''\rangle$ be states corresponding to the beam splitter being either left in place or moved out of the path of the particle, and assume a unitary time evolution

$$|Q'\rangle|B'\rangle \mapsto |Q'\rangle|B'\rangle, \quad |Q''\rangle|B'\rangle \mapsto |Q''\rangle|B''\rangle \qquad (19.2)$$

between t_1 and t_2. Finally, the motion of the particle from t_2 to t_3 is governed by

$$|2a\rangle|B'\rangle \mapsto (|3c\rangle + |3d\rangle)|B'\rangle/\sqrt{2}, \quad |2a\rangle|B''\rangle \mapsto |3d\rangle|B''\rangle, \qquad (19.3)$$

where $|2a\rangle$ is a wave packet on path a for the particle at time t_2, and a similar notation is used for wave packets on paths c and d in Fig. 19.1. The overall unitary time evolution of the system consisting of the particle, the quantum coin, and the apparatus during the time interval from t_0 until t_3 takes the form

$$
\begin{aligned}
|\Psi_0\rangle = |0a\rangle \otimes |Q\rangle|B'\rangle &\mapsto |1a\rangle \otimes \big(|Q'\rangle + |Q''\rangle\big)|B'\rangle/\sqrt{2} \\
&\mapsto |2a\rangle \otimes \big(|Q'\rangle|B'\rangle + |Q''\rangle|B''\rangle\big)/\sqrt{2} \\
&\mapsto \big(|3c\rangle + |3d\rangle\big) \otimes |Q'\rangle|B'\rangle/2 + |3d\rangle \otimes |Q''\rangle|B''\rangle/\sqrt{2}, \quad (19.4)
\end{aligned}
$$

where $\otimes$ helps to set the particle off from the rest of the quantum state.

There are reasons, discussed in Sec. 17.4, why macroscopic objects are best described not with individual kets but with macro projectors, or statistical distributions or density matrices. The use of kets is not misleading, however, and it makes the reasoning somewhat simpler. With a little effort — again, see Sec. 17.4 — one can reconstruct arguments of the sort we shall be considering so that macroscopic properties are represented by macro projectors. While we will continue to use the simpler arguments, projectors representing macroscopic properties will be denoted by symbols without square brackets, as in (19.5), so that the formulas remain unchanged in a more sophisticated analysis.

Consider the consistent family for the times $t_0 < t_1 < t_2 < t_3$ with support consisting of the two histories

$$\Psi_0 \odot \begin{cases} Q' \odot B' \odot [3\bar{a}], \\ Q'' \odot B'' \odot [3d], \end{cases} \qquad (19.5)$$

where

$$|3\bar{a}\rangle := \big(|3c\rangle + |3d\rangle\big)/\sqrt{2}. \qquad (19.6)$$

It allows one to say that if the quantum coin outcome is Q', then the particle is later in the coherent superposition state $|3\bar{a}\rangle$, a state which could be detected by bringing the beams back together again and passing them through a second beam splitter, as in Fig. 18.3(c). On the other hand, if the outcome is Q'', then the particle will later be in channel d in a wave packet $|3d\rangle$. As $[3\bar{a}]$ and $[3d]$ do not commute with each other, it is clear that these final states in (19.5) are dependent, in the sense discussed in Ch. 14, either upon the earlier beam splitter locations $|B'\rangle$ and $|B''\rangle$, or the still earlier outcomes $|Q'\rangle$ and $|Q''\rangle$ of the quantum coin toss.

The expressions in (19.4) are a bit cumbersome, and the same effect can be achieved with a somewhat simpler notation in which (19.1) and (19.2) are replaced by the single expression

$$|B_0\rangle \mapsto \big(|B'\rangle + |B''\rangle\big)/\sqrt{2}, \tag{19.7}$$

where $|B_0\rangle$ is the initial state of the entire apparatus, including the quantum coin and the beam splitter, whereas $|B'\rangle$ and $|B''\rangle$ are apparatus states in which the beam splitter is at the locations B' and B'' indicated in Fig. 19.1. The time development of the particle in interaction with the beam splitter is given, as before, by (19.3).

19.3 Stochastic counterfactuals

A workman falls from a scaffolding, but is caught by a safety net, so he is not injured. What *would* have happened *if* the safety net had not been present? This is an example of a *counterfactual* question, where one has to imagine something different from what actually exists, and then draw some conclusion. Answering it involves counterfactual reasoning, which is employed all the time in the everyday world, though it is still not entirely understood by philosophers and logicians. In essence it involves comparing two or more possible states-of-affairs, often referred to as "worlds", which are similar in certain respects and differ in others. In the example just considered, a world in which the safety net is present is compared to a world in which it is absent, while both worlds have in common the feature that the workman falls from the scaffolding.

We begin our study of counterfactual reasoning by looking at a scheme which is able to address a limited class of counterfactual questions in a *classical* but *stochastic* world, that is, one in which there is a random element added to classical dynamics. The world of everyday experience is such a world, since classical physics gives deterministic answers to some questions, but there are others, e.g., "What will the weather be two weeks from now?", for which only probabilistic answers are available.

Shall we play badminton or tennis this afternoon? Let us toss a coin: H (heads) for badminton, T (tails) for tennis. The coin turns up T, so we play tennis. What *would* have happened *if* the result of the coin toss had been H? It is useful to introduce a diagrammatic way of representing the question and deriving an answer, Fig. 19.2. The node at the left at time t_1 represents the situation before the coin toss, and the two nodes at t_2 are the mutually-exclusive possibilities resulting from that toss. The lower branch represents what actually occurred: the toss resulted in T and a game of tennis. To answer the question of what would have happened if the coin had turned up the other way, we start from the node representing what actually happened, go backwards in time to the node preceding the coin toss, which we shall call the *pivot*, and then forwards along the alternative branch to arrive at the badminton game. This type of counterfactual reasoning can be thought of as comparing histories in two "worlds" which are identical at all times up to and including the pivot point t_1 at which the coin is tossed. After that, one of these worlds contains the outcome H and the consequences which flow from this, including a game of badminton, while the other world contains the outcome T and its consequences.

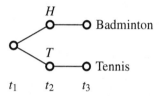

Fig. 19.2. Diagram for counterfactual analysis of a coin toss.

It is instructive to embed the preceding example in a slightly more complicated situation. Let us suppose that the choice between tennis or badminton was preceded by another: should we go visit the museum, or get some exercise? Once again, imagine the decision being made by tossing a coin at time t_0, with H leading to exercise and T to a museum visit. At the museum a choice between visiting one of two exhibits can also be carried out by tossing a coin. The set of possibilities is shown in Fig. 19.3. Suppose that the actual sequence of the two coins was H_1T_2, leading to tennis. If the first coin toss had resulted in T_1 rather than H_1, what would have happened? Start from the tennis node in Fig. 19.3, go back to the pivot node P_0 at t_0 preceding the first coin toss, and then forwards on the alternative, T_1 branch. This time there is not a unique possibility, for the second coin toss could have been either H_2 or T_2. Thus the appropriate answer would be: Had the first coin toss resulted in T_1, we would have gone to one or the other of the two exhibits at the museum, each possibility having probability $1/2$. That counterfactual questions have probabilistic answers is just what one would expect if the dynamics describing the situation is stochastic, rather than deterministic. The answer is deterministic only in the limiting cases of probabilities equal to 1 or 0.

However, a somewhat surprising feature of stochastic counterfactual reasoning comes to light if we ask the question, again assuming the afternoon was devoted to tennis, "What would have happened if the first coin had turned up H_1 (as it actually did)?", and attempt to answer it using the diagram in Fig. 19.3. Let us call this a *null counterfactual* question, since it asks not what would have happened if the world had been different in some way, but what would have happened if the world

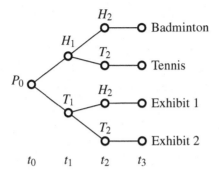

Fig. 19.3. Diagram for analyzing two successive coin tosses.

had been the same in this particular respect. The answer obtained by tracing from "tennis" backwards to P_0 in Fig. 19.3 and then forwards again along the upper, or H_1 branch, is not tennis, but it is badminton or tennis, each with probability $1/2$. We do not, in other words, reach the conclusion that what actually happened would have happened had the world been the same in respect to the outcome of the first coin toss. Is it reasonable to have a stochastic answer, with probability less than 1, for a null counterfactual question? Yes, because to have a deterministic answer would be to specify implicitly that the second coin toss turned out the way it actually did. But in a world which is not deterministic there is no reason why random events should not have turned out differently.

Counterfactual questions are sometimes ambiguous because there is more than one possibility for a pivot. For example, "What would we have done if we had not played tennis this afternoon?" will be answered in a different way depending upon whether H_1 or P_0 in Fig. 19.3 is used as the pivot. In order to make a counterfactual question precise, one must specify both a framework of possibilities, as in Fig. 19.3, and also a pivot, the point at which the actual and counterfactual worlds, identical at earlier times, "split apart".

This method of reasoning is useful for answering some types of counterfactual questions but not others. Even to use it for the case of a workman whose fall is broken by a safety net requires an exercise in imagination. Let us suppose that just after the workman started to fall (the pivot), the safety net was swiftly removed, or left in place, depending upon some rapid electronic coin toss, so that the situation could be represented in a diagram similar to Fig. 19.2. Is this an adequate, or at least a useful, way of thinking about this counterfactual question? At least it represents a way to get started, and we shall employ the same idea for quantum counterfactuals.

19.4 Quantum counterfactuals

Counterfactuals have played an important role in discussions of quantum measurements. Thus a perennial question in the foundations of quantum theory is whether measurements reveal pre-existing properties of a measured system, or whether they somehow "create" such properties. Suppose, to take an example, that a Stern–Gerlach measurement reveals the value $S_x = 1/2$ for a spin-half particle. Would the particle have had the same value of S_x even if the measurement had not been made? An interpretation of quantum theory which gives a "yes" answer to this counterfactual question can be said to be *realistic* in that it affirms the existence of certain properties or events in the world independent of whether measurements are made. (For some comments on realism in quantum theory, see Ch. 27.) Another similar counterfactual question is the following: Given that the S_x measurement outcome indicates, using an appropriate framework (see Ch. 17), that the value of S_x was $+1/2$ before the measurement, would this still have been the case if S_z had been measured instead of S_x?

The system of quantum counterfactual reasoning presented here is designed to answer these and similar questions. It is quite similar to that introduced in the previous section for addressing classical counterfactual questions. It makes use of quantum coins of the sort discussed in Sec. 19.2, and diagrams like those in Figs. 19.2 and 19.3. The nodes in these diagrams represent events in a consistent family of quantum histories, and nodes connected by lines indicate the histories with finite weight that form the support of the family. We require that the family be consistent, and that *all* the histories in the diagram belong to the *same* consistent family. This is a *single-framework* rule for quantum counterfactual reasoning comparable to the one discussed in Sec. 16.1 for ordinary quantum reasoning.

Let us see how this works in the case in which S_x is the component of spin actually measured for a spin-half particle, and we are interested in what would have been the case if S_z had been measured instead. Imagine a Stern–Gerlach apparatus of the sort discussed in Sec. 17.2 or Sec. 18.3, arranged so that it can be rotated about an axis (in the manner indicated in Sec. 18.3) to measure either S_x or S_z. When ready to measure S_x its initial state is $|X^\circ\rangle$, and its interaction with the particle results in the unitary time development

$$|x^+\rangle \otimes |X^\circ\rangle \mapsto |X^+\rangle, \quad |x^-\rangle \otimes |X^\circ\rangle \mapsto |X^-\rangle. \tag{19.8}$$

Similarly, when oriented to measure S_z the initial state is $|Z^\circ\rangle$, and the corresponding time development is

$$|z^+\rangle \otimes |Z^\circ\rangle \mapsto |Z^+\rangle, \quad |z^-\rangle \otimes |Z^\circ\rangle \mapsto |Z^-\rangle. \tag{19.9}$$

The symbols X°, etc., without square brackets will be used to denote the corresponding projectors. Because they refer to macroscopically distinct states, all the

Z projectors are orthogonal to all the X projectors: $X^+ Z^+ = 0$, etc. Without loss of generality we can consider the quantum coin and the associated servomechanism to be part of the Stern–Gerlach apparatus, which is initially in the state $|A\rangle$, with the coin toss corresponding to a unitary time development

$$|A\rangle \mapsto (|X^\circ\rangle + |Z^\circ\rangle)/\sqrt{2}. \tag{19.10}$$

Assume that the spin-half particle is prepared in an initial state $|w^+\rangle$, where the exact choice of w is not important for the following discussion, provided it is not $+x, -x, +z$, or $-z$. Suppose that X^+ is observed: the quantum coin resulted in the apparatus state X° appropriate for a measurement of S_x, and the outcome of the measurement corresponds to $S_x = +1/2$. What would have happened if the quantum coin toss had, instead, resulted in the apparatus state Z° appropriate for a measurement of S_z? To address this question we must adopt some consistent family and identify the event which serves as the pivot. As in other examples of quantum reasoning, there is more than one possible family, and the answer given to a counterfactual question can depend upon which family one uses. Let us begin with a family whose support consists of the four histories

$$\Psi_0 \odot I \odot \begin{cases} X^\circ \odot \begin{cases} X^+, \\ X^-, \end{cases} \\ Z^\circ \odot \begin{cases} Z^+, \\ Z^-, \end{cases} \end{cases} \tag{19.11}$$

at the times $t_0 < t_1 < t_2 < t_3$, where $|\Psi_0\rangle = |w^+\rangle \otimes |A\rangle$ is the initial state. It is represented in Fig. 19.4 in a diagram resembling those in Figs. 19.2 and 19.3. The quantum coin toss (19.10) takes place between t_1 and t_2. The particle reaches the Stern–Gerlach apparatus and the measurement occurs between t_2 and t_3, and at t_3 the outcome of the measurement is indicated by one of the four pointer states (end of Sec. 9.5) $X^\pm$, $Z^\pm$. Notice that only the first branching in Fig. 19.4, between t_1 and t_2, corresponds to the alternative outcomes of the quantum coin toss, while the later branching is due to other stochastic quantum processes.

Suppose S_x was measured with the result X^+. To answer the question of what would have occurred if S_z had been measured instead, start with the X^+ vertex in Fig. 19.4, trace the history back to I at t_1 (or Ψ_0 at t_0) as a pivot, and then go forwards on the lower branch of the diagram through the Z° node. The answer is that one of the two outcomes Z^+ or Z^- would have occurred, each possibility having a positive probability which depends on w, which seems reasonable. Rather than using the nodes in Fig. 19.4, one can equally well use the support of the consistent family written in the form (19.11), as there is an obvious correspondence between the nodes in the former and positions of the projectors in the latter. From

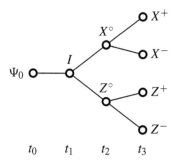

Fig. 19.4. Diagram for counterfactual analysis of the family (19.11).

now on we will base counterfactual reasoning on expressions of the form (19.11), interpreted as diagrams with nodes and lines in the fashion indicated in Fig. 19.4.

Now ask a different question. Assuming, once again, that X^+ was the actual outcome, what would have happened if the quantum coin had resulted (as it actually did) in X° and thus a measurement of S_x? To answer this null counterfactual question, we once again trace the actual history in (19.11) or Fig. 19.4 backwards from X^+ at t_3 to the I or the Ψ_0 node, and then forwards again along the upper branch through the X° node at t_2, since we are imagining a world in which the quantum coin toss had the same result as in the actual world. The answer to the question is that either X^+ or X^- would have occurred, each possibility having some positive probability. Since quantum dynamics is intrinsically stochastic in ways which are not limited to a quantum coin toss, there is no reason to suppose that what actually did occur, X^+, would necessarily have occurred, given only that we suppose the same outcome, X° rather than Z°, for the coin toss.

Nevertheless, it is possible to obtain a more definitive answer to this null counterfactual question by using a different consistent family with support

$$\Psi_0 \odot \begin{cases} [x^+] \odot \begin{cases} X^\circ \odot X^+, \\ Z^\circ \odot U^+, \end{cases} \\ [x^-] \odot \begin{cases} X^\circ \odot X^-, \\ Z^\circ \odot U^-, \end{cases} \end{cases} \tag{19.12}$$

where the nodes $[x^\pm]$ at t_1, a time which precedes the quantum coin toss, correspond to the spin states $S_x = \pm 1/2$, and U^+ and U^- are defined in the next paragraph. The history which results in X^+ can be traced back to the pivot $[x^+]$, and then forwards again along the same (upper) branch, since we are assuming that the quantum coin toss in the alternative (counterfactual) world did result in the X° apparatus state. The result is X^+ with probability 1. That this is reasonable can be seen in the following way. The actual measurement outcome X^+ shows that the particle had $S_x = +1/2$ at time t_1 before the measurement took place, since quantum measurements reveal pre-existing values if one employs a suitable framework. And by choosing $[x^+]$ at t_1 as the pivot, one is assuming that S_x had the same value at this time in both the actual and the counterfactual world. Therefore a later measurement of S_x in the counterfactual world would necessarily result in X^+.

However, we find something odd if we use (19.12) to answer our earlier counterfactual question of what would have happened if S_z had been measured rather than S_x. Tracing the actual history backwards from X^+ to $[x^+]$ and then forwards along the lower branch in the upper part of (19.12), through Z°, we reach U^+ at t_3 rather than the pair Z^+, Z^-, as in (19.11) or Fig. 19.4. Here U^+ is a projector on the state $|U^+\rangle$ obtained by *unitary* time evolution of $|x^+\rangle|Z^\circ\rangle$ using (19.9):

$$|x^+\rangle|Z^\circ\rangle = (|z^+\rangle + |z^-\rangle)|Z^\circ\rangle/\sqrt{2} \mapsto |U^+\rangle = (|Z^+\rangle + |Z^-\rangle)/\sqrt{2}. \quad (19.13)$$

Similarly, U^- in (19.12) projects on the state obtained by unitary time evolution of $|x^-\rangle|Z^\circ\rangle$. Both U^+ and U^- are macroscopic quantum superposition (MQS) states. The appearance of these MQS states in (19.12) reflects the need to construct a family satisfying the consistency conditions, which would be violated were we to use the pointer states Z^+ and Z^- at t_3 following the Z° nodes at t_2. The fact that consistency conditions sometimes require MQS states rather than pointer states is significant for analyzing certain quantum paradoxes, as we shall see in later chapters.

The contrasting results obtained using the families in (19.11) and (19.12) illustrate an important feature of quantum counterfactual reasoning of the type we are discussing: the outcome depends upon the family of histories which is used, and also upon the pivot. In order to employ the pivot $[x^+]$ rather than I at t_1, it is necessary to use a family in which the former occurs, and it cannot simply be added to the family (19.11) by a process of refinement. To be sure, this dependence upon the framework and pivot is not limited to the quantum case; it also arises for classical stochastic counterfactual reasoning. However, in a classical situation the framework is a classical sample space with its associated event algebra, and framework dependence is rather trivial. One can always, if necessary, refine the sample space, which corresponds to adding more nodes to a diagram such as Fig. 19.3, and there is never a problem with incompatibility or MQS states.

Consider a somewhat different question. Suppose the actual measurement outcome corresponds to $S_x = +1/2$. Would S_x have had the same value if no measurement had been carried out? To address this question, we employ an obvious modification of the previous gedanken experiment, in which the quantum coin leads either to a measurement of S_x, as actually occurred, or to no measurement at all, by swinging the apparatus out of the way before the arrival of the particle. Let

$|N\rangle$ denote the state of the apparatus when it has been swung out of the way. An appropriate consistent family is one with support

$$\Psi_0 \odot \begin{cases} [x^+] \odot \begin{cases} X^\circ \odot X^+, \\ N \odot [x^+], \end{cases} \\ [x^-] \odot \begin{cases} X^\circ \odot X^-, \\ N \odot [x^-]. \end{cases} \end{cases} \tag{19.14}$$

It resembles (19.12), but with Z° replaced by N, U^+ by $[x^+]$, and U^- by $[x^-]$, since if no measuring apparatus is present, the particle continues on its way in the same spin state.

We can use this family and the node $[x^+]$ at time t_1 to answer the question of what would have happened in a case in which the measurement result was $S_x = +1/2$ if, contrary to fact, no measurement had been made. Start with the X^+ node at t_3, trace it back to $[x^+]$ at t_1, and then forwards in time through the N node at t_2. The result is $[x^+]$, so the particle would have been in the state $S_x = +1/2$ at t_1 and at later times if no measurement had been made.

20

Delayed choice paradox

20.1 Statement of the paradox

Consider the Mach–Zehnder interferometer shown in Fig. 20.1. The second beam splitter can either be at its regular position B_{in} where the beams from the two mirrors intersect, as in (a), or moved out of the way to a position B_{out}, as in (b). When the beam splitter is in place, interference effects mean that a photon which enters the interferometer through channel a will always emerge in channel f to be measured by a detector F. On the other hand, when the beam splitter is out of the way, the probability is 1/2 that the photon will be detected by detector E, and 1/2 that it will be detected by detector F.

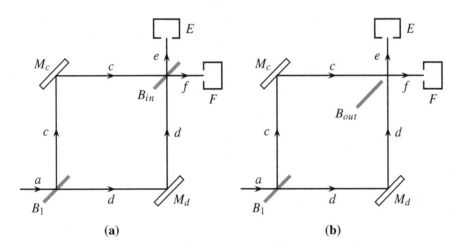

Fig. 20.1. Mach–Zehnder interferometer with the second beam splitter (a) in place, (b) moved out of the way.

The paradox is constructed in the following way. Suppose that the beam splitter is out of the way, Fig. 20.1(b), and the photon is detected in E. Then it seems

273

plausible that the photon was earlier in the d arm of the interferometer. For example, were the mirror M_d to be removed, no photons would arrive at E; if the length of the path in the d arm were doubled by using additional mirrors, the photon would arrive at E with a time delay, etc. On the other hand, when the beam splitter is in place, we understand the fact that the photon always arrives at F as due to an interference effect arising from a coherent superposition state of photon wave packets in both arms c and d. That this is the correct explanation can be supported by placing phase shifters in the two arms, Sec. 13.2, and observing that the phase *difference* must be kept constant in order for the photon to always be detected in F. Similarly, removing either of the mirrors will spoil the interference effect.

Suppose, however, that the beam splitter is in place until just before the photon reaches it, and is then suddenly moved out of the way. What will happen? Since the photon does not interact with the beam splitter, we conclude that the situation is the same as if the beam splitter had been absent all along. If the photon arrives at E, then it was earlier in the d arm of the interferometer. But this seems strange, because if the beam splitter had been left in place, the photon would surely have been detected by F, which requires, as noted above, that inside the interferometer it is in a superposition state between the two arms. Hence it would seem that a later event, the position or absence of the beam splitter as decided at the very last moment before its arrival must somehow influence the earlier state of the photon, when it was in the interferometer far away from the beam splitter, and determine whether it is in one of the individual arms or in a superposition state. How can this be? Can the future influence the past?

The reader may be concerned that given the dimensions of a typical laboratory Mach–Zehnder interferometer and a photon moving with the speed of light, it would be physically impossible to shift the beam splitter out of the way while the photon is inside the interferometer. But we could imagine a very large interferometer constructed someplace out in space so as to allow time for the mechanical motion. Also, modified forms of the delayed choice experiment can be constructed in the laboratory using tricks involving photon polarization and fast electronic devices.

It is possible to state the paradox in counterfactual terms. Suppose the beam splitter is not in place and the photon is detected by E, indicating that it was earlier in the d arm of the interferometer. What *would* have occurred *if* the beam splitter had been in place? On the one hand, it seems reasonable to argue that the photon would certainly have been detected by F; after all, it is always detected by F when the beam splitter is in place. On the other hand, experience shows that if a photon arrives in the d channel and encounters the beam splitter, it has a probability of $1/2$ of emerging in either of the two exit channels. This second conclusion is hard to reconcile with the first.

20.2 Unitary dynamics

Let $|0a\rangle$ be the photon state at t_0 when the photon is in channel a, Fig. 20.1, just before entering the interferometer through the first (immovable) beam splitter, and let the unitary evolution up to a time t_1 be given by

$$|0a\rangle \mapsto |1\bar{a}\rangle := \big(|1c\rangle + |1d\rangle\big)/\sqrt{2}, \qquad (20.1)$$

where $|1c\rangle$ and $|1d\rangle$ are photon wave packets in the c and d arms of the interferometer. These in turn evolve unitarily,

$$|1c\rangle \mapsto |2c\rangle, \quad |1d\rangle \mapsto |2d\rangle, \qquad (20.2)$$

to wave packets $|2c\rangle$ and $|2d\rangle$ in the c and d arms at a time t_2 just before the photon reaches the second (movable) beam splitter.

What happens next depends upon whether this beam splitter is in or out. If it is in, then

$$B_{in}: \quad |2c\rangle \mapsto |3\bar{c}\rangle, \quad |2d\rangle \mapsto |3\bar{d}\rangle, \qquad (20.3)$$

where

$$|3\bar{c}\rangle := \big(|3e\rangle + |3f\rangle\big)/\sqrt{2}, \quad |3\bar{d}\rangle := \big(-|3e\rangle + |3f\rangle\big)/\sqrt{2}, \qquad (20.4)$$

and $|3e\rangle$ and $|3f\rangle$ are photon wave packets at time t_3 in the e and the f output channels. If the beam splitter is out, the behavior is rather simple:

$$B_{out}: \quad |2c\rangle \mapsto |3f\rangle, \quad |2d\rangle \mapsto |3e\rangle. \qquad (20.5)$$

Finally, the detection of the photon during the time interval from t_3 to t_4 is described by

$$|3e\rangle|E^\circ\rangle \mapsto |E^*\rangle, \quad |3f\rangle|F^\circ\rangle \mapsto |F^*\rangle. \qquad (20.6)$$

Here $|E^\circ\rangle$ and $|F^\circ\rangle$ are the ready states of the two detectors, and $|E^*\rangle$ and $|F^*\rangle$ the states in which a photon has been detected.

The overall time development starting with an initial state

$$|\Psi_0\rangle = |0a\rangle|E^\circ\rangle|F^\circ\rangle \qquad (20.7)$$

at time t_0 leads to a succession of states $|\Psi_j\rangle$ at time t_j. These can be worked out by putting together the different transformations indicated in (20.1)–(20.6), assuming the detectors do not change except for the processes indicated in (20.6). For $j \geq 2$ the result depends upon whether the (second) beam splitter is in or out. At t_4 with the beam splitter in one finds

$$B_{in}: \quad |\Psi_4\rangle = |E^\circ\rangle|F^*\rangle, \qquad (20.8)$$

whereas if the beam splitter is out, the result is a macroscopic quantum superposition (MQS) state

$$B_{out}: \quad |\Psi_4\rangle = |S^+\rangle := (|E^*\rangle|F^\circ\rangle + |E^\circ\rangle|F^*\rangle)/\sqrt{2}. \qquad (20.9)$$

A second MQS state

$$|S^-\rangle := (-|E^*\rangle|F^\circ\rangle + |E^\circ\rangle|F^*\rangle)/\sqrt{2}, \qquad (20.10)$$

orthogonal to $|S^+\rangle$, will be needed later.

20.3 Some consistent families

Let us first consider the case in which the beam splitter is out. Unitary evolution leading to the MQS state $|S^+\rangle$, (20.9), at t_4 obviously does not provide a satisfactory way to describe the outcome of the final measurement. Consequently, we begin by considering the consistent family whose support consists of the two histories

$$B_{out}: \quad \Psi_0 \odot [1\bar{a}] \odot [2\bar{a}] \odot [3\bar{c}] \odot \{E^*, F^*\} \qquad (20.11)$$

at the times $t_0 < t_1 < t_2 < t_3 < t_4$. Here and later we use symbols without square brackets for projectors corresponding to macroscopic properties; see the remarks in Sec. 19.2 following (19.4). This family resembles ones used for wave function collapse, Sec. 18.2, in that there is unitary time evolution preceding the measurement outcomes. For this reason, however, it does not allow us to make the inference required in the statement of the paradox in Sec. 20.1, that if the photon is detected by E (final state E^*), it was earlier in the d arm of the interferometer. Such an assertion at t_1 or t_2 is incompatible with $[1\bar{a}]$ or $[2\bar{a}]$, as these projectors do not commute with the projectors C, D for the photon to be in the c or the d arm. (For toy versions of C and D, see (12.9) in Sec. 12.1.) In order to translate the paradox into quantum mechanical terms we need to use a different consistent family, such as the one with support

$$B_{out}: \quad \Psi_0 \odot \begin{cases} [1c] \odot [2c] \odot [3f] \odot F^*, \\ [1d] \odot [2d] \odot [3e] \odot E^*. \end{cases} \qquad (20.12)$$

Each of these histories has weight $1/2$, and using this family one can infer that

$$B_{out}: \quad \Pr([1d]_1 \mid E_4^*) = \Pr([2d]_2 \mid E_4^*) = 1, \qquad (20.13)$$

where, as usual, subscripts indicate the times of events. That is, if the photon is detected by E with the beam splitter out, then it was earlier in the d and not in the c

arm of the interferometer. Note, however, that using the consistent family (20.11) leads to the equally valid result

$$B_{out}: \quad \Pr([1\bar{a}]_1 \mid E_4^*) = \Pr([2\bar{a}]_2 \mid E_4^*) = 1. \qquad (20.14)$$

The single-framework rule prevents one from combining (20.13) and (20.14), because the families (20.11) and (20.12) are mutually incompatible.

Next, consider the situation in which the beam splitter is in place. In this case the unitary history

$$B_{in}: \quad \Psi_0 \odot [1\bar{a}] \odot [2\bar{a}] \odot [3f] \odot F^* \qquad (20.15)$$

allows one to discuss the outcome of the final measurement. It describes the photon using coherent superpositions of wave packets in the two arms at times t_1 and t_2, as suggested by the statement of the paradox. Based upon it one can conclude that

$$B_{in}: \quad \Pr([1\bar{a}]_1 \mid F_4^*) = \Pr([2\bar{a}]_2 \mid F_4^*) = 1, \qquad (20.16)$$

which is the analog of (20.14). (While (20.14) and (20.16) are correct as written, one should note that the conditions E^* and F^* at t_4 are not necessary, and the probabilities are still equal to 1 if one omits the final detector states from the condition. It is helpful to think of Ψ_0 as always present as a condition, even though it is not explicitly indicated in the notation.) On the other hand, it is also possible to construct the counterpart of (20.12) in which the photon is in a definite arm at t_1 and t_2, using the family with support

$$B_{in}: \quad \Psi_0 \odot \begin{cases} [1c] \odot [2c] \odot [3\bar{c}] \odot S^+, \\ [1d] \odot [2d] \odot [3\bar{d}] \odot S^-, \end{cases} \qquad (20.17)$$

where S^+ and S^- are projectors onto the MQS states defined in (20.9) and (20.10). Note that the MQS states in (20.17) cannot be replaced with pairs of pointer states $\{E^*, F^*\}$ as in (20.11), since the four histories would then form an inconsistent family. See the toy model example in Sec. 13.3.

It is worth emphasizing the fact that there is nothing "wrong" with MQS states from the viewpoint of fundamental quantum theory. If one supposes that the usual Hilbert space structure of quantum mechanics is the appropriate sort of mathematics for describing the world, then MQS states will be present in the theory, because the Hilbert space is a linear vector space, so that if it contains the states $|E^*\rangle|F^\circ\rangle$ and $|E^\circ\rangle|F^*\rangle$, it must also contain their linear combinations. However, if one is interested in discussing a situation in which a photon is detected by a detector, (20.17) is not appropriate, as within this framework the notion that one detector or the other has detected the photon makes no sense.

Let us summarize the results of our analysis as it bears upon the paradox stated in Sec. 20.1. No consistent families were actually specified in the initial statement of the paradox, and we have used four different families in an effort to analyze it: two with the beam splitter out, (20.11) and (20.12), and two with the beam splitter in, (20.15) and (20.17). In a sense, the paradox is based upon using only two of these families, (20.15) with B_{in} and the photon in a superposition state inside the interferometer, and (20.12) with B_{out} and the photon in a definite arm of the interferometer. By focusing on only these two families — they are, of course, only specified implicitly in the statement of the paradox — one can get the misleading impression that the difference between the photon states inside the interferometer in the two cases is somehow caused by the presence or absence of the beam splitter at a later time when the photon leaves the interferometer. But by introducing the other two families, we see that it is quite possible to have the photon either in a superposition state or in a definite arm of the interferometer both when the beam splitter is in place and when it is out of the way. Thus the difference in the type of photon state employed at t_1 and t_2 is not determined or caused by the location of the beam splitter; rather, it is a consequence of a choice of a particular type of quantum description for use in analyzing the problem.

One can, to be sure, object that (20.17) with the detectors in MQS states at t_4 is hardly a very satisfactory description of a situation in which one is interested in which detector detected the photon. It is true that if one wants a description in which no MQS states appear, then (20.15) is to be preferred to (20.17). But notice that what the physicist does in employing this altogether reasonable criterion is somewhat analogous to what a writer of a novel does when changing the plot in order to have the ending work out in a particular way. The physicist is selecting histories which at t_4 will be useful for addressing the question of which detector detected the photon, and not whether the detector system will end up in S^+ or S^-, and for this purpose (20.15), not (20.17) is appropriate. Were the physicist interested in whether the final state was S^+ or S^-, as could conceivably be the case — e.g., when trying to design some apparatus to measure such superpositions — then (20.17), not (20.15), would be the appropriate choice. Quantum mechanics as a fundamental theory allows either possibility, and does not determine the type of questions the physicist is allowed to ask.

If one does not insist that MQS states be left out of the discussion, then a comparison of the histories in (20.12) and (20.17), which are identical up to time t_2 while the photon is still inside the interferometer, and differ only at later times, shows the beam splitter having an ordinary causal effect upon the photon: events at a later time depend upon whether the beam splitter is or is not in place, and those at an earlier time do not. The relationship between these two families is then similar to that between (20.11) and (20.15), where again the presence or absence of the beam splitter when the photon leaves the interferometer can be said to be the cause of different behavior at later times. Causality is actually a rather subtle concept, which philosophers have been arguing about for a long time, and it seems unlikely that quantum theory by itself will contribute much to this discussion. However, the possibility of viewing the presence or absence of the beam splitter as influencing later events should at the very least make one suspicious of the alternative claim that its location influences earlier events.

20.4 Quantum coin toss and counterfactual paradox

Thus far we have worked out various consistent families for two quite distinct situations: the beam splitter in place, or moved out of the way. One can, however, include both possibilities in a single framework in which a quantum coin is tossed while the photon is still inside the interferometer, with the outcome of the toss fed to a servomechanism which moves the beam splitter out of the way or leaves it in place at the time when the photon leaves the interferometer. This makes it possible to examine the counterfactual formulation of the delayed choice paradox found at the end of Sec. 20.1.

The use of a quantum coin for moving a beam splitter was discussed in Sec. 19.2, and we shall use a simplified notation similar to (19.7). Let $|B_0\rangle$ be the state of the quantum coin, servomechanism, and beam splitter prior to the time t_1 when the photon is already inside the interferometer, and suppose that during the time interval from t_1 to t_2 the quantum coin toss occurs, leading to a unitary evolution

$$|B_0\rangle \mapsto (|B_{in}\rangle + |B_{out}\rangle)/\sqrt{2}, \tag{20.18}$$

with the states $|B_{in}\rangle$ and $|B_{out}\rangle$ corresponding to the beam splitter in place or removed from the path of the photon. The unitary time development of the photon from t_2 to t_3, in agreement with (20.3) and (20.5), is given by the expressions

$$
\begin{aligned}
|2c\rangle|B_{in}\rangle &\mapsto |3\bar{c}\rangle|B_{in}\rangle, \quad |2d\rangle|B_{in}\rangle \mapsto |3\bar{d}\rangle|B_{in}\rangle, \\
|2c\rangle|B_{out}\rangle &\mapsto |3f\rangle|B_{out}\rangle, \quad |2d\rangle|B_{out}\rangle \mapsto |3e\rangle|B_{out}\rangle.
\end{aligned}
\tag{20.19}
$$

The unitary time development of the initial state

$$|\Omega_0\rangle = |0a\rangle|B_0\rangle|E^\circ\rangle|F^\circ\rangle \tag{20.20}$$

can be worked out using the formulas in Sec. 20.2 combined with (20.18) and (20.19). In order to keep the notation simple, we assume that the apparatus states $|B_0\rangle$, $|B_{in}\rangle$, $|B_{out}\rangle$ do not change except during the time interval from t_1 to t_2, when the change is given by (20.18). The reader may find it helpful to work out $|\Omega_j\rangle = T(t_j, t_0)|\Omega_0\rangle$ at different times. At t_4, when the photon has been detected, it is given by

$$|\Omega_4\rangle = \Big(|B_{in}\rangle|E^\circ\rangle|F^*\rangle + |B_{out}\rangle|S^+\rangle \Big)/\sqrt{2}. \qquad (20.21)$$

Suppose the quantum coin toss results in the beam splitter being out of the way at the moment when the photon leaves the interferometer, and that the photon is detected by E. What would have occurred if the coin toss had, instead, left the beam splitter in place? As noted in Sec. 19.4, to address such a counterfactual question we need to use a particular consistent family, and specify a pivot. The answers to counterfactual questions are in general not unique, since one can employ more than one family, and more than one pivot within a single family.

Consider the family whose support consists of the three histories

$$\Omega_0 \odot [1\bar{a}] \odot \begin{cases} B_{in} \odot [3f] \odot F^*, \\ B_{out} \odot [3\bar{c}] \odot \{E^*, F^*\} \end{cases} \qquad (20.22)$$

at the times $t_0 < t_1 < t_2 < t_3 < t_4$. Note that B_{out} and E^* occur on the lower line, and we can trace this history back to $[1\bar{a}]$ at t_1 as the pivot, and then forwards again along the upper line corresponding to B_{in}, to conclude that if the beam splitter had been in place the photon would have been detected by F. This is not surprising and certainly not paradoxical. (Note that having the E detector detect the photon when the beam splitter is absent is quite consistent with the photon having been in a superposition state until just before the time of its detection; this corresponds to (20.11) in Sec. 20.3.) To construct a paradox we need to be able to infer from E^* at t_4 that the photon was earlier in the d arm of the interferometer. This suggests using the consistent family whose support is

$$\Omega_0 \odot \begin{cases} [1\bar{a}] \odot B_{in} \odot [3f] \odot F^*, \\ [1c] \odot B_{out} \odot [3f] \odot F^*, \\ [1d] \odot B_{out} \odot [3e] \odot E^*, \end{cases} \qquad (20.23)$$

rather than (20.22). (The consistency of (20.23) follows from noting that one of the two histories which ends in F^* is associated with B_{in} and the other with B_{out}, and these two states are mutually orthogonal, since they are macroscopically distinct.) The events at t_1 are contextual in the sense of Ch. 14, with $[1\bar{a}]$ dependent upon B_{in}, while $[1c]$ and $[1d]$ depend on B_{out}.

The family (20.23) does allow one to infer that the photon was earlier in the d arm if it was later detected by E, since E^* occurs only in the third history, preceded by $[1d]$ at t_1. However, since this event precedes B_{out} but not B_{in}, it cannot serve as a pivot for answering a question in which the actual B_{out} is replaced by the counterfactual B_{in}. The only event in (20.23) which can be used for this purpose is Ω_0. Using Ω_0 as a pivot, we conclude that had the beam splitter been in, the photon would surely have arrived at detector F, which is a sensible result. However, the null counterfactual question, "What would have happened if the beam splitter had been out of the way (as in fact it was)?", receives a rather indefinite, probabilistic answer. Either the photon would have been in the d arm and detected by E, or it would have been in the c arm and detected by F. Thus using Ω_0 as the pivot means, in effect, answering the counterfactual question after erasing the information that the photon was detected by E rather than by F, or that it was in the d arm rather than the c arm. Hence if we use the family (20.23) with Ω_0 as the pivot, the original counterfactual paradox, with its assumption that detection by E implied that the photon was earlier in d, and then asking what would have occurred if this photon had encountered the beam splitter, seems to have disappeared, or at least it has become rather vague.

To be sure, one might argue that there is something paradoxical in that the superposition state $[1\bar{a}]$ in (20.23) is present in the B_{in} history, whereas nonsuperposition states $[1c]$ and $[1d]$ precede B_{out}. Could this be a sign of the future influencing the past? That is not very plausible, for, as noted in Ch. 14, the sort of contextuality we have here, with the earlier photon state depending on the later B_{in} and B_{out}, reflects the way in which the quantum description has been constructed. If there is an influence of the future on the past, it is rather like the influence of the end of a novel on its beginning, as noted in the previous section. Or, to put it in somewhat different terms, this influence manifests itself in the theoretical physicist's notebook rather than in the experimental physicist's laboratory.

What might come closer to representing the basic idea behind the delayed choice paradox is a family in which $[1d]$ at t_1 can serve as a pivot for a counterfactual argument, rather than having to rely on Ω_0 at t_0. Here is such a family:

$$
\Omega_0 \odot
\begin{cases}
[1c] \odot
\begin{cases}
B_{in} \odot [3\bar{c}] \odot S^+, \\
B_{out} \odot [3f] \odot F^*,
\end{cases} \\
[1d] \odot
\begin{cases}
B_{in} \odot [3\bar{d}] \odot S^-, \\
B_{out} \odot [3e] \odot E^*.
\end{cases}
\end{cases}
\tag{20.24}
$$

If we use $[1d]$ at t_1 as the pivot for a case in which the beam splitter is out and the photon is detected in E, it gives a precise answer to the null counterfactual question of what would have happened had the beam splitter been out (as it actually was): the photon would have been detected by E and not by F. But now when we ask what would have happened had the beam splitter been left in place, the answer is that the system of detectors would later have been in the MQS state S^-. In the same way, if the photon is detected in F when the beam splitter is out, a counterfactual argument using $[1c]$ at t_1 as the pivot leads to the conclusion that had the beam splitter been in, the detectors would later have been in the MQS state S^+, which is orthogonal to, and hence quite distinct from S^-. Thus detection in F rather than E when the beam splitter is out leads to a different counterfactual conclusion, in contrast with what we found earlier when using Ω_0 as the pivot. That the answers to our counterfactual questions involve MQS states is hardly surprising, given the discussion in Sec. 20.3. And, as in the case of (20.17), the MQS states in (20.24) cannot be replaced with ordinary pointer states (as defined at the end of Sec. 9.5) E^* and F^* of the detectors, for doing so would result in an inconsistent family. Also note the analogy with the situation considered in Sec. 19.4, where looking for a framework which could give a more precise answer to a counterfactual question involving a spin measurement led to a family (19.12) containing MQS states.

Let us summarize the results obtained by using a quantum coin and studying various consistent families related to the counterfactual statement of the delayed choice paradox. We have looked at three different frameworks, (20.22), (20.23), and (20.24), and found that they give somewhat different answers to the question of what would have happened if the beam splitter had been left in place, when what actually happened was that the photon was detected in E with the beam splitter out. (Such a multiplicity of answers is typical of quantum and — to a lesser degree — classical stochastic counterfactual questions; see Sec. 19.4.) In the end, none of the frameworks supports the original paradox, but each framework evades it for a somewhat different reason. Thus (20.22) does not have photon states localized in the arms of the interferometer, (20.23) has such states, but they cannot be used as a pivot for the counterfactual argument, and remedying this last problem by using (20.24) results in the counterfactual question being answered in terms of MQS states, which were certainly not in view in the original statement of the paradox.

20.5 Conclusion

The analysis of the delayed choice paradox given above provides some useful lessons on how to analyze quantum paradoxes of this general sort. Perhaps the first and most important lesson is that a paradox must be turned into an explicit quantum mechanical model, complete with a set of unitary time transformations. The model should be kept as simple as possible: there is no point in using long expressions and extensive calculations when the essential elements of the paradox and the ideas for untangling it can be represented in a simple way. Indeed, the simpler the representation, the easier it will be to spot the problematic reasoning underlying the paradox. In the interests of simplicity we used single states, rather than macroscopic projectors or density matrices, for the measuring apparatus, and for discussing the outcomes of a quantum coin toss. A more sophisticated approach is available, see Sec. 17.4, but it leads to the same conclusions.

A second lesson is that in order to discuss a paradox, it is necessary to introduce an appropriate framework, which will be a consistent family if the paradox involves time development. There will, typically, be more than one possible framework, and it is a good idea to look at several, since different frameworks allow one to investigate different aspects of a situation.

A third lesson has to do with MQS states. These are usually not taken into account when stating a paradox, and this is not surprising: most physicists do not have any intuitive idea as to what they mean. Nevertheless, families containing MQS states may be very useful for understanding where the reasoning underlying a paradox has gone astray. For example, a process of implicitly (and thus unconsciously) choosing families which contain no MQS states, and then inferring from this that the future influences the past, or that there are mysterious nonlocal influences, lies behind a number of paradoxes. This becomes evident when one works out various alternative families of histories and sees what is needed in order to satisfy the consistency conditions.

21

Indirect measurement paradox

21.1 Statement of the paradox

The paradox of indirect measurement, often called interaction-free measurement, can be put in a form very similar to the delayed choice paradox discussed in Ch. 20. Consider a Mach–Zehnder interferometer, Fig. 21.1, with two beam splitters, which are always present. A mirror M can be placed either (a) in the c arm of the interferometer, where it reflects the photon out of this arm into channel g, thus preventing it from reaching the second beam splitter, or (b) outside the c arm, in a place where it has no effect. The two positions of M are denoted by M_{in} and M_{out}. Detectors E, F, and G detect the photon when it emerges from the apparatus in channels e, f, or g. With M out of the way, the path differences inside the interferometer are such that a photon which enters through channel a will always emerge in channel f, so the photon will always be detected by F. With M in place, a photon which passes into the c channel cannot reach the second beam splitter B_2. However, a photon which reaches B_2 by passing through the d arm can emerge in either the e or the f channel, with equal probability. As a consequence, for M_{in} the probabilities for detection by E, F, and G are 1/4, 1/4, and 1/2, respectively.

Detection of a photon by G can be thought of as a measurement indicating that the mirror was in the position M_{in} rather than M_{out}. It is a *partial* measurement of the mirror's position in that while a photon detected by G implies the mirror is in place, the converse is not true: the mirror can be in place without the photon being detected by G, since it might have passed through the d arm of the interferometer. Detection of the photon by E can likewise be thought of as a measurement indicating that M is in the c arm, since when M is not there the photon is always detected by F. Detection by E is an indirect measurement that M is in place, in contrast to the direct measurement which occurs when G detects the photon. And detection by E is also a partial measurement: it can only occur, but does not always occurs when M is in the c arm.

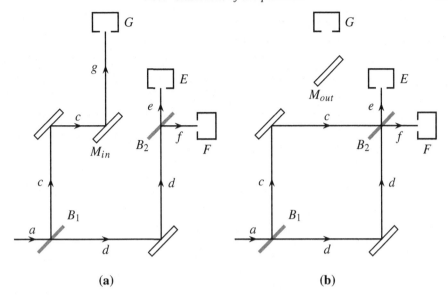

Fig. 21.1. Mach–Zehnder interferometer with extra mirror M located (a) in the c arm, (b) outside the interferometer.

The indirect measurement using E seems paradoxical for the following reason. In order to reach E, the photon must have passed through the d arm of the interferometer, since the c arm was blocked by M. Hence the photon was never anywhere near M, and could not have interacted with M. How, then, could the photon have been affected by the presence or absence of the mirror in the c arm, that is, by the difference between M_{in} and M_{out}? How could it "know" that the c arm was blocked, and that therefore it was allowed to emerge (with a certain probability) in the e channel, an event not possible had M been outside the c arm?

The paradox becomes even more striking in a delayed choice version analogous to that used in Ch. 20. Suppose the mirror M is initially not in the c arm. However, just before the time of arrival of the photon — that is, the time the photon would arrive were it to pass through the c arm — M is either left outside or rapidly moved into place inside the arm by a servomechanism actuated by a quantum coin flip which took place when the photon had already passed the first beam splitter B_1. In this case one can check, see the analysis in Sec. 21.4, that if the photon was later detected in E, M must have been in place blocking the c arm at the instant when the photon would have struck it had the photon been in the c arm. That is, despite the fact that the photon arriving in E was earlier in the d arm it seems to have been sensitive to the state of affairs existing far away in the c arm at just the instant when it would have encountered M! Is there any way to explain this apart from some mysterious nonlocal influence of M upon the photon?

A yet more striking version of the delayed choice version comes from contemplating an extremely large interferometer located somewhere out in space, in which one can arrange that the entire decision process as to whether or not to place M in the c arm occurs during a time when the photon, later detected in E, is at a point on its trajectory through the d arm of the interferometer which is space-like separated (in the sense of relativity theory) from the relevant events in the c arm. Not only does one need nonlocal influences; in addition, they must travel faster than the speed of light! One way to avoid invoking superluminal signals is by assuming a message is carried, at the speed of light, from M to the second beam splitter B_2 in time to inform the photon arriving in the d arm that it is allowed to leave B_2 in the e channel, rather than having to use the f channel, the only possibility for M_{out}. The problem, of course, is to find a way of getting the message from M to B_2, given that the photon is in the d arm and hence unavailable for this task.

A counterfactual version of this paradox is readily constructed. Suppose that with M in the location M_{in} blocking the c arm, the photon was detected in E. What would have occurred if M_{out} had been the case rather than M_{in}? In particular, if the position of M was decided by a quantum coin toss after the photon was already inside the interferometer, what would have happened to the photon — which must have been in the d arm given that it later was detected by E — if the quantum coin had resulted in M remaining outside the c arm? Would the photon have emerged in the f channel to be detected by F? — this seems the only plausible possibility. But then we are back to asking the same sort of question: how could the photon "know" that the c arm was unblocked?

21.2 Unitary dynamics

The unitary dynamics for the system shown in Fig. 21.1 is in many respects the same as for the delayed choice paradox of Ch. 20, and we use a similar notation for the unitary time development. Let $|0a\rangle$ be a wave packet for the photon at t_0 in the input channel a just before it enters the interferometer, $|1c\rangle$ and $|1d\rangle$ be wave packets in the c and d arms of the interferometer at time t_1, and $|2c\rangle$ and $|2d\rangle$ their counterparts at a time t_2 chosen so that if M is in the c arm the photon will have been reflected by it into a packet $|2g\rangle$ in the g channel. At time t_3 the photon will have emerged from the second beam splitter in channel e or f — the corresponding wave packets are $|3e\rangle$ and $|3f\rangle$ — or will be in a wave packet $|3g\rangle$ in the g channel. Finally, at t_4 the photon will have been detected by one of the three detectors in Fig. 21.1. Their ready states are $|E^\circ\rangle$, $|F^\circ\rangle$, and $|G^\circ\rangle$, with $|E^*\rangle$, $|F^*\rangle$, and $|G^*\rangle$ the corresponding states when a photon has been detected.

The unitary time development from t_0 to t_1 is given by

$$|0a\rangle \mapsto |1\bar{a}\rangle := (|1c\rangle + |1d\rangle)/\sqrt{2}. \tag{21.1}$$

For t_1 to t_2 it depends on the location of M:

$$\begin{aligned} M_{out} : & \quad |1c\rangle \mapsto |2c\rangle, \quad |1d\rangle \mapsto |2d\rangle, \\ M_{in} : & \quad |1c\rangle \mapsto |2g\rangle, \quad |1d\rangle \mapsto |2d\rangle, \end{aligned} \tag{21.2}$$

with no difference between M_{in} and M_{out} if the photon is in the d arm of the interferometer. For the time step from t_2 to t_3 the relevant unitary transformations are independent of the mirror position:

$$\begin{aligned} |2c\rangle &\mapsto |3\bar{c}\rangle := \left(+|3e\rangle + |3f\rangle\right)/\sqrt{2}, \\ |2d\rangle &\mapsto |3\bar{d}\rangle := \left(-|3e\rangle + |3f\rangle\right)/\sqrt{2}, \\ |2g\rangle &\mapsto |3g\rangle. \end{aligned} \tag{21.3}$$

The detector states remain unchanged from t_0 to t_3, and the detection events between t_3 and t_4 are described by:

$$|3e\rangle|E^\circ\rangle \mapsto |E^*\rangle, \quad |3f\rangle|F^\circ\rangle \mapsto |F^*\rangle, \quad |3g\rangle|G^\circ\rangle \mapsto |G^*\rangle. \tag{21.4}$$

If the photon is not detected, the detector remains in the ready state; thus (21.4) is an abbreviated version of

$$|3e\rangle|E^\circ\rangle|F^\circ\rangle|G^\circ\rangle \mapsto |E^*\rangle|F^\circ\rangle|G^\circ\rangle, \tag{21.5}$$

etc. One could also use macro projectors or density matrices for the detectors, see Sec. 17.4, but this would make the analysis more complicated without altering any of the conclusions.

21.3 Comparing M_{in} and M_{out}

In Sec. 20.3 we considered separate consistent families depending upon whether the second beam splitter was in or out. That approach could also be used here, but for the sake of variety we adopt one which is slightly different: a single family with *two* initial states at time t_0, one with the mirror in and one with the mirror out,

$$|\Psi_0\rangle|M_{out}\rangle, \quad |\Psi_0\rangle|M_{in}\rangle, \tag{21.6}$$

each with a positive (nonzero) probability, where

$$|\Psi_0\rangle = |0a\rangle|E^\circ\rangle|F^\circ\rangle|G^\circ\rangle. \tag{21.7}$$

Here the mirror is treated as an inert object, so $|M_{in}\rangle$ and $|M_{out}\rangle$ do not change with time. They do, however, influence the time development of the photon state as indicated in (21.2). Thus unitary time development of the first of the two states in (21.6) leads at time t_4 to

$$|E^\circ\rangle|F^*\rangle|G^\circ\rangle|M_{out}\rangle, \tag{21.8}$$

while the second results in a macroscopic quantum superposition (MQS) state

$$\tfrac{1}{2}\Big(-|E^*\rangle|F^\circ\rangle|G^\circ\rangle + |E^\circ\rangle|F^*\rangle|G^\circ\rangle + \sqrt{2}|E^\circ\rangle|F^\circ\rangle|G^*\rangle\Big)|M_{in}\rangle. \tag{21.9}$$

Consider the consistent family with support given by the four histories

$$\begin{aligned}
\Psi_0 M_{out} &\odot [1\bar{a}] \odot [2\bar{a}] \odot [3f] \odot F^*, \\
\Psi_0 M_{in} &\odot [1\bar{a}] \odot [2s] \odot [3s] \odot \{E^*, F^*, G^*\},
\end{aligned} \tag{21.10}$$

at the times $t_0 < t_1 < t_2 < t_3 < t_4$, where

$$|2s\rangle := \big(|2d\rangle + |2g\rangle\big)/\sqrt{2}, \quad |3s\rangle := \tfrac{1}{2}\big(-|3e\rangle + |3f\rangle + \sqrt{2}|3g\rangle\big) \tag{21.11}$$

are superposition states in which the photon is not located in a definite channel. This corresponds to unitary time development until the photon is detected, and then pointer states (as defined at the end of Sec. 9.5) for the detectors. It shows that E^* and G^* can only occur with M_{in}, and in this sense either of these events constitutes a measurement indicating that the mirror was in the c arm. There is nothing paradoxical about the histories in (21.10), because an important piece of the paradox stated in Sec. 21.1 was the notion that the photon detected by E must at an earlier time have been in the d arm of the interferometer. But since the projectors C and D for the particle to be in the c or the d arm do not commute with $[1\bar{a}]$, the assertion that the photon was in one or the other arm of the interferometer at time t_1 makes no sense when one uses (21.10), and the same is true at t_2.

Therefore let us consider a different consistent family with support

$$\begin{aligned}
\Psi_0 M_{out} &\odot \quad [1\bar{a}] \odot [2\bar{a}] \odot \quad [3f] \odot F^*, \\
\Psi_0 M_{in} &\odot \begin{cases} [1c] \odot [2g] \odot & [3g] \odot G^*, \\ [1d] \odot [2d] \odot & \begin{cases} [3e] \odot E^*, \\ [3f] \odot F^*. \end{cases} \end{cases}
\end{aligned} \tag{21.12}$$

Using this family allows us to assert that if the photon was later detected by E, then the mirror was in the c arm, and the photon itself was in the d arm while inside the interferometer, and thus far away from M. The photon states at time t_1 in this family are contextual in the sense discussed in Ch. 14, since $[1c]$ and $[1d]$ do not commute with $[1\bar{a}]$, and the same is true for $[2d]$ at t_2. Thus $[1d]$ and $[2d]$ depend, in the sense of Sec. 14.1, on M_{in}, and it makes no sense to talk about whether the photon is in the c or the d arm if the mirror is out of the way, M_{out}. For this reason it is not possible to use (21.12) in order to investigate what effect replacing M_{in} with M_{out} has on the photon while it is in arm d. Hence while (21.12) represents some advance over (21.10) in stating the paradox, it cannot be used to infer the existence of nonlocal effects.

As noted in Ch. 14, the fact that certain events are contextual should not be thought of as something arising from a physical cause; in particular, it is misleading to think of contextual events as "caused" by the events on which they depend, in the technical sense defined in Sec. 14.1. Thinking that the change from M_{out} to M_{in} in (21.12) somehow "collapses" the photon from a superposition into one localized in one of the arms is quite misleading. Instead, the appearance of a superposition in the M_{out} case and not in the M_{in} case reflects our decision to base a quantum description upon (21.12) rather than, for example, the family (21.10), where the photon is in a superposition both for M_{out} and M_{in}.

One can also use a consistent family in which the photon is in a definite arm while inside the interferometer both when M is in and when it is out of the c arm, so that the c and d states are not contextual:

$$\Psi_0 M_{out} \odot \begin{cases} [1c] \odot [2c] \odot \quad [3\bar{c}] \odot S^+, \\ [1d] \odot [2d] \odot \quad [3\bar{d}] \odot S^-, \end{cases}$$
$$\Psi_0 M_{in} \odot \begin{cases} [1c] \odot [2g] \odot [3g] \odot G^*, \\ [1d] \odot [2d] \odot \begin{cases} [3e] \odot E^*, \\ [3f] \odot F^*. \end{cases} \end{cases} \tag{21.13}$$

The states S^+ and S^- are the macroscopic quantum superposition (MQS) states of detectors E and F as defined in (20.9) and (20.10). Just as in the case of the family (20.17) in Sec. 20.3, the MQS states in the last two histories in (21.13) cannot be eliminated by replacing them with E^* and F^*, as that would violate the consistency conditions. And since the projectors S^+ and S^- do not commute with E^* and F^*, contextuality has not really disappeared when (21.12) is replaced by (21.13): it has been removed from the events at t_1 and t_2, but reappears in the events at t_3 and t_4. In particular, it would make no sense to look at the events at the final time t_4 in (21.13) and conclude that a detection of the photon by E^* was evidence that the mirror M was in rather than out of the c arm. While such a conclusion would be valid using (21.10) or (21.12), it is not supported by (21.13) since in the latter E^* only makes sense in the case M_{in}, and is meaningless with M_{out}.

The preceding analysis has uncovered a very basic problem. Using E^* as a way of determining that M_{in} is the case rather than M_{out} is incompatible with using E^* as an indication that the photon was earlier in the d rather than the c arm. For the former, (21.10) is perfectly adequate, as is (21.12). However, when we try to construct a family in which $[1c]$ vs. $[1d]$ makes sense whether or not the mirror is blocking the c arm, the result, (21.13), is unsatisfactory, both because of the appearance of MQS states at t_4 and also because E^* is now contextual in a way which makes it depend on M_{in}, so that it is meaningless in the case M_{out}.

Hence the detection of the photon by E cannot be used to distinguish M_{in} from M_{out} if one uses (21.13). If this were a problem in classical physics, one could try combining results from (21.10), (21.12), and (21.13) in order to complete the argument leading to the paradox. But these are incompatible quantum frameworks, so the single-framework rule means that the results obtained using one of them cannot be combined with results obtained using the others. From this perspective the paradox stated in Sec. 21.1 arises from using rules of reasoning which work quite well in classical physics, but do not always function properly when imported into the quantum domain.

21.4 Delayed choice version

In order to construct a delayed choice version of the paradox, we suppose that a quantum coin is connected to a servomechanism, and during the time interval between t_1 and $t_{1.5}$, while the photon is inside the interferometer but before it reaches the mirror M, the coin is tossed and the outcome fed to the servomechanism. The servomechanism then places the mirror M in the c arm or leaves it outside, as determined by the outcome of the quantum coin toss. Using the abbreviated notation at the end of Sec. 19.2, the corresponding unitary time development from t_1 to $t_{1.5}$ can be written in the form

$$|M_0\rangle \mapsto (|M_{in}\rangle + |M_{out}\rangle)/\sqrt{2}, \qquad (21.14)$$

the counterpart of (20.18) for the delayed choice paradox of Ch. 20. Here $|M_0\rangle$ is the initial state of the quantum coin, servomechanism, and mirror. The kets $|M_{in}\rangle$ and $|M_{out}\rangle$ in (21.14) include the mirror and the rest of the apparatus (coin and servomechanism), and thus they have a slightly different physical interpretation from those in (21.6). However, since the photon dynamics which interests us depends only on where the mirror M is located, this distinction makes no difference for the present analysis. Combining (21.14) with (21.2) gives an overall unitary time development of the photon and the mirror (and associated apparatus) from t_1 to t_2 in the form:

$$
\begin{aligned}
|1c\rangle|M_0\rangle &\mapsto (|2c\rangle|M_{out}\rangle + |2g\rangle|M_{in}\rangle)/\sqrt{2}, \\
|1d\rangle|M_0\rangle &\mapsto |2d\rangle(|M_{out}\rangle + |M_{in}\rangle)/\sqrt{2}.
\end{aligned}
\qquad (21.15)
$$

What is important is the location of the mirror at the time $t_{1.5}$ when the photon interacts with it — assuming both the mirror and the photon are in the c arm of the interferometer — and not its location in the initial state $|M_0\rangle$; the latter could be either the M_{in} or M_{out} position, or someplace else.

Let the initial state of the entire system at t_0 be

$$|\Omega_0\rangle = |0a\rangle|M_0\rangle|E^\circ\rangle|F^\circ\rangle|G^\circ\rangle, \qquad (21.16)$$

and let $|\Omega_j\rangle$ be the state which results at time t_j from unitary time development. We assume that $|M_0\rangle$, $|M_{out}\rangle$, and $|M_{in}\rangle$ do not change outside the time interval where (21.14) and (21.15) apply. At t_3 one has

$$|\Omega_3\rangle = \left[(-|3e\rangle + |3f\rangle + \sqrt{2}|3g\rangle)|M_{in}\rangle + 2|3f\rangle|M_{out}\rangle \right]$$
$$\otimes |E^\circ\rangle|F^\circ\rangle|G^\circ\rangle/2\sqrt{2}. \qquad (21.17)$$

We leave to the reader the task of working out $|\Omega_j\rangle$ at other times, a useful exercise if one wants to check the properties of the various consistent families described below.

Corresponding to (21.10) there is a family, now based on the single initial state Ω_0, with support

$$\Omega_0 \odot [1\bar{a}] \odot \begin{cases} M_{out} \odot [3f] \odot F^*, \\ M_{in} \odot [3s] \odot \{E^*, F^*, G^*\}, \end{cases} \qquad (21.18)$$

where $|3s\rangle$ is defined in (21.11). This confirms the fact that whether or not a photon arrives at E^* depends on the position of the mirror M at the time when the photon reaches the corresponding position in the c arm of the interferometer, not on where M was at an earlier time, in accordance with what was stated in Sec. 21.1. Suppose that the photon has been detected in E^*. From (21.18) it is evident that the quantum coin toss resulted in M_{in}. What would have happened if, instead, the result had been M_{out}? If we use $[1\bar{a}]$ at t_1 as a pivot, the answer is that the photon would have been detected by F. This is reasonable, but as noted in our discussion of (21.10), not at all paradoxical, since it is impossible to use this family to discuss whether or not the photon was in the d arm.

The counterpart of (21.12) is the family with support

$$\Omega_0 \odot \begin{cases} [1\bar{a}] \odot M_{out} \odot [3f] \odot F^*, \\ [1c] \odot M_{in} \odot [3g] \odot G^*, \\ [1d] \odot M_{in} \odot [3\bar{c}] \odot \{E^*, F^*\}. \end{cases} \qquad (21.19)$$

Just as in (21.12), the photon states at t_1 are contextual; $[1c]$ and $[1d]$ depend on M_{in}, while $[1\bar{a}]$ depends on M_{out}. The only difference is that here the dependence is on the later, rather than earlier, position of the mirror M. Note once again that dependence, understood in the sense defined in Ch. 14, does not refer to a physical cause, and there is no reason to think that the future is influencing the past — see the discussion in Secs. 20.3 and 20.4. We can use (21.19) to conclude that the detection of the photon by E means that the photon was earlier in the d and not the c arm of the interferometer. However, due to the contextuality just mentioned, $[1d]$ at t_1 cannot serve as a pivot in a counterfactual argument which tells what would have happened had M_{out} occurred rather than M_{in}. The only pivot available in (21.19) is the initial state Ω_0. But the corresponding counterfactual assertion is too vague to serve as a satisfactory basis of a paradox, for precisely the same reasons given in Sec. 20.4 in connection with the analogous (20.23).

The counterpart of (21.13) is the family with support

$$\Omega_0 \odot \begin{cases} [1c] \odot \begin{cases} M_{out} \odot [3\bar{c}] \odot S^+, \\ M_{in} \odot [3g] \odot G^*, \end{cases} \\ [1d] \odot \begin{cases} M_{out} \odot [3\bar{d}] \odot S^-, \\ M_{in} \odot [3\bar{d}] \odot S^-. \end{cases} \end{cases} \tag{21.20}$$

Here $[1c]$ and $[1d]$ are no longer contextual. Also note that in this family there is not the slightest evidence of any nonlocal influence by the mirror on the photon: the later time development if the photon is in the d arm at t_1 is exactly the same for M_{in} and for M_{out}. However, (21.20) is clearly not a satisfactory formulation of the paradox, despite the fact that $[1d]$ at t_1 can serve as a pivot. Among other things, E^* does not appear at t_4. This can be remedied in part by replacing the fourth history in (21.20) with the two histories

$$\Omega_0 \odot [1d] \odot M_{in} \odot [3\bar{d}] \odot \{E^*, F^*\}. \tag{21.21}$$

The resulting family, now supported on five histories of nonzero weight, remains consistent. But E^* is a contextual event dependent on M_{in}, and if we use this family, E^* makes no sense in the case M_{out}. Thus, as noted above in connection with (21.13), we cannot when using this family employ E^* as evidence that the mirror was in rather than outside of the c arm. In addition to the difficulty just mentioned, (21.20) has MQS states at t_4. While one can modify the fourth history by replacing it with (21.21), the same remedy will not work in the other two cases, for it would violate the consistency conditions.

Let us summarize what we have learned from considering a situation in which a quantum coin toss at a time when the photon is already inside the interferometer determines whether or not the c arm will be blocked by the mirror M. For the photon to later be detected by E, it is necessary that M be in the c arm at the time when the photon arrives at this point, and in this respect E^* does, indeed, provide a (partial) measurement indicating M_{in} is the case rather than M_{out}. However, the attempt to infer from this that there is some sort of nonlocal influence between M and the photon fails, for reasons which are quite similar to those summarized at the end of Sec. 21.3: one needs to find a consistent family in which the photon is in the d arm both for M_{in} and for M_{out}. This is obviously not the case for (21.18) and

(21.19), whereas (21.20) — with or without the fourth line replaced with (21.21) — is unsatisfactory because of the states which appear at t_4. Thus by the time one has constructed a family in which $[1d]$ at t_1 can serve as a pivot, the counterfactual analysis runs into difficulty because of what happens at later times. Just as in Sec. 21.3, one can construct various pieces of a paradox by using different consistent families. But the fact that these families are mutually incompatible prevents putting the pieces together to complete the paradox.

21.5 Interaction-free measurement?

It is sometimes claimed that the determination of whether M is blocking the c arm by means of a photon detected in E is an "interaction-free measurement": The photon did not actually interact with the mirror, but nonetheless provided information about its location. The term "interact with" is not easy to define in quantum theory, and we will want to discuss two somewhat different reasons why one might suppose that such an indirect measurement involves no interaction. The first is based on the idea that detection by E implies that the photon was earlier in the d arm of the interferometer, and thus far from the mirror and unable to interact with it — unless, of course, one believes in the existence of some mysterious long-range interaction. The second comes from noting that when it is in the c arm, Fig. 21.1(a), the mirror M is oriented in such a way that any photon hitting it will later be detected by G. Obviously a photon detected by E was not detected by G, and thus, according to this argument, could not have interacted with M. The consistent families introduced earlier are useful for discussing both of these ideas.

Let us begin with (21.10), or its counterpart (21.18) if a quantum coin is used. In these families the time development of the photon state is given by unitary transformations until it has been detected. As one would expect, the photon state is different, at times t_2 and later, depending upon whether M is in or out of the c arm. Hence if unitary time development reflects the presence or absence of some interaction, these families clearly do not support the idea that during the process which eventually results in E^* the photon does not interact with M. Indeed, one comes to precisely the opposite conclusion.

Suppose one considers families of histories in which the photon state evolves in a stochastic, rather than a unitary, fashion preceding the final detection. Are the associated probabilities affected by the presence or absence of M in the c arm? In particular, can one find cases in which certain probabilities are the same for both M_{in} and M_{out}? Neither (21.12) nor its quantum coin counterpart (21.19) provide examples of such invariant probabilities, but (21.20) does supply an example: if the photon is in the d arm at t_1, then it will certainly be in the superposition state

[$3\bar{d}$] just after leaving the interferometer, and at a slightly later time the detector system will be in the MQS state S^-. (One would have the same thing in (21.13) if the last two histories were collapsed into a single history representing unitary time development after t_1, ending in S^- at t_4.) So in this case we have grounds to say that there was no interaction between the photon and the mirror if the photon was in the d arm at t_1. However, (21.20), for reasons noted in Sec. 21.4, cannot be used if one wants to speak of photon detection by E as representing a measurement of M_{in} as against M_{out}. Thus we have found a case which is "interaction-free", but it cannot be called a "measurement".

Finally, let us consider the argument for noninteraction based upon the idea that, had it interacted with the mirror, the photon would surely have been scattered into channel g to be detected by G. This argument would be plausible if we could be sure that the photon was in or not in the c arm of the interferometer at the time when it (might have) interacted with M. However, if the photon was in a superposition state at the relevant time, as is the case in the families (21.10) and (21.18), the argument is no longer compelling. Indeed, one could say that the M_{in} histories in these families provide a counterexample showing that when a quantum particle is in a delocalized state, a local interaction can produce effects which are contrary to the sort of intuition one builds up by using examples in classical physics, where particles always have well-defined positions.

In conclusion, there seems to be no point of view from which one can justify the term "interaction-free measurement". The one that comes closest might be that based on the family (21.20), in which the photon can be said to be definitely in the c or d arm of the interferometer, and when in the d arm it is not influenced by whether M is or is not in the c arm. But while this family can be used to argue for the absence of any mysterious long-range influences of the mirror on the photon, it is incompatible with using detection of the photon by E as a measurement of M_{in} in contrast to M_{out}.

It is worthwhile comparing the indirect measurement situation considered in this chapter with a different type of "interaction-free" measurement discussed in Sec. 12.2 and in Secs. 18.1 and 18.2: A particle (photon or neutron) passes through a beam splitter, and because it is *not* detected by a detector in one of the two output channels, one can infer that it left the beam splitter through the other channel. In this situation there actually is a consistent family, see (12.31) or the analogous (18.7), containing the measurement outcomes, and in which the particle is far away from the detector in the case in which it is not detected. Thus one might have some justification for referring to this as "interaction-free". However, since such a situation can be understood quite simply in classical terms, and because "interaction-free" has generally been associated with confused ideas of wave function collapse, see Sec. 18.2, even in this case the term is probably not very helpful.

21.6 Conclusion

The paradox stated in Sec. 21.1 was analyzed by assuming, in Sec. 21.3, that the mirror positions M_{in} and M_{out} are specified at the initial time t_0, before the photon enters the interferometer, and then in Sec. 21.4 by assuming these positions are determined by a quantum coin toss which takes place when the photon is already inside the interferometer. Both analyses use several consistent families, and come to basically similar conclusions. In particular, while various parts of the argument leading to the paradoxical result — e.g., the conclusion that detection by E means the photon was earlier in the d arm of the interferometer — can be supported by choosing an appropriate framework, it is not possible to put all the pieces together within a single consistent family. Thus the reasoning which leads to the paradox, when restated in a way which makes it precise, violates the single-framework rule.

This indicates a fourth lesson on how to analyze quantum paradoxes, which can be added to the three in Sec. 20.5. Very often quantum paradoxes rely on reasoning which violates the single-framework rule. Sometimes such a violation is already evident in the way in which a paradox is stated, but in other instances it is more subtle, and analyzing several different frameworks may be necessary in order to discover where the difficulty lies.

The idea of a mysterious nonlocal influence of the position of mirror M (M_{in} vs. M_{out}) on the photon when the latter is far away from M in the d arm of the interferometer is not supported by a consistent quantum analysis. In the family (21.20) the absence of any influence is quite explicit. In the family (21.19) the fact that the photon states inside the interferometer are contextual events indicates that the difference between the photon states arises not from some physical influence of the mirror position, but rather from the physicist's choice of one form of description rather than another. (We found a very similar sort of "influence" of B_{in} and B_{out} in the delayed choice paradox of Ch. 20, and the remarks made there in Secs. 20.3 and 20.4 also apply to the indirect measurement paradox.) It is, of course, important to distinguish differences arising simply because one employs a different way of describing a situation from those which come about due to genuine physical influences.

22

Incompatibility paradoxes

22.1 Simultaneous values

There is never any difficulty in supposing that a classical mechanical system possesses, at a particular instant of time, precise values of each of its physical variables: momentum, kinetic energy, potential energy of interaction between particles 3 and 5, etc. Physical variables, see Sec. 5.5, correspond to real-valued functions on the classical phase space, and if at some time the system is described by a point γ in this space, the variable A has the value $A(\gamma)$, B has the value $B(\gamma)$, etc.

In quantum theory, where physical variables correspond to observables, that is, Hermitian operators on the Hilbert space, the situation is very different. As discussed in Sec. 5.5, a physical variable A has the value a_j provided the quantum system is in an eigenstate of A with eigenvalue a_j or, more generally, if the system has a property represented by a nonzero projector P such that

$$AP = a_j P. \tag{22.1}$$

It is very often the case that two quantum observables have no eigenvectors in common, and in this situation it is impossible to assign values to both of them for a single quantum system at a single instant of time. This is the case for S_x and S_z for a spin-half particle, and as was pointed out in Sec. 4.6, even the assumption that "$S_z = 1/2$ AND $S_x = 1/2$" is a false (rather than a meaningless) statement is enough to generate a paradox if one uses the usual rules of classical logic. This is perhaps the simplest example of an *incompatibility paradox* arising out of the assumption that quantum properties behave in much the same way as classical properties, so that one can ignore the rules of quantum reasoning summarized in Ch. 16, in particular the rule which forbids combining incompatible properties and families. By contrast, if two Hermitian operators A and B

commute, there is at least one orthonormal basis $\{|j\rangle\}$, Sec. 3.7, in which both are diagonal,

$$A = \sum_j a_j |j\rangle\langle j|, \quad B = \sum_j b_j |j\rangle\langle j|. \tag{22.2}$$

If the quantum system is described by this framework, there is no difficulty with supposing that A has (to take an example) the value a_2 at the same time as B has the value b_2.

The idea that *all* quantum variables should simultaneously possess values, as in classical mechanics, has a certain intuitive appeal, and one can ask whether there is not some way to extend the usual Hilbert space description of quantum mechanics, perhaps by the addition of some hidden variables, in order to allow for this possibility. For this to be an extension rather than a completely new theory, one needs to place some restrictions upon which values will be allowed, and the following are reasonable requirements: (i) The value assigned to a particular observable will always be one of its eigenvalues. (ii) Given a collection of *commuting observables*, the values assigned to them will be eigenvalues corresponding to a single eigenvector. For example, with reference to A and B in (22.2), assigning a_2 to A and b_2 to B is a possibility, but assigning a_2 to A and b_3 to B (assuming $a_2 \neq a_3$ and $b_2 \neq b_3$) is not. That condition (ii) is reasonable if one intends to assign values to *all* observables can be seen by noting that the projector $|2\rangle\langle 2|$ in (22.2) is itself an observable with eigenvalues 0 and 1. If it is assigned the value 1, then it seems plausible that A should be assigned the value a_2 and B the value b_2.

Bell and Kochen and Specker have shown that in a Hilbert space of dimension 3 or more, assigning values to all quantum observables in accordance with (i) and (ii) is not possible. In Sec. 22.3 we shall present a simple example due to Mermin which shows that such a value assignment is not possible in a Hilbert space of dimension 4 or more. Such a counterexample is a paradox in the sense that it represents a situation that is surprising and counterintuitive from the perspective of classical physics. Section 22.2 is devoted to introducing the notion of a value functional, a concept which is useful for discussing the two-spin paradox of Sec. 22.3. A truth functional, Sec. 22.4, is a special case of a value functional, and is useful for understanding how the concept of "truth" is used in quantum descriptions. The three-box paradox in Sec. 22.5 employs incompatible frameworks of histories in a manner similar to the way in which the two-spin paradox uses incompatible frameworks of properties at one time, and Sec. 22.6 extends the results of Sec. 22.4 on truth functionals to the case of histories.

22.2 Value functionals

A *value functional* v assigns to all members $A, B, \ldots$ of some collection $\mathcal{C}$ of physical variables numerical values of a sort which could be appropriate for describing a single system at a single instant of time. For example, with γ a fixed point in the classical phase space, the value functional v_γ assigns to each physical variable C the value

$$v_\gamma(C) = C(\gamma) \tag{22.3}$$

of the corresponding function at the point γ. In this case $\mathcal{C}$ could be the collection of all physical variables, or some more restricted set. If there is some algebraic relationship among certain physical variables, as in the formula

$$E = p^2/2m + V \tag{22.4}$$

for the total energy in terms of the momentum and potential energy of a particle in one dimension, this relationship will also be satisfied by the values assigned by v_γ:

$$v_\gamma(E) = [v_\gamma(p)]^2/2m + v_\gamma(V). \tag{22.5}$$

To define a value functional for a quantum system, let $\{D_j\}$ be some fixed decomposition of the identity, and let the collection $\mathcal{C}$ consist of all operators of the form

$$C = \sum_j c_j D_j, \tag{22.6}$$

with real eigenvalues c_j. The value functional v_k defined by

$$v_k(C) = c_k \tag{22.7}$$

assigns to each physical variable C its value on the subspace D_k. Note that there are as many distinct value functionals as there are members in the decomposition $\{D_j\}$. As in the classical case, if there is some algebraic relationship among the observables belonging to $\mathcal{C}$, such as

$$F = 2I - A + B^2, \tag{22.8}$$

it will be reflected in the values assigned by v_k:

$$v_k(F) = 2 - v_k(A) + [v_k(B)]^2. \tag{22.9}$$

It is important to note that the class $\mathcal{C}$ on which a quantum value functional is defined is a collection of *commuting* observables, since the decomposition of the identity is held fixed and only the eigenvalues in (22.6) are allowed to vary. Conversely, given a collection of commuting observables, one can find an orthonormal basis in which they are simultaneously diagonal, Sec. 3.7, and the corresponding decomposition of the identity can be used to define value functionals which assign values simultaneously to all of the observables in the collection.

The problem posed in Sec. 22.1 of defining values for *all* quantum observables can be formulated as follows: Find a *universal value functional* v_u defined on the collection of *all* observables or Hermitian operators on a quantum Hilbert space, and satisfying the conditions:

U1. For any observable A, $v_u(A)$ is one of its eigenvalues.

U2. Given any decomposition of the identity $\{D_j\}$, with C the corresponding collection of observables of the form (22.6), there is some D_k from the decomposition such that

$$v_u(C) = c_k \qquad (22.10)$$

for every C in C, where c_k is the coefficient in (22.6).

Conditions U1 and U2 are the counterparts of the requirements (i) and (ii) stated in Sec. 22.1. Note that any algebraic relationship, such as (22.8), among the members of a collection of *commuting* observables will be reflected in the values assigned to them by v_u, as in (22.9). The reason is that there will be an orthonormal basis in which these observables are simultaneously diagonal, and (22.10) will hold for the corresponding decomposition of the identity.

22.3 Paradox of two spins

There are various examples which show explicitly that a universal value functional satisfying conditions U1 and U2 in Sec. 22.2 cannot exist. One of the simplest is the following two-spin paradox due to Mermin. For a spin-half particle let σ_x, σ_y, and σ_z be the operators $2S_x$, $2S_y$, and $2S_z$, with eigenvalues ± 1. The corresponding matrices using a basis of $|z^+\rangle$ and $|z^-\rangle$ are the familiar Pauli matrices:

$$\sigma_x = \begin{pmatrix} 0 & 1 \\ 1 & 0 \end{pmatrix}, \quad \sigma_y = \begin{pmatrix} 0 & -i \\ i & 0 \end{pmatrix}, \quad \sigma_x = \begin{pmatrix} 1 & 0 \\ 0 & -1 \end{pmatrix}. \qquad (22.11)$$

The Hilbert space for two spin-half particles a and b is the tensor product

$$\mathcal{H} = \mathcal{A} \otimes \mathcal{B}, \qquad (22.12)$$

and we define the corresponding spin operators as

$$\sigma_{ax} = \sigma_x \otimes I, \quad \sigma_{by} = I \otimes \sigma_y, \qquad (22.13)$$

etc.

The nine operators on $\mathcal{H}$ in the 3×3 square

$$
\begin{array}{ccc}
\sigma_{ax} & \sigma_{bx} & \sigma_{ax}\sigma_{bx} \\[4pt]
\sigma_{by} & \sigma_{ay} & \sigma_{ay}\sigma_{by} \\[4pt]
\sigma_{ax}\sigma_{by} & \sigma_{ay}\sigma_{bx} & \sigma_{az}\sigma_{bz}
\end{array}
\tag{22.14}
$$

have the following properties:

    M1. Each operator is Hermitian, with two eigenvalues equal to $+1$ and two equal to -1.

    M2. The three operators in each row commute with each other, and likewise the three operators in each column.

    M3. The product of the three operators in each row is equal to the identity I.

    M4. The product of the three operators in both of the first two columns is I, while the product of those in the last column is $-I$.

These statements can be verified by using the well-known properties of the Pauli matrices:

$$
(\sigma_x)^2 = I, \quad \sigma_x\sigma_y = i\sigma_z,
\tag{22.15}
$$

etc. Note that σ_{ax} and σ_{by} commute with each other, as they are defined on separate factors in the tensor product, see (22.13), whereas σ_{ax} and σ_{ay} do not commute with each other. Statement M1 is obvious when one notes that the trace of each of the nine operators in (22.14) is 0, whereas its square is equal to I.

A universal value functional v_u will assign one of its eigenvalues, $+1$ or -1, to each of the nine observables in (22.14). Since the product of the operators in the first row is I and an assignment of values preserves algebraic relations among commuting observables, as in (22.9), it must be the case that

$$
v_u(\sigma_{ax})\, v_u(\sigma_{bx})\, v_u(\sigma_{ax}\sigma_{bx}) = 1.
\tag{22.16}
$$

The products of the values in the other rows and in the first two columns is also 1, whereas for the last column the product is

$$
v_u(\sigma_{ax}\sigma_{bx})\, v_u(\sigma_{ay}\sigma_{by})\, v_u(\sigma_{az}\sigma_{bz}) = -1.
\tag{22.17}
$$

The set of values

$$
\begin{array}{ccc}
-1 & -1 & +1 \\[4pt]
+1 & -1 & -1 \\[4pt]
-1 & -1 & +1
\end{array}
\tag{22.18}
$$

for the nine observables in (22.14) satisfies (22.16), (22.17), and all of the other product conditions except that the product of the integers in the center column is

−1 rather than +1. This seems like a small defect, but there is no obvious way to remedy it, since changing any −1 in this column to +1 will result in a violation of the product condition for the corresponding row. In fact, a value assignment simultaneously satisfying all six product conditions is impossible, because the three product conditions for the rows imply that the product of all nine numbers is +1, while the three product conditions for the columns imply that this same product must be −1, an obvious contradiction. To be sure, we have only looked at a rather special collection of observables in (22.14), but this is enough to show that there is no *universal* value functional capable of assigning values to *every* observable in a manner which satisfies conditions U1 and U2 of Sec. 22.2.

It should be emphasized that the two-spin paradox is not a paradox for quantum mechanics as such, because quantum theory provides no mechanism for assigning values simultaneously to noncommuting observables (except for special cases in which they happen to have a common eigenvector). Each of the nine observables in (22.14) commutes with four others: two in the same row, and two in the same column. However, it does *not* commute with the other four observables. Hence there is no reason to expect that a single value functional can assign sensible values to all nine, and indeed it cannot. The motivation for thinking that such a function might exist comes from the analogy provided by classical mechanics, as noted in Sec. 22.1. What the two-spin paradox shows is that at least in this respect there is a profound difference between quantum and classical physics.

This example shows that a universal value functional is not possible in a four-dimensional Hilbert space, or in any Hilbert space of higher dimension, since one could set up the same example in a four-dimensional subspace of the larger space. The simplest known examples showing that universal value functionals are impossible in a three-dimensional Hilbert space are much more complicated. Universal value functionals are possible in a two-dimensional Hilbert space, a fact of no particular physical significance, since very little quantum theory can be carried out if one is limited to such a space.

22.4 Truth functionals

Additional insight into the difference between classical and quantum physics comes from considering *truth functionals*. A truth functional is a value functional defined on a collection of indicators (in the classical case) or projectors (in the quantum case), rather than on a more general collection of physical variables or observables. A classical truth functional θ_γ can be defined by choosing a fixed point γ in the phase space, and then for every indicator P belonging to some collection $\mathcal{L}$

writing

$$\theta_\gamma(P) = P(\gamma), \tag{22.19}$$

which is the same as (22.3). Since an indicator can only take the values 0 or 1, $\theta_\gamma(P)$ will either be 1, signifying that system in the state γ possesses the property P, and thus that P is *true*; or 0, indicating that P is *false*. (Recall that the indicator for a classical property, (4.1), takes the value 1 on the set of points in the phase space where the system has this property, and 0 elsewhere.)

A quantum truth functional is defined on a Boolean algebra $\mathcal{L}$ of projectors of the type

$$P = \sum_j \pi_j D_j, \tag{22.20}$$

where each π_j is either 0 or 1, and $\{D_j\}$ is a decomposition of the identity. It has the form

$$\theta_k(P) = \begin{cases} 1 & \text{if } PD_k = D_k, \\ 0 & \text{if } PD_k = 0, \end{cases} \tag{22.21}$$

for some choice of k. This is a special case of (22.7), with (22.20) and π_k playing the role of (22.6) and c_k. If one thinks of the decomposition $\{D_j\}$ as a sample space of mutually exclusive events, one and only one of which occurs, then the truth functional θ_k assigns the value 1 to all properties P which are true, in the sense that $\Pr(P \mid D_k) = 1$, when D_k is the event which actually occurs, and 0 to all properties P which are false, in the sense that $\Pr(P \mid D_k) = 0$. Thus as long as one only considers a single decomposition of the identity the situation is analogous to the classical case: the projectors in $\{D_j\}$ constitute what is, in effect, a discrete phase space. The difference between classical and quantum physics lies in the fact that θ_γ in (22.19) can be applied to as large a collection of indicators as one pleases, whereas the definition θ_k in (22.21) will not work for an arbitrary collection of projectors; in particular, if P does not commute with D_k, PD_k is not a projector.

For a given decomposition $\{D_j\}$ of the identity, the truth functional θ_k is simply the value functional v_k of (22.7) restricted to projectors belonging to $\mathcal{L}$ rather than to more general operators; that is,

$$\theta_k(P) = v_k(P) \tag{22.22}$$

for all P in $\mathcal{L}$. Conversely, v_k is determined by θ_k in the sense that for any operator C of the form (22.6) one has

$$v_k(C) = \sum_j c_j \theta_k(D_j). \tag{22.23}$$

An alternative approach to defining a truth functional is the following. Let $\theta(P)$ assign the value 0 or 1 to every projector in the Boolean algebra $\mathcal{L}$ generated by the decomposition of the identity $\{D_j\}$, subject to the following conditions:

$$\theta(I) = 1,$$
$$\theta(I - P) = 1 - \theta(P), \qquad (22.24)$$
$$\theta(PQ) = \theta(P)\theta(Q).$$

One can think of these as a special case of a value functional preserving algebraic relations, as discussed in Sec. 22.2. Thus it is evident that θ_k as defined in (22.21), since it is derived from a value functional, (22.22), will satisfy (22.24). It can also be shown that a functional θ taking the values 0 and 1 and satisfying (22.24) must be of the form (22.21) for some k.

We shall define a *universal truth functional* to be a functional θ_u which assigns 0 or 1 to *every* projector P on the Hilbert space, not simply those associated with a particular Boolean algebra $\mathcal{L}$, in such a way that the relations in (22.24) are satisfied whenever they make sense. In particular, the third relation in (22.24) makes no sense if P and Q do not commute, for then PQ is not a projector, so we modify it to read:

$$\theta_u(PQ) = \theta_u(P)\theta_u(Q) \quad \text{if} \quad PQ = QP. \qquad (22.25)$$

When P and Q both belong to the same Boolean algebra they commute with each other, so θ_u when restricted to a particular Boolean algebra $\mathcal{L}$ satisfies (22.24). Consequently, when θ_u is thought of as a function on the projectors in $\mathcal{L}$, it coincides with an "ordinary" truth functional θ_k for this algebra, for some choice of k.

Given a universal value functional v_u, we can define a corresponding universal truth functional θ_u by letting $\theta_u(P) = v_u(P)$ for every projector P. Conversely, given a universal truth functional one can use it to construct a universal value functional satisfying conditions U1 and U2 of Sec. 22.2 by using the counterpart of (22.23):

$$v_u(C) = \sum_j c_j \theta_u(D_j). \qquad (22.26)$$

That is, given any Hermitian operator C there is a decomposition of the identity $\{D_j\}$ such that C can be written in the form (22.6). On this decomposition of the identity θ_u must agree with θ_k for some k, so the right side of (22.23) makes sense, and can be used to define $v_u(C)$. It then follows that U1 and U2 of Sec. 22.2 are satisfied. If the eigenvalues of C are degenerate there is more than one way of writing it in the form (22.6), but it can be shown that the properties we are assuming for θ_u imply that these different possibilities lead to the same $v_u(C)$.

This close connection between universal value functionals and universal truth functionals means that arguments for the existence or nonexistence of one immediately apply to the other. Thus neither of these universal functionals can be constructed in a Hilbert space of dimension 3 or more, and the two-spin paradox of Sec. 22.3, while formulated in terms of a universal value functional, also demonstrates the nonexistence of a universal truth functional in a four (or higher) dimensional Hilbert space. It is, indeed, somewhat disappointing that there is nothing very significant to which the formulas (22.25) and (22.26) actually apply!

The nonexistence of universal quantum truth functionals is not very surprising. It is simply another manifestation of the fact that quantum incompatibility makes it impossible to extend certain ideas associated with the classical notion of truth into the quantum domain. Similar problems were discussed earlier in Sec. 4.6 in connection with incompatible properties, and in Sec. 16.4 in connection with incompatible frameworks.

22.5 Paradox of three boxes

The three-box paradox of Aharonov and Vaidman resembles the two-spin paradox of Sec. 22.3 in that it is a relatively simple example which is incompatible with the existence of a universal truth functional. Whereas the two-spin paradox refers to properties of a quantum system at a single instant of time, the three-box paradox employs histories, and the incompatibility of the different frameworks reflects a violation of consistency conditions rather than the fact that projectors do not commute with each other. The paradox is discussed in this section, and the connection with truth functionals for histories is worked out in Sec. 22.6.

Consider a three-dimensional Hilbert space spanned by an orthonormal basis consisting of three states $|A\rangle$, $|B\rangle$, and $|C\rangle$. As in the original statement of the paradox, we shall think of these states as corresponding to a particle being in one of three separate boxes, though one could equally well suppose that they are three orthogonal states of a spin-one particle, or the states $m = -1$, 0, and 1 in a toy model of the type introduced in Sec. 2.5. The dynamics is trivial, $T(t', t) = I$: if the particle is in one of the boxes, it stays there. We shall be interested in quantum histories involving three times $t_0 < t_1 < t_2$, based upon an initial state

$$|D\rangle = (|A\rangle + |B\rangle + |C\rangle)/\sqrt{3} \qquad (22.27)$$

at t_0, and ending at t_2 in one of the two events F or $\tilde{F} = I - F$, where F is the projector corresponding to

$$|F\rangle = (|A\rangle + |B\rangle - |C\rangle)/\sqrt{3}. \qquad (22.28)$$

In the first consistent family $\mathcal{A}$ the events at the intermediate time t_1 are A and

$\tilde{A} = I - A$, with A the projector $|A\rangle\langle A|$. The support of this family consists of the three histories

$$D \odot \begin{cases} A \odot F, \\ A \odot \tilde{F}, \\ \tilde{A} \odot \tilde{F}, \end{cases} \tag{22.29}$$

since $D \odot \tilde{A} \odot F$ has zero weight. Checking consistency is straightforward. The chain operator for the first history in (22.29) is obviously orthogonal to the other two because of the final states. The orthogonality of the chain operators for the second and third histories can be worked out using chain kets, or by replacing the final $\tilde{F}$ with F and employing the trick discussed in connection with (11.5). Because $D \odot \tilde{A} \odot F$ has zero weight, if the event F occurs at t_2, then A rather than $\tilde{A}$ must have been the case at t_1; that is, $\tilde{A}$ at t_1 is never followed by F at t_2. Thus one has

$$\Pr(A_1 \mid D_0 \wedge F_2) = 1, \tag{22.30}$$

with our usual convention of a subscript indicating the time of an event.

Now consider a second consistent family $\mathcal{B}$ with events B and $\tilde{B} = I - B$ at t_1; B is the projector $|B\rangle\langle B|$. In this case the support consists of the histories

$$D \odot \begin{cases} B \odot F, \\ B \odot \tilde{F}, \\ \tilde{B} \odot \tilde{F}, \end{cases} \tag{22.31}$$

from which one can deduce that

$$\Pr(B_1 \mid D_0 \wedge F_2) = 1, \tag{22.32}$$

the obvious counterpart of (22.30) given the symmetry between $|A\rangle$ and $|B\rangle$ in the definition of $|D\rangle$ and $|F\rangle$.

The paradox arises from noting that from the same initial data D and F ("initial" refers to position in a logical argument, not temporal order in a history; see Sec. 16.1) one is able to infer by using $\mathcal{A}$ that A occurred at time t_1, and by using $\mathcal{B}$ that B was the case at t_1. However, A and B are mutually exclusive properties, since $BA = 0$. That is, we seem to be able to conclude with probability 1 that the particle was in box A, and also that it was in box B, despite the fact that the rules of quantum theory indicate that it cannot simultaneously be in both boxes! Thus it looks as if the rules of quantum reasoning have given rise to a contradiction.

However, these rules, as summarized in Ch. 16, require that both the initial data and the conclusions be embedded in a *single framework*, whereas we have employed two different consistent families, $\mathcal{A}$ and $\mathcal{B}$. In addition, in order to reach a contradiction we used the assertion that A and B are mutually exclusive, and this requires a third framework C, since A does not include B and B does not include A at t_1. If the frameworks A, B, and C were compatible with each other, as is always the case in classical physics, there would be no problem, for the inferences carried out in the separate frameworks would be equally valid in the common refinement. But, as we shall show, these frameworks are mutually incompatible, despite the fact that the history projectors commute with one another.

Any common refinement of A and B would have to contain, among other things, the first history in (22.29) and the first one in (22.31):

$$D \odot A \odot F, \quad D \odot B \odot F. \tag{22.33}$$

The product of these two history projectors is zero, since $AB = 0$, but the chain operators are *not* orthogonal to each other. If one works out the chain kets one finds that they are both equal to a nonzero constant times $|F\rangle$. Thus having the two histories in (22.33) in the same family will violate the consistency conditions. A convenient choice for C is the family whose support is the three histories

$$D \odot \begin{cases} A \odot I, \\ B \odot I, \\ C \odot I. \end{cases} \tag{22.34}$$

Note that this is, in effect, a family of histories defined at only two times, t_0 and t_1, as I provides no information about what is going on at t_2, and for this reason it is automatically consistent, Sec. 11.3. It is incompatible with A because a common refinement would have to include the two histories

$$D \odot A \odot F, \quad D \odot B \odot I, \tag{22.35}$$

whose projectors are orthogonal, but whose chain kets are not, and it is likewise incompatible with B.

Thus the paradox arises because of reasoning in a way which violates the single-framework rule, and in this respect it resembles the two-spin paradox of Sec. 22.3. An important difference is that the incompatibility between frameworks in the case of two spins results from the fact that some of the nine operators in (22.14) do not commute with each other, whereas in the three-box paradox the projectors for the histories commute with each other, and incompatibility arises because the consistency conditions are not fulfilled in a common refinement.

Rewording the paradox in a slightly different way may assist in understanding why some types of inference which seem quite straightforward in terms of ordinary reasoning are not valid in quantum theory. Let us suppose that we have used the family A together with the initial data of D at t_0 and F at t_2 to reach the conclusion that at time t_1 A is true and $\tilde{A} = I - A$ is false. Since $\tilde{A} = B + C$, it seems natural to conclude that both B and C are false, contradicting the result (from framework

$\mathcal{B}$) that B is true. The step from the falsity of $\tilde{A}$ to the falsity of B as a consequence of $\tilde{A} = B + C$ would be justified in classical mechanics by the following rule: If $P = Q + R$ is an indicator which is the sum of two other indicators, and P is false, meaning $P(\gamma) = 0$ for the phase point γ describing the physical system, then $Q(\gamma) = 0$ and $R(\gamma) = 0$, so both Q and R are false. For example, if the energy of a system is not in the range between 10 and 20 J, then it is not between 10 and 15 J, nor is it between 15 and 20 J.

The corresponding rule in quantum physics states that if the projector P is the sum of two projectors Q and R, and P is known to be false, then *if Q and R are part of the Boolean algebra of properties entering into the logical discussion,* both Q and R are false. The words in italics apply equally to the case of classical reasoning, but they are usually ignored, because if Q is not among the list of properties available for discussion, it can always be added, and $R = P - Q$ added at the same time to ensure that the properties form a Boolean algebra. In classical physics there is never any problem with adding a property which has not previously come up in the discussion, and therefore the rule in italics can safely be relegated to the dusty books on formal logic which scientists put off reading until after they retire. However, in quantum theory it is by no means the case that Q (and therefore $R = P - Q$) can always be added to the list of properties or events under discussion, and this is why the words in italics are extremely important. If by using the family $\mathcal{A}$ we have come to the conclusion that $\tilde{A} = B + C$ is false, and, as is in fact the case, B at t_1 *cannot* be added to this family while maintaining consistency, then B has to be regarded as meaningless from the point of view of the discussion based upon $\mathcal{A}$, and something which is meaningless cannot be either true or false.

One can also think about it as follows. The physicist who first uses the initial data and framework $\mathcal{A}$ to conclude that A was true at t_1, and then inserts B at t_1 into the discussion has, in effect, changed the framework to something other than $\mathcal{A}$. In classical physics such a change in framework causes no problems, and it certainly does not alter the correctness of a conclusion reached earlier in a framework which made no mention of B. But in the quantum case, adding B means that something else must be changed in order to ensure that one still has a consistent framework. Since A occurs at the end of the previous step of the argument and is thus still at the center of attention, the physicist who introduces B is unconsciously (which is what makes the move so dangerous!) shifting to a framework, such as (22.34), in which either D at t_0 or F at t_2 has been forgotten. But as the new framework does not include the initial data, it is no longer possible to derive the truth of A. Hence adding B to the discussion in this manner is, relative to the truth of A, rather like sawing off the branch on which one is seated, and the whole argument comes crashing to the ground.

22.6 Truth functionals for histories

The notion of a truth functional can be applied to histories as well as to properties of a quantum system at a single time, and makes perfectly good sense as long as one considers a *single* framework or consistent family, based upon a sample space consisting of some decomposition of the history identity into elementary histories, as discussed in Sec. 8.5. Given this framework, one and only one of its elementary histories will actually occur, or be true, for a single quantum system during a particular time interval or run. A truth functional is then a function which assigns 1 (true) to a particular elementary history, 0 (false) to the other elementary histories, and 1 or 0 to other members of the Boolean algebra of histories using a formula which is the obvious analog of (22.21). The number of distinct truth functionals will typically be less than the number of elementary histories, since one need not count histories with zero weight — they are dynamically impossible, so they never occur — and certain elementary histories will be excluded by the initial data, such as an initial state.

A universal truth functional θ_u for histories can be defined in a manner analogous to a universal truth functional for properties, Sec. 22.4. We assume that θ_u assigns a value, 1 or 0, to every projector representing a history which is not intrinsically inconsistent (Sec. 11.8), that is, any history which is a member of at least one consistent family, and that this assignment satisfies the first two conditions of (22.24) and the third condition whenever it makes sense. That is, (22.25) should hold when P and Q are two histories belonging to the same consistent family (which implies, among other things, that $PQ = QP$). For the purposes of the following discussion it will be convenient to denote by $\mathcal{T}$ the collection of all true histories, the histories to which θ_u assigns the value 1. Given that θ_u satisfies these conditions, it is not hard to see that when it is restricted to a particular consistent family or framework $\mathcal{F}$, that is, regarded as a function on the histories belonging to this family, it will coincide with one of the "ordinary" truth functionals for this family, and therefore $\mathcal{T} \cap \mathcal{F}$, the subset of all true histories belonging to $\mathcal{F}$, will consist of one elementary history and all compound histories which contain this particular elementary history. In particular, θ_u can never assign the value 1 to two distinct elementary histories belonging to the same framework.

Since a decomposition of the identity at a single time is an example, albeit a rather trivial one, of a consistent family of one-time "histories", it follows that there can be no truly universal truth functional for histories of a quantum system whose Hilbert space is of dimension 3 or more. Nonetheless, it interesting to see how the three-box paradox of the previous section provides an explicit example, with non-trivial histories, of a circumstance in which there is no universal truth functional. Imagine it as an experiment which is repeated many times, always starting with the same initial state D. The universal truth functional and the corresponding list T of true histories will vary from one run to the next, since different histories will occur in different runs. Think of a run (it will occur with a probability of 1/9) in which the final state is F, so the history $D \odot I \odot F$ is true, and therefore an element of T. What other histories belong to T, and are thus assigned the value 1 by the universal truth functional θ_u?

Consider the consistent family $\mathcal{A}$ whose sample space is shown in (22.29), aside from histories of zero weight to which θ_u will always assign the value 0. One and only one of these histories must be true, so it is the history $D \odot A \odot F$, as the other two terminate in $\tilde{F}$. From this we can conclude, using the counterpart of (22.21), that $D \odot A \odot I$, which belongs to the Boolean algebra of $\mathcal{A}$, is true, a member of T. Following the same line of reasoning for the consistent family $\mathcal{B}$, we conclude that $D \odot B \odot F$ and $D \odot B \odot I$ are elements of T. But now consider the consistent family $\mathcal{C}$ with sample space (22.34). One and only one of these three elementary histories can belong to T, and this contradicts the conclusion we reached previously using $\mathcal{A}$ and $\mathcal{B}$, that *both $D \odot A \odot I$ and $D \odot B \odot I$* belong to T.

Our analysis does not by itself rule out the possibility of a universal truth functional which assigns the value 0 to the history $D \odot I \odot F$, and could be used in a run in which $D \odot I \odot \tilde{F}$ occurs. But it shows that the concept can, at best, be of rather limited utility in the quantum domain, despite the fact that it works without any difficulty in classical physics. Note that quantum truth functionals form a perfectly valid procedure for analyzing histories (and properties at a single time) as long as one restricts one's attention to a *single framework*, a single consistent family. With this restriction, quantum truth as it is embodied in a truth functional behaves in much the same way as classical truth. It is only when one tries to extend this concept of truth to something which applies simultaneously to different incompatible frameworks that problems arise.

23

Singlet state correlations

23.1 Introduction

This and the following chapter can be thought of as a single unit devoted to discussing various issues raised by a famous paper published by Einstein, Podolsky, and Rosen in 1935, in which they claimed to show that quantum mechanics, as it was understood at that time, was an incomplete theory. In particular, they asserted that a quantum wave function cannot provide a complete description of a quantum system. What they were concerned with was the problem of assigning simultaneous values to noncommuting operators, a topic which has already been discussed to some extent in Ch. 22. Their strategy was to consider an entangled state (see the definition in Sec. 6.2) of two spatially separated systems, and they argued that by carrying out a measurement on one system it was possible to determine a property of the other.

A simple example of an entangled state of spatially separated systems involves the spin degrees of freedom of two spin-half particles that are in different regions of space. In 1951 Bohm pointed out that the claim of the Einstein, Podolsky, and Rosen paper, commonly referred to as EPR, could be formulated in a simple way in terms of a singlet state of two spins, as defined in (23.2) below. Much of the subsequent discussion of the EPR problem has followed Bohm's lead, and that is the approach adopted in this and the following chapter. In this chapter we shall discuss various histories for two spin-half particles initially in a singlet state, and pay particular attention to the statistical correlations between the two spins. The basic correlation function which enters many discussions of the EPR problem is evaluated in Sec. 23.2 using histories involving just two times. A number of families of histories involving three times are considered in Sec. 23.3, while Sec. 23.4 discusses what happens when a spin measurement is carried out on one particle, and Sec. 23.5 the case of measurements of both particles.

The results found in this chapter may seem a bit dull and repetitious, and the reader who finds them so should skip ahead to the next chapter where the EPR problem itself, in Bohm's formulation, is stated in Sec. 24.1 in the form of a paradox, and the paradox is explored using various results derived in the present chapter. An alternative way of looking at the paradox using counterfactuals is discussed in Sec. 24.2. The remainder of Ch. 24 deals with an alternative approach to the EPR problem in which one adds an additional mathematical structure, usually referred to as "hidden variables", to the standard quantum Hilbert space of wave functions. A simple example of hidden variables in the context of measurements on particles in a spin singlet state, due to Mermin, is the topic of Sec. 24.3. It disagrees with the predictions of quantum theory for the spin correlation function, and this disagreement is not a coincidence, for Bell has shown by means of an inequality that *any* hidden variables theory of this sort *must* disagree with the predictions of quantum theory. The derivation of this inequality is taken up in Sec. 24.4, which also contains some remarks on its significance for the (non)existence of mysterious nonlocal influences in the quantum world.

23.2 Spin correlations

Imagine two spin-half particles a and b traveling away from each other in a region of zero magnetic field (so the spin direction of each particle will remain fixed), which are described by a wave function

$$|\chi_t\rangle = |\psi_0\rangle \otimes |\omega_t\rangle, \tag{23.1}$$

where $|\omega_t\rangle$ is a wave packet $\omega(\mathbf{r}_a, \mathbf{r}_b, t)$ describing the positions of the two particles, while

$$|\psi_0\rangle = \left(|z_a^+\rangle|z_b^-\rangle - |z_a^-\rangle|z_b^+\rangle\right)/\sqrt{2} \tag{23.2}$$

is the singlet state of the spins of the two particles, the state with total angular momentum equal to 0. Hereafter we shall ignore $|\omega_t\rangle$, as it plays no essential role in the following arguments, and concentrate on the spin state $|\psi_0\rangle$.

Rather than using eigenstates of S_{az} and S_{bz}, $|\psi_0\rangle$ can be written equally well in terms of eigenstates of S_{aw} and S_{bw}, where w is some direction in space described by the polar angles ϑ and φ. The states $|w^+\rangle$ and $|w^-\rangle$ are given as linear combinations of $|z^+\rangle$ and $|z^-\rangle$ in (4.14), and using these expressions one can rewrite $|\psi_0\rangle$ in the form

$$|\psi_0\rangle = (|w_a^+\rangle|w_b^-\rangle - |w_a^-\rangle|w_b^+\rangle)/\sqrt{2}, \tag{23.3}$$

or as

$$|\psi_0\rangle = \sin(\vartheta/2)\left(e^{-i\varphi/2}|z_a^+\rangle|w_b^+\rangle + e^{i\varphi/2}|z_a^-\rangle|w_b^-\rangle\right)/\sqrt{2}$$
$$+ \cos(\vartheta/2)\left(e^{-i\varphi/2}|z_a^+\rangle|w_b^-\rangle - e^{i\varphi/2}|z_a^-\rangle|w_b^+\rangle\right)/\sqrt{2}, \qquad (23.4)$$

where ϑ and φ are the polar angles for the direction w, with w the positive z axis when $\vartheta = 0$. The fact that $|\psi_0\rangle$ has the same functional form in (23.3) as in (23.2) reflects the fact that this state is spherically symmetrical, and thus does not single out any particular direction in space.

Consider the consistent family whose support is a set of four histories at the two times $t_0 < t_1$:

$$\psi_0 \odot \{z_a^+, z_a^-\}\{w_b^+, w_b^-\}, \qquad (23.5)$$

where the product of the two curly brackets stands for the set of four projectors $z_a^+ w_b^+, z_a^+ w_b^-, z_a^- w_b^+$, and $z_a^- w_b^-$. The time development operator $T(t_1, t_0)$ is equal to I, since we are only considering the spins and not the spatial wave function $\omega(\mathbf{r}_a, \mathbf{r}_b, t)$. Thus one can calculate the probabilities of these histories, or of the events at t_1 given ψ_0 at t_0, by thinking of $|\psi_0\rangle$ in (23.4) as a pre-probability and using the absolute squares of the corresponding coefficients. The result is:

$$\Pr(z_a^+, w_b^+) = \Pr(z_a^-, w_b^-) = \tfrac{1}{2}\sin^2(\vartheta/2) = (1 - \cos\vartheta)/4,$$
$$\Pr(z_a^+, w_b^-) = \Pr(z_a^-, w_b^+) = \tfrac{1}{2}\cos^2(\vartheta/2) = (1 + \cos\vartheta)/4, \qquad (23.6)$$

where one could also write $\Pr(z_a^+ \wedge w_b^+)$ in place of $\Pr(z_a^+, w_b^+)$ for the probability of $S_{az} = +1/2$ and $S_{bw} = +1/2$. Using these probabilities one can evaluate the *correlation function*

$$C(z, w) = \langle(2S_{az})(2S_{bw})\rangle = 4\langle\psi_0|S_{az}S_{bw}|\psi_0\rangle =$$
$$\Pr(z_a^+, w_b^+) + \Pr(z_a^-, w_b^-) - \Pr(z_a^+, w_b^-) - \Pr(z_a^-, w_b^+) = -\cos\vartheta. \qquad (23.7)$$

Because $|\psi_0\rangle$ is spherically symmetrical, one can immediately generalize these results to the case of a family of histories in which the directions z and w in (23.5) are replaced by arbitrary directions w_a and w_b, which can conveniently be written in the form of unit vectors $\mathbf{a}$ and $\mathbf{b}$. Since the cosine of the angle between $\mathbf{a}$ and $\mathbf{b}$ is equal to the dot product $\mathbf{a} \cdot \mathbf{b}$, the generalization of (23.6) is

$$\Pr(\mathbf{a}^+, \mathbf{b}^+) = \Pr(\mathbf{a}^-, \mathbf{b}^-) = (1 - \mathbf{a} \cdot \mathbf{b})/2,$$
$$\Pr(\mathbf{a}^+, \mathbf{b}^-) = \Pr(\mathbf{a}^-, \mathbf{b}^+) = (1 + \mathbf{a} \cdot \mathbf{b})/2, \qquad (23.8)$$

while the correlation function (23.7) is given by

$$C(\mathbf{a}, \mathbf{b}) = -\mathbf{a} \cdot \mathbf{b}. \qquad (23.9)$$

As will be shown in Sec. 23.5, $C(\mathbf{a}, \mathbf{b})$ is also the correlation function for the outcomes, expressed in a suitable way, of measurements of the spin components of particles a and b in the directions **a** and **b**.

23.3 Histories for three times

Let us now consider various families of histories for the times $t_0 < t_1 < t_2$, assuming an initial state ψ_0 at t_0. One possibility is a unitary history with ψ_0 at all three times, but in addition there are various stochastic histories. As a first example, consider the consistent family whose support consists of the two histories

$$\psi_0 \odot \begin{cases} z_a^+ z_b^- \odot z_a^+ z_b^-, \\ z_a^- z_b^+ \odot z_a^- z_b^+. \end{cases} \tag{23.10}$$

Each history carries a weight of $1/2$ and describes a situation in which $S_{bz} = -S_{az}$, with values which are independent of time for $t > t_0$. In particular, one has conditional probabilities

$$\Pr(z_{a1}^+ \mid z_{a2}^+) = \Pr(z_{b1}^- \mid z_{a2}^+) = \Pr(z_{b2}^- \mid z_{a2}^+) = 1, \tag{23.11}$$

$$\Pr(z_{a1}^- \mid z_{b1}^+) = \Pr(z_{a2}^- \mid z_{b1}^+) = \Pr(z_{b2}^+ \mid z_{b1}^+) = 1, \tag{23.12}$$

among others, where the time, t_1 or t_2, at which an event occurs is indicated by a subscript 1 or 2. Thus if $S_{az} = +1/2$ at t_2, then it had this same value at t_1, and one can be certain that S_{bz} has the value $-1/2$ at both t_1 and t_2.

Because of spherical symmetry, the same sort of family can be constructed with z replaced by an arbitrary direction w. In particular, with $w = x$, we have a family with support

$$\psi_0 \odot \begin{cases} x_a^+ x_b^- \odot x_a^+ x_b^-, \\ x_a^- x_b^+ \odot x_a^- x_b^+. \end{cases} \tag{23.13}$$

Again, each history has a weight of $1/2$, and now it is the values of S_{ax} and S_{bx} which are of opposite sign and independent of time, and the results in (23.11) and (23.12) hold with z replaced by x. The two families (23.10) and (23.13) are obviously incompatible with each other because the projectors for one family do not commute with those of the other. There is no way in which they can be combined in a single description, and the corresponding conditional probabilities cannot be related to one another, since they are defined on separate sample spaces.

One can also consider a family in which a stochastic branching takes place between t_1 and t_2 instead of between t_0 and t_1; thus (23.10) can be replaced with

$$\psi_0 \odot \psi_0 \odot \{z_a^+ z_b^-, z_a^- z_b^+\}. \tag{23.14}$$

In this case the last equality in (23.11) remains valid, but the other conditional probabilities in (23.11) and (23.12) are undefined, because (23.14) does not contain projectors corresponding to values of S_{az} and S_{bz} at time t_1, and they cannot be added to this family, as they do not commute with ψ_0.

One need not limit oneself to families in which the same component of spin angular momentum is employed for both particles. The four histories

$$\psi_0 \odot \begin{cases} z_a^+ x_b^+ \odot z_a^+ x_b^+, \\ z_a^+ x_b^- \odot z_a^+ x_b^-, \\ z_a^- x_b^+ \odot z_a^- x_b^+, \\ z_a^- x_b^- \odot z_a^- x_b^- \end{cases} \qquad (23.15)$$

form the support of a consistent family. Since they all have equal weight, one has conditional probabilities

$$\Pr(x_b^+ \,|\, z_a^+) = 1/2 = \Pr(x_b^- \,|\, z_a^+), \qquad (23.16)$$

and others of a similar type which hold for events at both t_1 and t_2, which is why subscripts 1 and 2 have been omitted. In addition, the values of S_{az} and S_{bx} do not change with time:

$$\begin{aligned} \Pr(z_{a2}^+ \,|\, z_{a1}^+) &= 1 = \Pr(z_{a2}^- \,|\, z_{a1}^-), \\ \Pr(x_{b2}^+ \,|\, x_{b1}^+) &= 1 = \Pr(x_{b2}^- \,|\, x_{b1}^-). \end{aligned} \qquad (23.17)$$

Yet another consistent family, with support

$$\psi_0 \odot \begin{cases} z_a^+ z_b^- \odot z_a^+ \{x_b^+, x_b^-\}, \\ z_a^- z_b^+ \odot z_a^- \{x_b^+, x_b^-\}, \end{cases} \qquad (23.18)$$

where $z_a^+ \{x_b^+, x_b^-\}$ denotes the pair of projectors $z_a^+ x_b^+$ and $z_a^+ x_b^-$, combines features of (23.10) and (23.15): values of S_{az} are part of the description at both t_1 and t_2, but in the case of particle b, two separate components S_{bz} and S_{bx} are employed at t_1 and t_2. It is important to notice that this change is *not* brought about by any dynamical effect; instead, it is simply a consequence of using (23.18) rather than (23.10) or (23.15) as the framework for constructing the stochastic description. In particular, one can have a history in which $S_{bz} = +1/2$ at t_1 and $S_{bx} = -1/2$ at t_2. This does not mean that some torque is present which rotates the direction of the spin from the $+z$ to the $-x$ direction, for there is nothing which could produce such a torque. See the discussion following (9.33) in Sec. 9.3.

The families of histories considered thus far all satisfy the consistency conditions, as is clear from the fact that the final projectors are mutually orthogonal. Given that three times are involved, inconsistent families are also possible. Here is one which will be discussed later from the point of view of measurements. It contains the sixteen histories which can be represented in the compact form

$$\psi_0 \odot \{x_a^+, x_a^-\} \{z_b^+, z_b^-\} \odot \{z_a^+, z_a^-\} \{x_b^+, x_b^-\}, \qquad (23.19)$$

where the product of curly brackets at each of the two times stands for a collection of four projectors, as in (23.5). Each history makes use of one of the four projectors at each of the two times; for example,

$$\psi_o \odot x_a^+ z_b^- \odot z_a^- x_b^- \tag{23.20}$$

is one of the sixteen histories. Each of these histories has a finite weight, and the chain kets of the four histories ending in $z_a^- x_b^-$, to take an example, are all proportional to $|z_a^-\rangle |x_b^-\rangle$, so cannot be orthogonal to each other.

23.4 Measurements of one spin

Suppose that the z-component S_{az} of the spin of particle a is measured using a Stern–Gerlach apparatus as discussed in Ch. 17. The initial state of the apparatus is $|Z_a^\circ\rangle$, and its interaction with the particle during the time interval from t_1 to t_2 gives rise to a unitary time evolution

$$|z_a^+\rangle |Z_a^\circ\rangle \mapsto |Z_a^+\rangle, \quad |z_a^-\rangle |Z_a^\circ\rangle \mapsto |Z_a^-\rangle, \tag{23.21}$$

where $|Z_a^+\rangle$ and $|Z_a^-\rangle$ are apparatus states ("pointer positions") indicating the two possible outcomes of the measurement. Note that the spin states no longer appear on the right side; we are assuming that at t_2 the spin-half particle has become part of the measuring apparatus. (Thus (23.21) represents a destructive measurement in the terminology of Sec. 17.1. One could also consider nondestructive measurements in which the value of S_{az} is the same after the measurement as it is before, by using (18.17) in place of (23.21), but these will not be needed for the following discussion.) The b particle has no effect on the apparatus, and vice versa. That is, one can place an arbitrary spin state $|w_b^+\rangle$ for the b particle on both sides of the arrows in (23.21).

Consider the consistent family with support

$$\Psi_0^z \odot \begin{cases} z_a^+ z_b^- \odot Z_a^+ z_b^-, \\ z_a^- z_b^+ \odot Z_a^- z_b^+, \end{cases} \tag{23.22}$$

where the initial state is $|\Psi_0^z\rangle = |\psi_0\rangle |Z_a^\circ\rangle$. The conditional probabilities

$$\Pr(z_{a1}^+ \mid Z_{a2}^+) = 1 = \Pr(z_{a1}^- \mid Z_{a2}^-), \tag{23.23}$$

$$\Pr(z_{b1}^- \mid Z_{a2}^+) = 1 = \Pr(z_{b1}^+ \mid Z_{a2}^-), \tag{23.24}$$

$$\Pr(z_{b2}^- \mid Z_{a2}^+) = 1 = \Pr(z_{b2}^+ \mid Z_{a2}^-), \tag{23.25}$$

$$\Pr(z_{b2}^+ \mid z_{b1}^+) = 1 = \Pr(z_{b2}^- \mid z_{b1}^-) \tag{23.26}$$

are an obvious consequence of (23.22). The first pair, (23.23), tell us that the measurement is, indeed, a measurement: the outcomes $Z^\pm$ actually reveal values of S_{az} before the measurement took place. Those in (23.24) and (23.25) tell us that the measurement is also an *indirect* measurement of S_{bz} for particle b, even though this particle never interacts with the apparatus that measures S_{az}, since the measurement outcomes Z_a^+ and Z_a^- are correlated with the properties z_b^- and z_b^+.

There is nothing very surprising about carrying out an indirect measurement of the property of a distant object in this way, and the ability to do so does not indicate any sort of mysterious long-range or nonlocal influence. Consider the following analogy. Two slips of paper, one red and one green, are placed in separate opaque envelopes. One envelope is mailed to a scientist in Atlanta and the other to a scientist in Boston. When the scientist in Atlanta opens the envelope and looks at the slip of paper, he can immediately infer the color of the slip in the envelope in Boston, and for this reason he has, in effect, carried out an indirect measurement. Furthermore, this measurement indicates the color of the slip of paper in Boston not only at the time the measurement is carried out, but also at earlier and later times, assuming the slip in Boston does not undergo some process which changes its color. In the same way, the outcome, Z_a^+ or Z_a^-, for the measurement of S_{az} allows one to infer the value of S_{bz} both at t_1 and at t_2, and at later times as well if one extends the histories in (23.22) in an appropriate manner. In order for this inference to be correct, it is necessary that particle b not interact with anything, such as a measuring device or magnetic field, which could perturb its spin.

The conditional probabilities in (23.26) tell us that S_{bz} is the same at t_2 as at t_1, consistent with our assumption that particle b has not interacted with anything during this time interval. Note, in particular, that carrying out a measurement on S_{az} has no influence on S_{bz}, which is just what one would expect, since particle b is isolated from particle a, and from the measuring apparatus, at all times later than t_0.

A similar discussion applies to a measurement carried out on some other component of the spin of particle a. To measure S_{ax}, what one needs is an apparatus initially in the state $|X_a^\circ\rangle$, which during the time interval from t_1 to t_2 interacts with particle a in such a way as to give rise to the unitary time transformation

$$|x_a^+\rangle|X_a^\circ\rangle \mapsto |x_a^+\rangle|X_a^+\rangle, \quad |x_a^-\rangle|X_a^\circ\rangle \mapsto |x_a^-\rangle|X_a^-\rangle. \tag{23.27}$$

The counterpart of (23.22) is the consistent family with support

$$\Psi_0^x \odot \begin{cases} x_a^+ x_b^- \odot X_a^+ x_b^-, \\ x_a^- x_b^+ \odot X_a^- x_b^+, \end{cases} \tag{23.28}$$

where the initial state is now $|\Psi_0^x\rangle = |\psi_0\rangle|X_a^\circ\rangle$. Using this family, one can calculate probabilities analogous to those in (23.23)–(23.26), with z and Z replaced by x and X. Thus in this framework a measurement of S_{ax} is an indirect measurement of S_{bx}, and one can show that the measurement has no effect upon S_{bx}.

Comparing (23.22) with (23.10), or (23.28) with (23.13) shows that the families which describe measurement results are close parallels of those describing the system of two spins in the absence of any measurements. To include the measurement, one simply introduces an appropriate initial state at t_0, and replaces one of the lower case letters at t_2 with the corresponding capital to indicate a measurement outcome. This should come as no surprise: apparatus designed to measure some property will, if it is working properly, measure that property. Once one knows how to describe a quantum system in terms of its microscopic properties, the addition of a measurement apparatus of an appropriate type will simply confirm the correctness of the microscopic description.

Replacing lower case with capital letters can also be used to construct measurement counterparts of other consistent families in Sec. 23.3. The counterpart of (23.14) when S_{az} is measured is the family with support

$$\Psi_0^z \odot \Psi_0^z \odot \{Z_a^+ z_b^-, \ Z_a^- z_b^+\}. \tag{23.29}$$

Using this family one can deduce the conditional probabilities in (23.25) referring to the values of S_{bz} at t_2, and thus the measurement of S_{az}, viewed within this framework, is again an indirect measurement of S_{bz} at t_2. However, the results in (23.23), (23.24), and (23.26) are not valid for the family (23.29), because values of S_{az} and S_{bz} cannot be defined at t_1: the corresponding projectors do not commute with the ψ_0 part of Ψ_0^z.

One reason for introducing (23.29) is that it is the family which comes closest to representing the idea that a measurement is associated with a collapse of the wave function of the measured system. In the case at hand, the measured system can be thought of as the spin state of the two particles, but since particle a is no longer relevant to the discussion at t_2, collapse should be thought of as resulting in a state $|z_b^-\rangle$ or $|z_b^+\rangle$ for particle b, depending upon whether the measurement outcome is Z_a^+ or Z_a^-. (In the case of a nondestructive measurement on particle a the states resulting from the collapse would be $|z_a^+\rangle|z_b^-\rangle$ and $|z_a^-\rangle|z_b^+\rangle$.) As pointed out in Sec. 18.2, wave function collapse is basically a mathematical procedure for computing certain types of conditional probabilities. Regarding it as some sort of physical process gives rise to a misleading picture of instantaneous influences which can travel faster than the speed of light. The remarks in Sec. 18.2 with reference to the beam splitter in Fig. 18.1 apply equally well to spatially separated systems of spin-half particles, or of photons, etc.

One way to see that the measurement of S_{az} is not a process which somehow brings S_{bz} into existence at t_2 is to note that the change between t_1 and the final time t_2 in (23.29) is similar to the change which occurs in the family (23.14), where there is no measurement. Another way to see this is to consider the family whose support consists of the four histories

$$\Psi_0^z \odot \Psi_0^z \odot \{Z_a^+, Z_a^-\}\{x_b^+, x_b^-\} \qquad (23.30)$$

in the compact notation used earlier in (23.5). This resembles (23.29), except that the components of S_{bx} rather than S_{bz} appear at t_2. Were the measurement having some physical effect on particle b, it would be just as sensible to suppose that it produces random values of S_{bx}, as that it results in a value of S_{bz} correlated with the outcome of the measurement!

It was noted earlier that (23.26) implies that measuring S_{az} has no effect upon S_{bz}. Nor does such a measurement influence any other component of the spin of particle b, as can be seen by constructing an appropriate consistent family in which this component enters the description at both t_1 and t_2. Thus in the case of S_{bx} one can use the measurement counterpart of (23.15), a family with support

$$\Psi_0^z \odot \begin{cases} z_a^+ x_b^+ \odot Z_a^+ x_b^+, \\ z_a^+ x_b^- \odot Z_a^+ x_b^-, \\ z_a^- x_b^+ \odot Z_a^- x_b^+, \\ z_a^- x_b^- \odot Z_a^- x_b^-. \end{cases} \qquad (23.31)$$

It is then evident by inspection that S_{bx} is the same at t_1 and t_2. Using this family one obtains the conditional probabilities

$$\begin{aligned} \Pr(x_b^+ \mid Z_{a2}^+) &= 1/2 = \Pr(x_b^- \mid Z_{a2}^+), \\ \Pr(x_b^+ \mid Z_{a2}^-) &= 1/2 = \Pr(x_b^- \mid Z_{a2}^-), \end{aligned} \qquad (23.32)$$

where the subscript indicating the time has been omitted from $x_b^{\pm}$, since these results apply equally at t_1 and t_2. Of course (23.32) is nothing but the measurement counterpart of (23.16). It tells one that a measurement of S_{az} can in no way be regarded as an indirect measurement of S_{bx}. Similar results are obtained if the projectors corresponding to S_{bx} in (23.31) are replaced by those corresponding to S_{bw} for some other direction w, except that the conditional probabilities for w_b^+ and w_b^- in the expression corresponding to (23.32) will depend upon w. If w is close to z, a measurement of S_{az} is an approximate indirect measurement of S_{bw} in the sense that $S_{bw} = -S_{az}$ for most experimental runs, with occasional errors.

The family

$$\Psi_0^z \odot \begin{cases} z_a^+ z_b^- \odot Z_a^+ \{x_b^+, x_b^-\}, \\ z_a^- z_b^+ \odot Z_a^- \{x_b^+, x_b^-\} \end{cases} \qquad (23.33)$$

is the counterpart of (23.18) when S_{az} is measured. Here the events involving the spin of particle b are different at t_2 from what they are at t_1. However, just as in the case of (23.18), for which no measurement occurs, one should not think of this change as a physical consequence of the measurement. See the discussion following (23.18).

23.5 Measurements of two spins

Thus far we have only considered measurements on particle a. One can also imagine carrying out measurements on the spins of both particles. All that is needed is a second measuring device of a type appropriate for whatever component of the spin of particle b is of interest. If, for example, this is S_{bx}, then the unitary time transformation from t_1 to t_2 will be the same as (23.27) except for replacing the subscript a with b. In what follows it will be convenient to assume that measurements are carried out on both particles at the same time. However, this is not essential; analogous results are obtained if measurements are carried out at different times. The properties of a particle will, in general, be different before and after it is measured, but the time at which a measurement is carried out on the other particle is completely irrelevant.

For the combined system of two particles and two measuring devices a typical unitary transformation from t_1 to t_2 takes the form:

$$|z_a^+\rangle |x_b^-\rangle |Z_a^\circ\rangle |X_b^\circ\rangle \mapsto |Z_a^+\rangle |X_b^-\rangle. \qquad (23.34)$$

Once again, one can generate consistent families for measurements by starting off with any of the consistent families in Sec. 23.3, replacing ψ_0 with an appropriate initial state which includes each apparatus in its ready state, and then replacing lower case letters at the final time t_2 with corresponding capitals. For example

$$\Psi_0^{zz} \odot \begin{cases} z_a^+ z_b^- \odot Z_a^+ Z_b^-, \\ z_a^- z_b^+ \odot Z_a^- Z_b^+, \end{cases} \qquad (23.35)$$

with $|\Psi_0^{zz}\rangle = |\psi_0\rangle |Z_a^\circ\rangle |Z_b^\circ\rangle$, is the counterpart of (23.10), and it shows that the outcomes of measurements of S_{az} and S_{bz} will be perfectly anticorrelated:

$$\Pr(Z_b^- \mid Z_a^+) = 1 = \Pr(Z_b^+ \mid Z_a^-). \qquad (23.36)$$

Not only does one obtain consistent families by this process of "capitalizing" those in Sec. 23.3, the *weights* for histories involving measurements are also precisely the same as their counterparts that involve only particle properties. This means that the correlation function $C(\mathbf{a}, \mathbf{b})$ introduced in Sec. 23.1 can be applied to measurement outcomes as well as to microscopic properties. To do this, let $\alpha(w)$ be $+1$ if the apparatus designed to measure S_{aw} is in the state $|W_a^+\rangle$ at t_2, and -1 if it is in the state $|W_a^-\rangle$, and define $\beta(w)$ in the same way for measurements on particle b. Then we can write

$$C(\mathbf{a}, \mathbf{b}) = \langle \alpha(w_a)\beta(w_b)\rangle = -\mathbf{a} \cdot \mathbf{b} \qquad (23.37)$$

as the average over a large number of experimental runs of the product $\alpha(w_a)\beta(w_b)$ when w_a is **a** and w_b is **b**.

The physical significance of C in (23.37) is, of course, different from that in (23.9). The former refers to measurement outcomes and the latter to properties of the two particles. However, they are identical functions of **a** and **b**, and given that the measurements accurately reflect previous values of the corresponding spin components, no confusion will arise from using the same symbol in both cases. One could also, to be sure, define the same sort of correlation for a case in which a spin component is measured for only one particle, using the product of the outcome of that measurement, understood as ±1, with twice the value (in units of $\hbar$) of the appropriate spin component for the other particle; for example

$$C(w, w') = \langle \alpha(w)2S_{bw'} \rangle. \tag{23.38}$$

As noted in Sec. 23.4, the outcome of a measurement of the z component of the spin of particle a can be used to infer the value of S_{az} before the measurement, and the value of S_{bz} for particle b as long as that particle remains isolated. The roles of particles a and b can be interchanged: a measurement of S_{bz} for particle b allows one to infer the value of S_{az}. And because of the spherical symmetry of ψ_0, the same results hold if z is replaced by any other direction w. How are these results modified, or extended, if the spins of *both* particles are measured? If the same component of spin is measured for particle b as for particle a, the results are just what one would expect. Suppose it is the z component, then (23.35) shows that one can infer both z_a^+ and z_b^- on the basis of the outcome Z_a^+, or of the outcome Z_b^-, a result which is not surprising since one outcome implies the other, (23.36).

Things become more complicated if the a and b measurements involve different components, and in this case it is necessary to pay careful attention to the framework one is using for inferring microscopic properties from the outcomes of the measurements. To illustrate this, let us suppose that S_{az} is measured for particle a and S_{bx} for particle b. One consistent family that can be used for analyzing this situation is the counterpart of (23.15):

$$\Psi_0^{zx} \odot \begin{cases} z_a^+ x_b^+ \odot Z_a^+ X_b^+, \\ z_a^+ x_b^- \odot Z_a^+ X_b^-, \\ z_a^- x_b^+ \odot Z_a^- X_b^+, \\ z_a^- x_b^- \odot Z_a^- X_b^-. \end{cases} \tag{23.39}$$

Here the initial state is $|\Psi_0^{zx}\rangle = |\psi_0\rangle|Z_a^\circ\rangle|X_b^\circ\rangle$. Using this family allows one to infer from the outcome of each measurement something about the spin of the same particle at an earlier time, but nothing about the spin of the other particle. Thus one has

$$\Pr(z_{a1}^+ \mid Z_{a2}^+) = 1 = \Pr(z_{a1}^- \mid Z_{a2}^-), \tag{23.40}$$

$$\Pr(x_{b1}^+ \mid X_{b2}^+) = 1 = \Pr(x_{b1}^- \mid X_{b2}^-), \tag{23.41}$$

but there is no counterpart of (23.24) relating S_{bz} to $Z_a^\pm$, nor a way to relate S_{ax} to $X_b^\pm$, because the relevant projectors, such as z_b^+, are not present in (23.39) at t_1, nor can they be added, since they do not commute with the projectors which are already there.

On the other hand, the family with support

$$\Psi_0^{zx} \odot \begin{cases} z_a^+ z_b^- \odot Z_a^+ \{X_b^+, X_b^-\}, \\ z_a^- z_b^+ \odot Z_a^- \{X_b^+, X_b^-\}, \end{cases} \tag{23.42}$$

which is the counterpart of (23.18) and (23.33), can be used to infer values of S_{bz} from the outcomes $Z_a^\pm$. By using it, one obtains the conditional probabilities

$$\Pr(z_{b1}^- \mid Z_{a2}^+) = 1 = \Pr(z_{b1}^+ \mid Z_{a2}^-) \tag{23.43}$$

in addition to (23.40). However, if one uses (23.42) the outcome of the b measurement tells one nothing about S_{bx} at t_1. It is worth noting that a refinement of (23.42) in which additional events are added at a time $t_{1.5}$, so that the histories

$$\Psi_0^{zx} \odot \begin{cases} z_a^+ z_b^- \odot \begin{cases} z_a^+ x_b^+ \odot Z_a^+ X_b^+, \\ z_a^+ x_b^- \odot Z_a^+ X_b^-, \end{cases} \\ z_a^- z_b^+ \odot \begin{cases} z_a^- x_b^+ \odot Z_a^- X_b^+, \\ z_a^- x_b^- \odot Z_a^- X_b^- \end{cases} \end{cases} \tag{23.44}$$

are defined at $t_0 < t_1 < t_{1.5} < t_2$, is the support of a consistent family in which one can infer from X_b^+ or X_b^- at t_2 the value of S_{bx} at $t_{1.5}$, but not at an earlier time. As this is a refinement of (23.42), both (23.40) and (23.43) remain valid.

The consistent family with support

$$\Psi_0^{zx} \odot \begin{cases} x_a^+ x_b^- \odot \{Z_a^+, Z_a^-\} X_b^-, \\ x_a^- x_b^+ \odot \{Z_a^+, Z_a^-\} X_b^+ \end{cases} \tag{23.45}$$

is the counterpart of (23.42) with x rather than z-components at t_1. One can use it to infer the x-component of the spin of either particle at t_1 from the outcome of the S_{bx} measurement:

$$\Pr(x_{b1}^+ \mid X_{b2}^+) = 1 = \Pr(x_{b1}^- \mid X_{b2}^-), \tag{23.46}$$

$$\Pr(x_{a1}^- \mid X_{b2}^+) = 1 = \Pr(x_{a1}^+ \mid X_{b2}^-). \tag{23.47}$$

Given the conditional probabilities in (23.43) and (23.47), and no indication of the consistent families from which they were obtained, one might be tempted to combine them and draw the conclusion that for a run in which the measurement outcomes are, say, Z_a^+ and X_b^+ at t_2, both S_{ax} and S_{bz} had the value $-1/2$ at t_1:

$$\Pr(x_{a1}^- \wedge z_{b1}^- \mid Z_{a2}^+ \wedge X_{b2}^+) = 1. \tag{23.48}$$

This, however, is not correct. To begin with, the frameworks (23.42) and (23.45) are mutually incompatible because of the projectors at t_1, so they cannot be used to derive (23.48) by combining (23.43) with (23.47). Next, if one tries to construct a single consistent family in which it might be possible to derive (23.48), one runs into the following difficulty. A description which ascribes values to both S_{ax} and S_{bz} at t_1 requires a decomposition of the identity which includes the four projectors $x_a^+ z_b^+$, $x_a^+ z_b^-$, $x_a^- z_b^+$, and $x_a^- z_b^-$. This by itself is not a problem, but when combined with the four measurement outcomes, the result is the *inconsistent* family

$$\Psi_0^{zx} \odot \{x_a^+, x_a^-\}\{z_b^+, z_b^-\} \odot \{Z_a^+, Z_a^-\}\{X_b^+, X_b^-\} \tag{23.49}$$

obtained by replacing ψ_0 with Ψ_0^{zx} and capitalizing x and z at t_2 in (23.19). The same arguments used to show that (23.19) is inconsistent apply equally to (23.49); adding measurements does not improve things. Consequently, because it cannot be obtained using a consistent family, (23.48) is not a valid result.

24

EPR paradox and Bell inequalities

24.1 Bohm version of the EPR paradox

Einstein, Podolsky, and Rosen (EPR) were concerned with the following issue. Given two spatially separated quantum systems A and B and an appropriate initial entangled state, a measurement of a property on system A can be an indirect measurement of B in the sense that from the outcome of the A measurement one can infer with probability 1 a property of B, because the two systems are correlated. There are cases in which either of two properties of B represented by noncommuting projectors can be measured indirectly in this manner, and EPR argued that this implied that system B could possess two incompatible properties at the same time, contrary to the principles of quantum theory.

In order to understand this argument, it is best to apply it to a specific model system, and we shall do so using Bohm's formulation of the EPR paradox in which the systems A and B are two spin-half particles a and b in two different regions of space, with their spin degrees of freedom initially in a spin singlet state (23.2). As an aid to later discussion, we write the argument in the form of a set of numbered assertions leading to a paradox: a result which seems plausible, but contradicts the basic principles of quantum theory. The assertions E1–E4 are not intended to be exact counterparts of statements in the original EPR paper, even when the latter are translated into the language of spin-half particles. However, the general idea is very similar, and the basic conundrum is the same.

E1. Suppose S_{az} is measured for particle a. The result allows one to predict S_{bz} for particle b, since $S_{bz} = -S_{az}$.

E2. In the same way, the outcome of a measurement of S_{ax} allows one to predict S_{bx}, since $S_{bx} = -S_{ax}$.

E3. Particle b is isolated from particle a, and therefore it cannot be affected by measurements carried out on particle a.

E4. Consequently, particle b must simultaneously possess values for both S_{bz}

and S_{bx}, namely the values revealed by the corresponding measurements on particle a, either of which could be carried out in any given experimental run.

E5. But this contradicts the basic principles of quantum theory, since in the two-dimensional spin space one cannot simultaneously assign values of both S_z and S_x to particle b.

Let us explore the paradox by asking how each of these assertions is related to a precise quantum mechanical description of the situation. We begin with E1, and employ the notation in Sec. 23.4, with the particles initially in a spin singlet state $|\psi_0\rangle$, and an apparatus designed to measure S_{az} initially in the state $|Z_a^{\circ}\rangle$ at time t_0. The interaction of particle a with the apparatus during the time interval from t_1 to t_2 gives rise to the unitary time transformation (23.21). We then need a consistent family which includes the possible outcomes Z_a^+ and Z_a^- of the measurement, corresponding to $S_{az} = +1/2$ and $-1/2$, together with the values of S_{bz}.

It is useful to begin with the family in (23.29), since it comes the closest among all the families in Sec. 23.4 to representing how physicists would have thought about the problem in 1935, when the EPR paper was published. In this family the initial state evolves unitarily until after the measurement has occurred, when there is a split (or "collapse") into the two possibilities $Z_a^+ z_b^-$ and $Z_a^- z_b^+$. Using this family one can deduce $S_{bz} = -1/2$ from the measurement outcome Z_a^+, and $S_{bz} = +1/2$ from Z_a^-; the results can be expressed formally as conditional probabilities, (23.25). This means that E1 is in agreement with the principles of quantum theory.

Even stronger results can be obtained using the family (23.22) in which the stochastic split takes place at an earlier time. In this family it is possible to view the measurement of S_{az} as revealing a pre-existing property of particle a at a time before the measurement took place, a value which was already the opposite of S_{bz}. In addition, the value of S_{bz} was unaffected by the measurement of S_{az}, a fact expressed formally by the conditional probabilities in (23.26). Thus this family both confirms E1 and lends support to E3. Additional support for E3 comes from the family (23.31), which shows that a measurement of S_{az} does not have any effect upon S_{bx}, and of course one could set up an analogous family using any other component of spin of particle b, and reach the same conclusion.

Next we come to E2. It is nothing but E1 with S_z replaced by S_x for both particles, so the preceding discussion of E1 will apply to E2, with obvious modifications. The family (23.28) with its apparatus for measuring S_{ax} must be used in place of (23.22), and from it one can deduce the counterparts of (23.23)–(23.26) with z and Z replaced by x and X. And of course the S_{ax} measurement will not alter any component of the spin of particle b, which confirms E3.

Assertion E4 would seem to be an immediate consequence of those preceding it were it not for the requirement that quantum reasoning employ a *single* framework in order to reach a sound conclusion, Sec. 16.1. Assertions E1 and E2 have been justified on the basis of two distinct consistent families, (23.22) and (23.28). Are these families compatible, that is, can they be combined in a single framework? One's first thought is that they cannot be combined, because the projectors for the properties associated with S_{az} and S_{bz} at t_1 (the intermediate time) in (23.22) obviously do not commute with those in (23.28), which are associated with S_{ax} and S_{bx}, and the same is true of the projectors at t_2. However, the situation is not so simple. The projectors representing the complete histories in (23.22) are orthogonal to, and hence commute with, the history projectors in (23.28), because the initial states $|Z_a^\circ\rangle$ and $|X_a^\circ\rangle$ for the apparatus will be orthogonal. This follows from the fact that an apparatus designed to measure S_z will differ in a visible (macroscopic) way from one designed to measure S_x; see the discussion following (17.10).

Consequently, (23.22) and (23.28) can be combined in a single consistent family with two distinct initial states: the spin singlet state of the particles combined with either of the measuring apparatuses. However, the resulting framework does *not* support E4. The reason is that the two initial states are mutually exclusive, so that only one or the other will occur in a particular experimental run. Consequently, the conclusion that S_{bz} will have a particular value, at t_1 or t_2, as determined by the measurement outcome, is only correct for a run in which the apparatus is set up to measure S_{az}, and the corresponding conclusion for S_{bx} only holds for runs in which the apparatus is set up to measure S_{ax}. But E4 asserts that particle b *simultaneously* possesses values of S_z and S_x, and this conclusion obviously cannot be reached using the framework under consideration.

To put the matter in a different way, E1 is correct in a situation in which S_{az} is measured, and E2 in a situation in which S_{ax} is measured. But there is no way to measure S_{az} and S_{ax} simultaneously for a single particle, and therefore no situation in which E1 and E2 can be applied to the same particle. Einstein, Podolsky and Rosen were aware of this type of objection, as they mention it towards the end of their paper, and they respond in a fashion which can be translated into the language of spin-half particles in the following way. If one allows that an S_{az} measurement can be used to predict S_{bz} and an S_{ax} measurement to predict S_{bx}, but then asserts that S_{bx} does not exist when S_{az} is measured, and S_{bz} does not exist when S_{ax} is measured, this makes the properties of particle b depend upon which measurement is carried out on particle a, and no reasonable theory could allow this sort of thing.

There is nothing in the analysis presented in Sec. 23.4 to suggest that the properties of particle b depend in any way upon the type of measurement carried out on particle a. However, the type of property considered for particle b, S_{bz} as against S_{bx}, depends upon the choice of framework. There are frameworks, such as (23.22) and (23.29), in which a measurement of S_{az} is combined with values of S_{bz}, and other frameworks, such as (23.30) and (23.31), in which a measurement of S_{az} is combined with values for S_{bx}. Quantum theory does not specify which framework is to be used for a situation in which S_{az} is measured. However, only a framework which includes S_{bz} can be used to correlate the outcome of an S_{az} measurement with some property of the spin of particle b in a way which constitutes an indirect measurement of the latter.

Thus implicit in the analysis given in the EPR paper is the assumption that quantum theory is limited to a single framework in the case of an S_{az} measurement, one corresponding to a wave function collapse picture, (23.29), for this particular measurement. Once one recognizes that there are many possible frameworks, the argument no longer works. One can hardly fault Einstein and his colleagues for making such an assumption, as they were seeking to point out an inadequacy of quantum mechanics as it had been developed up to that time, with measurement and wave function collapse essential features of its physical interpretation. One can see in retrospect that they had, indeed, located a severe shortcoming of the principal interpretation of quantum theory then available, though they themselves did not know how to remedy it.

24.2 Counterfactuals and the EPR paradox

An alternative way of thinking about assertion E4 in the previous section is to consider a case in which S_{az} is measured (and thus S_{bz} is indirectly measured), and ask what would have been the case, in this particular experimental run, if S_{ax} had been measured instead, e.g., by rotating the direction of the field gradient in the Stern–Gerlach apparatus just before the arrival of particle a. This requires a counterfactual analysis, which can be carried out with the help of a quantum coin toss in the manner indicated in Sec. 19.4. Let the total quantum system be described by an initial state

$$|\Phi_0\rangle = |\psi_0\rangle|Q\rangle, \tag{24.1}$$

where $|\psi_0\rangle$ is the spin singlet state (23.2), and $|Q\rangle$ the initial state of the quantum coin, servomechanism, and the measuring apparatus. (As it is not important for the following discussion, the center of mass wave function $|\omega_t\rangle$, (23.1), has been omitted, just as in Ch. 23.) It will be convenient to assume that the quantum coin toss corresponds to a unitary time development

$$|Q\rangle \mapsto \left(|X_a^\circ\rangle + |Z_a^\circ\rangle\right)/\sqrt{2}, \tag{24.2}$$

during the interval from t_1 to t_2, and that the measurement of S_{ax} or S_{az} takes place during the time interval from t_2 to t_3, rather than between t_1 and t_2 as in Ch. 23.

Here $|X_a^\circ\rangle$ and $|Z_a^\circ\rangle$ are states of the apparatus in which it is ready to measure S_{ax} and S_{az}, respectively, and the servomechanism, etc., is thought of as included in these states. Thus the overall unitary time development from the initial time t_0 to the final time t_3 is given by

$$|\Phi_0\rangle \mapsto |\Phi_0\rangle \mapsto |\psi_0\rangle(|X_a^\circ\rangle + |Z_a^\circ\rangle)/\sqrt{2} \mapsto$$
$$\left(|x_b^-\rangle|X_a^+\rangle - |x_b^+\rangle|X_a^-\rangle + |z_b^-\rangle|Z_a^+\rangle - |z_b^+\rangle|Z_a^-\rangle\right)/2. \qquad (24.3)$$

The final step from t_2 to t_3 is obtained by assuming that (23.27) applies when the apparatus is in the state $|X_a^\circ\rangle$, and (23.21) when it is in the state $|Z_a^\circ\rangle$ at t_2.

A consistent family $\mathcal{F}_1$ which provides one way of analyzing the counterfactual question posed at the beginning of this section has for its support six histories for times $t_0 < t_1 < t_2 < t_3$. It is convenient to arrange them in two groups of three:

$$\Phi_0 \odot z_a^+ z_b^- \odot \begin{cases} Z_a^\circ \odot & Z_a^+ z_b^-, \\ X_a^\circ \odot & \begin{cases} X_a^+ z_b^-, \\ X_a^- z_b^-, \end{cases} \end{cases} \quad \Phi_0 \odot z_a^- z_b^+ \odot \begin{cases} Z_a^\circ \odot & Z_a^- z_b^+, \\ X_a^\circ \odot & \begin{cases} X_a^+ z_b^+, \\ X_a^- z_b^+. \end{cases} \end{cases} \qquad (24.4)$$

Suppose the coin toss resulted in S_{az} being measured, and the outcome was Z_a^+, implying $S_{bz} = -1/2$. To answer the question of what would have happened if S_{ax} had been measured instead, use the procedure of Sec. 19.4 and trace the outcome $Z_a^+ z_b^-$ in the first set of histories in (24.4) backwards to the pivot $z_a^+ z_b^-$ and then forwards through the X_a° node to the corresponding events at t_3. One concludes that had the quantum coin toss resulted in a measurement of S_{ax}, the outcome would have been X_a^+ or X_a^-, each with probability $1/2$, but *in either case* S_{bz} would have had the value $-1/2$, corresponding to z_b^-, that is to say, the same value it had in the actual world in which S_{az}, not S_{ax}, was measured. This conclusion seems very reasonable on physical grounds, for one would not expect a last minute choice to measure S_x rather than S_z for particle a to have any influence on the distant particle b, since the measuring apparatus does not interact in any way with particle b. To put the matter in another way, the conclusion of this counterfactual analysis agrees with the discussion of E3 in Sec. 24.1.

On the other hand, (24.4) by itself provides no immediate support for E4, for it supplies no information at all about S_{bx}. Of course, this is only one consistent family, and one might hope to do better using some other framework. One possibility might be the consistent family $\mathcal{F}_2$ with support

$$\Phi_0 \odot \Phi_0 \odot \begin{cases} Z_a^\circ \odot & \begin{cases} Z_a^+ z_b^-, \\ Z_a^- z_b^+, \end{cases} \\ X_a^\circ \odot & \begin{cases} X_a^+ x_b^-, \\ X_a^- x_b^+, \end{cases} \end{cases} \qquad (24.5)$$

which corresponds pretty closely to the notion of wave function collapse. Once again assume that the quantum coin toss leads to an S_{az} measurement, and that the outcome of this measurement is Z_a^+. Using Φ_0 at t_0 or t_1 as the pivot, one concludes that had S_{ax} been measured instead, S_{bx} would have been $-1/2$ for the outcome X_a^+, and $+1/2$ for the outcome X_a^-.

This result seems encouraging, for we have found a consistent family in which both S_{bz} and S_{bx} values appear, correlated in the expected way with S_{az} and S_{ax} measurements. However the S_{bz} states $z_b^{\pm}$ and the S_{bx} states $x_b^{\pm}$ in (24.5) are *contextual* properties in the sense of Ch. 14: z_b^+ and z_b^- both depend on Z_a°, and x_b^+ and x_b^- both depend on X_a°. This means — see the discussion in Ch. 14 — that when using (24.5), one cannot think of S_{bz} and S_{bx} as having values independent of the quantum coin toss. Only if the toss results in Z_a° is it meaningful to talk about S_{bz}, and only if it results in X_a° can one talk about S_{bx}. And since the two outcomes of the quantum coin toss are mutually-exclusive possibilities, one and only one of which will occur in any given experimental run, we have again failed to establish E4, and for basically the same reason pointed out in Sec. 24.1 when discussing the family with two initial states that combines (23.22) and (23.28). Indeed, in the latter family S_{bz} and S_{bx} are contextual properties which depend upon the corresponding initial states — something we did not bother to point out in Sec. 24.1 because dependence (in the technical sense used in Ch. 14) on *earlier* events never poses much of an intuitive problem. But does this contextuality mean that there is some mysterious long-range influence in that a last minute choice to measure S_{ax} rather than S_{az} would somehow determine whether particle b has a definite value of S_x rather that S_z? No, for dependence or contextuality in the technical sense used in Ch. 14 denotes a logical relationship brought about by choosing a framework in a particular way, and does not indicate any sort of physical causality. Thus there is no contradiction with the arguments presented in Sec. 24.1 in support of E3.

The reader with the patience to follow the analysis in this and the previous section may with some justification complain that the outcome was already certain at the outset: if E4 really does contradict the basic principles of quantum theory, as asserted by E5, then it is evident that it can never be obtained by an analysis based upon those principles. True enough, but there are various reasons why working out the details is still worthwhile. First, there is no way to establish with absolute certainty the consistency of the basic principles of a physical theory, as it is always something more than a piece of abstract mathematics or logic; one has to apply these principles to various examples and see what they predict. Second, it is of some interest to find out where and why the seemingly plausible chain of arguments from E1 to E5 comes apart, for this tells us something about the difference between quantum and classical physics. The preceding analysis shows that it is basically violations of the single-framework rule which cause the trouble, and in this respect the EPR paradox has quite a bit in common with the paradoxes discussed in previous chapters. But the nonclassical behavior of contextual events can also play a role, depending on how one analyzes the paradox.

Third, the analysis supports the correctness of the basic *locality* assumption of EPR as expressed in E3, an assertion which is confirmed by the analysis in Ch. 23. Given that the EPR paradox has sometimes been cited to support the claim that there are mysterious nonlocal influences in the quantum world, it is worth emphasizing that the analysis given here does not show any evidence of such influences. On the other hand, certain *modifications* of quantum mechanics in which the Hilbert space is supplemented by "hidden variables" of a particular sort will necessarily involve peculiar nonlocal influences if they are to reproduce the spin correlations (23.9) of standard quantum theory, and these are the subject of the remaining sections in this chapter.

24.3 EPR and hidden variables

A *hidden variable* theory is an alternative approach to quantum mechanics in which the Hilbert space of the standard theory is either replaced by or supplemented with a set of "hidden" (the name is not particularly apt) variables which behave like those one is accustomed to in classical mechanics. One of the best-known examples was proposed in 1952 by Bohm, using an approach similar to one employed earlier by de Broglie, in which at any instant of time all particles have precise positions, and these positions constitute the new (hidden) variables.

The simplest hidden variable model of a spin-half particle is one in which the different components of its spin angular momentum simultaneously possess well-defined values, something which is not true if one uses a quantum Hilbert space, for reasons discussed in Sec. 4.6. A measurement of some component of spin using a Stern–Gerlach apparatus will then reveal the value that the corresponding (hidden) variable had just before the measurement took place. More complicated models are possible, but the general idea is that measurement outcomes are determined by variables that behave classically in the sense that they simultaneously possess definite values. John Bell pointed out in 1964 that hidden variable models of this kind cannot reproduce the correlation function $C(\mathbf{a}, \mathbf{b})$, (23.9) or (23.37), for spin-half particles in an initial singlet state, if one makes the reasonable assumption that no mysterious long-range influences link the particles and the measuring apparatuses. This result led to a number of experimental measurements of the spin correlation function. Most of the experiments have used the polarizations of correlated photons rather than spin-half particles, but the principles are the same, and the results are in good agreement with the predictions of quantum mechanics. Note that one can think of this correlation function as referring to particle spins in the absence of any measurement when one uses the framework (23.5), or as the correlation function between outcomes of measurements of the spins of both particles, (23.37). In line with most discussions of Bell's result, we shall think of $C(\mathbf{a}, \mathbf{b})$ as referring to measurement outcomes.

Before exhibiting one version of Bell's argument in Sec. 24.4, it is useful to look at a specific setup discussed by Mermin. Imagine two apparatuses, one to measure the spin of particle a and the other the spin of particle b, each of which can measure the component of spin angular momentum in one of three directions in space, u, v, and x, lying in the x, y plane, with an angle of 120° between every pair of directions, Fig. 24.1. The component of spin which will be measured is determined by a switch setting on the apparatus, and these settings will also be denoted by u, v, and x. Let $\alpha(w) = \pm 1$ denote the two possible outcomes of the measurement when the switch setting of the a apparatus is w: $+1$ if the spin is found to be in the $+w$ direction, $S_{aw} = +1/2$, and -1 if it is in the opposite direction, $S_{aw} = -1/2$. Let $\beta(w) = \pm 1$ be the possible outcomes of the b apparatus measurement when its switch setting is w. In any given experiment these results will be random, but if they are averaged over a large number of runs, the averages of $\alpha(w)$ and of $\beta(w)$ will be zero for any choice of w, whereas the correlation function (23.37) will be given by:

$$C(w_a, w_b) = \langle \alpha(w_a)\beta(w_b) \rangle = \begin{cases} -1 & \text{if } w_a = w_b, \\ +1/2 & \text{if } w_a \neq w_b, \end{cases} \tag{24.6}$$

since if the switch settings w_a and w_b for the a and b apparatuses are unequal, the angle between the two directions is 120°, and the inner product of the two corresponding unit vectors is $-1/2$.

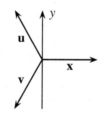

Fig. 24.1. Directions u, v, x in the x, y plane.

Let us try to construct a hidden variable model which can reproduce the correlation function (24.6). Suppose that particle a when it leaves the source which prepares the two particles in a singlet state contains an "instruction set" which will determine the outcomes of the measurements in each of the three directions u, v, and x. For example, if the particle carries the instruction set $(+1, +1, -1)$, a measurement of S_{au} will yield the result $+1/2$, a measurement of S_{av} will also yield

$+1/2$, and one of S_{ax} will yield $-1/2$. Of course, only one of these measurements will actually be carried out, the one determined by the switch setting on the apparatus when particle a arrives. Whichever measurement it may be, the result is determined ahead of time by the particle's instruction set. One can think of the instruction set as a list of the components of spin angular momentum in each of the three directions, in units of $\hbar/2$. This is what is called a "deterministic hidden variable" model because the instruction set, which constitutes the hidden variables in this model, determines the later measurement outcome without any extra element of randomness. It is possible to construct stochastic hidden variable models, but they turn out to be no more successful than deterministic models in reproducing the correlations predicted by standard quantum theory.

There are eight possible instruction sets for particle a and eight for particle b, thus a total of sixty-four possibilities for the two particles together. However, the perfect anticorrelation when $w_a = w_b$ in (24.6) can only be achieved if the instruction set for b is the *complementary* set to that of a, obtained by changing the sign of each instruction. If the a set is $(+1, +1, -1)$, the b set must be $(-1, -1, +1)$. For were the b set something else, say $(+1, -1, +1)$, then there would be identical switch settings, in this case $w_a = w_b = u$, leading to $\alpha(u) = \beta(u)$, which is not possible. Similarly, perfect anticorrelations for equal switch settings means that the instruction sets, once prepared at the source which produces the singlet state, cannot change in a random manner as a particle moves from the source to the measuring apparatus.

We will assume that the source produces singlet pairs with one of the eight instruction sets for a, and the complementary set for b, chosen randomly with a certain probability. Let $P_a(+ + -)$ denote the probability that the instruction set for a is $(+1, +1, -1)$. The correlation functions can be expressed in terms of these probabilities; for example,

$$
\begin{aligned}
C(u, v) = C(v, u) = &-P_a(+ + +) - P_a(+ + -) \\
&+ P_a(+ - +) + P_a(- + +) + P_a(+ - -) \\
&+ P_a(- + -) - P_a(- - +) - P_a(- - -).
\end{aligned} \tag{24.7}
$$

Consider the following sum of correlation functions calculated in this way:

$$
\begin{aligned}
C(u, v) + C(u, x) + C(x, v) = \\
-3P_a(+ + +) - 3P_a(- - -) + P_a(+ + -) + P_a(+ - +) \\
+ P_a(- + +) + P_a(+ - -) + P_a(- + -) + P_a(- - +).
\end{aligned} \tag{24.8}
$$

Since the probabilities of the different instruction sets add to 1, this quantity has a value lying between -3 and $+1$. However, if we use the quantum mechanical values (24.6) for the correlation functions, the left side of (24.8) is $3/2$, substantially greater than 1. Thus our hidden variable model cannot reproduce the correlation functions predicted by quantum theory. As we shall see in the next section, this failure is not an accident; it is something which one must expect in hidden variable models of this sort.

24.4 Bell inequalities

The inequality (24.10) was derived in 1969 by Clauser, Horne, Shimony, and Holt. As it is closely related to Bell's original result in 1964, this CHSH inequality is nowadays also referred to as a "Bell inequality", and by studying it one can learn the essential ideas behind such inequalities. We assume that when the a apparatus measures a spin component in the direction w_a, the outcome is given by a function $\alpha(w_a, \lambda) = \pm 1$ which depends on both w_a and a hidden variable, or collection of hidden variables, denoted by λ. Similarly, the outcome of the b measurement for a spin component in the direction w_b is given by a function $\beta(w_b, \lambda) = \pm 1$. In the example in Sec. 24.3, w_a and w_b can take on any of the three values u, v, or x, and λ should be thought of as the pair of instruction sets for *both* particles a and b. Hence λ could take on sixty-four different values, though we argued in Sec. 24.3 that the probabilities of all but eight of these must be 0. For the purpose of deriving the inequality, one need not think of w_a as a direction in space; it can simply be some sort of switch setting on the a apparatus, which, together with the value of the hidden variable λ associated with the particle, determines the outcome of the measurement through the function $\alpha(w_a, \lambda)$. The same remark applies to the b apparatus and the function $\beta(w_b, \lambda)$. Also, the derivation makes no use of the fact that the two spin-half particles are initially in a spin singlet state.

The source which produces the correlated particles produces different possible values of λ with a probability $\rho(\lambda)$, so the correlation function is given by

$$C(w_a, w_b) = \sum_\lambda \rho(\lambda)\, \alpha(w_a, \lambda)\, \beta(w_b, \lambda). \qquad (24.9)$$

(If λ is a continuous variable, $\sum_\lambda \rho(\lambda)$ should be replaced by $\int \rho(\lambda)\, d\lambda$.) Let a, a' be any two possible values for w_a, and b and b' any two possible values for w_b. Then as long as $\alpha(w_a, \lambda)$ and $\beta(w_b, \lambda)$ are functions which take only the two values $+1$ or -1, the correlations defined by (24.9) satisfy the inequality

$$|C(a, b) + C(a, b') + C(a', b) - C(a', b')| \leq 2. \qquad (24.10)$$

To see that this is so, consider the quantity

$$\alpha(a, \lambda)\, \beta(b, \lambda) + \alpha(a, \lambda)\, \beta(b', \lambda) + \alpha(a', \lambda)\, \beta(b, \lambda) - \alpha(a', \lambda)\, \beta(b', \lambda)$$
$$= \big[\alpha(a, \lambda) + \alpha(a', \lambda)\big]\, \beta(b, \lambda) + \big[\alpha(a, \lambda) - \alpha(a', \lambda)\big]\, \beta(b', \lambda). \qquad (24.11)$$

It can take on only two values, $+2$ and -2, because each of the four quantities

$\alpha(a, \lambda)$, $\alpha(a', \lambda)$, $\beta(b, \lambda)$ and $\beta(b', \lambda)$ is either $+1$ or -1. Thus either $\alpha(a, \lambda) = \alpha(a', \lambda)$, so that the right side of (24.11) is $2\alpha(a, \lambda)\beta(b, \lambda)$, or else $\alpha(a, \lambda) = -\alpha(a', \lambda)$, in which case it is $2\alpha(a, \lambda)\beta(b', \lambda)$. If one multiplies (24.11) by $\rho(\lambda)$ and sums over λ, the result of this weighted average is

$$C(a, b) + C(a, b') + C(a', b) - C(a', b'). \tag{24.12}$$

A weighted average of a quantity which takes on only two values must lie between them, so (24.12) lies somewhere between -2 and $+2$, which is what (24.10) asserts.

Consider the example in Sec. 24.3, and set $a = u$, $b = v$, $a' = b' = x$. If one inserts the quantum values (24.6) for these correlation functions in (24.12), the result is $3 \times 1/2 + 1 = 2.5$, which obviously violates the inequality (24.10). On the other hand, the hidden variable model in Sec. 24.3 assigns to the sum $C(u, v) + C(u, x) + C(x, v)$, see (24.8), a value between -3 and $+1$, and since $C(x, x) = -1$, the inequality (24.10) will be satisfied.

If quantum theory is a correct description of the world, then since it predicts correlation functions which violate (24.10), one or more of the assumptions made in the derivation of this inequality must be wrong. The first and most basic of these assumptions is the *existence of hidden variables* with a mathematical structure which differs from the Hilbert space used in standard quantum mechanics. This assumption is plausible from the perspective of classical physics if measurements reveal pre-existing properties of the measured system. In quantum physics it is also the case that a measurement reveals a pre-existing property *provided* this property is part of the framework which is being used to construct the quantum description. If S_{az} is measured for particle a, the outcome of a suitable (ideal) measurement will be correlated with the value of this component of spin angular momentum before the measurement in a framework which includes $|z_a^+\rangle$ and $|z_a^-\rangle$. However, there is no framework which includes the eigenstates of *both* S_{az} and S_{aw} for a direction w not equal to z or $-z$.

Thus the point at which the derivation of (24.10) begins to deviate from quantum principles is in the assumption that a function $\alpha(w_a, \lambda)$ exists *for different directions* w_a. As long as only a single choice for w_a is under consideration there is no problem, for then the "hidden" variable λ can simply be the value of S_{aw} at some earlier time. But when two (excluding the trivial case of w_a and $-w_a$) or even more possibilities are allowed, the assumption that $\alpha(w_a, \lambda)$ exists is in conflict with basic quantum principles. Precisely the same comments apply to the function $\beta(w_b, \lambda)$.

Of course, if postulating hidden variables is itself in error, there is no need to search for problems with the other assumptions having to do with the nature of these hidden variables. Nonetheless, let us see what can be said about them. A second assumption entering the derivation of (24.10) is that the hidden variable theory is *local*. Locality appears in the assumption that the outcome $\alpha(w_a, \lambda)$ of the *a* measurement depends on the setting w_a of this piece of apparatus, but not the setting w_b for the *b* apparatus, and that $\beta(w_b, \lambda)$ does not depend upon w_a. These assumptions are plausible, especially if one supposes that the particles *a* and *b* and the corresponding apparatuses are far apart at the time when the measurements take place. For then the settings w_a and w_b could be chosen at the very last moment before the measurements take place, and it is hard to see how either value could have any influence on the outcome of the measurement made by the other apparatus. Indeed, for a sufficiently large separation, an influence of this sort would have to travel faster than the speed of light, in violation of relativity theory.

The claim is sometimes made that quantum theory must be nonlocal simply because its predictions violate (24.10). But this is not correct. First, what follows logically from the violation of this inequality is that hidden variable theories, if they are to agree with quantum theory, must be nonlocal or embody some other peculiarity. But hidden variable theories by definition employ a different mathematical structure from (or in addition to) the quantum Hilbert space, so this tells us nothing about standard quantum mechanics. Second, the detailed quantum analysis of a spin singlet system in Ch. 23 shows no evidence of nonlocality; indeed, it demonstrates precisely the opposite: the spin of particle *b* is not influenced in any way by the measurements carried out on particle *a*. (To be sure, in Ch. 23 we did not discuss how a measurement on particle *a* might influence the outcome of a *measurement* on particle *b*, but the argument can be easily extended to include that case, and the conclusion is exactly the same.) Hidden variable theories, on the other hand, can indeed be nonlocal. The Bohm theory mentioned in Sec. 24.3 is known to be nonlocal in a rather thorough-going way, and this is one reason why it has been difficult to construct a relativistic version of it.

A third assumption which was made in deriving the inequality (24.10) is that the probability distribution $\rho(\lambda)$ for the hidden variable(s) λ does not depend upon either w_a or w_b. This seems plausible if there is a significant interval between the time when the two particles are prepared in some singlet state by a source which sets the value of λ, and the time when the spin measurements occur. For w_a and w_b could be chosen just before the measurements take place, and this choice should not affect the value of λ determined earlier, unless the future can influence the past.

In summary, the basic lesson to be learned from the Bell inequalities is that it is difficult to construct a plausible hidden variable theory which will mimic the sorts of correlations predicted by quantum theory and confirmed by experiment. Such a theory must either exhibit peculiar nonlocalities which violate relativity theory, or else incorporate influences which travel backwards in time, in contrast to everyday experience. This seems a rather high price to pay just to have a theory which is more "classical" than ordinary quantum mechanics.

25

Hardy's paradox

25.1 Introduction

Hardy's paradox resembles the Bohm version of the Einstein–Podolsky–Rosen paradox, discussed in Chs. 23 and 24, in that it involves two correlated particles, each of which can be in one of two states. However, Hardy's initial state is chosen in such a way that by following a plausible line of reasoning one arrives at a logical contradiction: something is shown to be true which one knows to be false. This makes this paradox in some respects more paradoxical than the EPR paradox as stated in Sec. 24.1. A paradox of a somewhat similar nature involving three spin-half particles was discovered (or invented) by Greenberger, Horne, and Zeilinger a few years earlier. The basic principles behind this GHZ paradox are very similar to those involved in Hardy's paradox. We shall limit our analysis to Hardy's paradox, as it is a bit simpler, but the same techniques can be used to analyze the GHZ paradox.

Hardy's paradox can be discussed in the language of spin-half particles, but we will follow the original paper, though with some minor modifications, in thinking of it as involving two particles, each of which can move through one of two arms (the two arms are analogous to the two states of a spin-half particle) of an interferometer, as indicated in Fig. 25.1. These are particles without spin, or for which the spin degree of freedom plays no role in the gedanken experiment. The source S at the center of the diagram produces two particles a and b moving to the left and right, respectively, in an initial state

$$|\psi_0\rangle = (|c\bar{c}\rangle + |c\bar{d}\rangle + |d\bar{c}\rangle)/\sqrt{3}. \qquad (25.1)$$

Here $|c\bar{d}\rangle$ stands for $|c\rangle \otimes |\bar{d}\rangle$, a state in which particle a is in the c arm of the left interferometer, and particle b in the $\bar{d}$ arm of the interferometer on the right. The other kets are defined in the same way. One can think of the two particles as two photons, but other particles will do just as well. In Hardy's original paper one

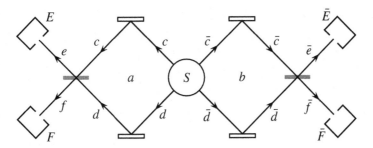

Fig. 25.1. Double interferometer for Hardy's paradox.

particle was an electron and the other a positron, and the absence of a $d\bar{d}$ term in (25.1) was due to their meeting and annihilating each other.

Suppose that S produces the state (25.1) at the time t_0. The unitary time development from t_0 to a time t_1, which is before either of the particles passes through the beam splitter at the output of its interferometer, is trivial: each particle remains in the same arm in which it starts out. We shall denote the states at t_1 using the same symbol as at t_0: $|c\rangle$, $|\bar{d}\rangle$, etc. One could change c to c', etc., but this is really not necessary. In this simplified notation the time development operator for the time interval from t_0 to t_1 is simply the identity I. During the time interval from t_1 to t_2, each particle passes through the beam splitter at the exit of its interferometer, and these beam splitters produce unitary transformations

$$B : |c\rangle \mapsto (|e\rangle + |f\rangle)/\sqrt{2}, \quad |d\rangle \mapsto (-|e\rangle + |f\rangle)/\sqrt{2},$$
$$\bar{B} : |\bar{c}\rangle \mapsto (|\bar{e}\rangle + |\bar{f}\rangle)/\sqrt{2}, \quad |\bar{d}\rangle \mapsto (-|\bar{e}\rangle + |\bar{f}\rangle)/\sqrt{2},$$

(25.2)

where $|e\rangle$, etc., denote wave packets in the output channels, and the phases are chosen to agree with those used for the toy model in Sec. 12.1. Combining the transformations in (25.2) results in the unitary transformations

$$|c\bar{c}\rangle \mapsto (+|e\bar{e}\rangle + |e\bar{f}\rangle + |f\bar{e}\rangle + |f\bar{f}\rangle)/2,$$
$$|c\bar{d}\rangle \mapsto (-|e\bar{e}\rangle + |e\bar{f}\rangle - |f\bar{e}\rangle + |f\bar{f}\rangle)/2,$$
$$|d\bar{c}\rangle \mapsto (-|e\bar{e}\rangle - |e\bar{f}\rangle + |f\bar{e}\rangle + |f\bar{f}\rangle)/2,$$
$$|d\bar{d}\rangle \mapsto (+|e\bar{e}\rangle - |e\bar{f}\rangle - |f\bar{e}\rangle + |f\bar{f}\rangle)/2,$$

(25.3)

for the combined states of the two particles during the time interval from t_0 or t_1 to t_2. Adding up the appropriate terms in (25.3), one finds that the initial state (25.1) is transformed into

$$B\bar{B} : |\psi_0\rangle \mapsto \left(-|e\bar{e}\rangle + |e\bar{f}\rangle + |f\bar{e}\rangle + 3|f\bar{f}\rangle\right)/\sqrt{12}$$

(25.4)

by the beam splitters.

We will later need to know what happens if one or both of the beam splitters has been taken out of the way. Let O and $\bar{O}$ denote situations in which the left and the right beam splitters, respectively, have been removed. Then (25.2) is to be replaced with

$$
\begin{aligned}
O &: |c\rangle \mapsto |f\rangle, \quad |d\rangle \mapsto |e\rangle, \\
\bar{O} &: |\bar{c}\rangle \mapsto |\bar{f}\rangle, \quad |\bar{d}\rangle \mapsto |\bar{e}\rangle,
\end{aligned}
\tag{25.5}
$$

in agreement with what one would expect from Fig. 25.1. The time development of $|\psi_0\rangle$ from t_0 to t_2 if one or both of the beam splitters is absent can be worked out using (25.5) together with (25.2):

$$
\begin{aligned}
B\bar{O} &: |\psi_0\rangle \mapsto \left(|e\bar{e}\rangle + |f\bar{e}\rangle + 2|f\bar{f}\rangle\right)/\sqrt{6}, \\
O\bar{B} &: |\psi_0\rangle \mapsto \left(|e\bar{e}\rangle + |e\bar{f}\rangle + 2|f\bar{f}\rangle\right)/\sqrt{6}, \\
O\bar{O} &: |\psi_0\rangle \mapsto \left(|e\bar{f}\rangle + |f\bar{e}\rangle + |f\bar{f}\rangle\right)/\sqrt{3}.
\end{aligned}
\tag{25.6}
$$

When they emerge from the beam splitters, the particles are detected, see Fig. 25.1. In order to have a compact notation, we shall use $|M\rangle$ for the initial state of the two detectors for particle a, and $|\bar{M}\rangle$ that of the detectors for particle b, and assume that the process of detection corresponds to the following unitary transformations for the time interval from t_2 to t_3:

$$
\begin{aligned}
|e\rangle|M\rangle &\mapsto |E\rangle, \quad |f\rangle|M\rangle \mapsto |F\rangle, \\
|\bar{e}\rangle|\bar{M}\rangle &\mapsto |\bar{E}\rangle, \quad |\bar{f}\rangle|\bar{M}\rangle \mapsto |\bar{F}\rangle.
\end{aligned}
\tag{25.7}
$$

Thus $|E\rangle$ means that particle a was detected by the detector located on the e channel. We are now ready to consider the paradox, which can be formulated in two different ways. Both of these are found in Hardy's original paper, though in the opposite order.

25.2 The first paradox

For this paradox we suppose that both beam splitters are in place. Consider the consistent family of histories at the times t_0, t_1, and t_2 whose support consists of the four histories

$$
\mathcal{F}_1 : \ \psi_0 \odot \{\bar{c}, \bar{d}\} \odot \{e, f\},
\tag{25.8}
$$

with the same initial state ψ_0, and one of the two possibilities $\bar{c}$ or $\bar{d}$ at t_1, followed by e or f at t_2. Here the symbols stand for projectors associated with the corresponding kets: $\bar{c} = |\bar{c}\rangle\langle\bar{c}|$, etc. That this family is consistent can be seen by noting that the unitary dynamics for particle a is independent of that for particle b at all times after t_0: the time development operator factors. Thus the Heisenberg operators (see Sec. 11.4) for $\bar{c}$ and $\bar{d}$, which refer to the b particle, commute with those for e and f, which refer to the a particle, so that for purposes of checking consistency, (25.8) is the same as a history involving only two times: t_0 and one later time. Hence one can apply the rule that a family of histories involving only two times is automatically consistent, Sec. 11.3. Of course, one can reach the same conclusion by explictly calculating the chain kets (Sec. 11.6) and showing that they are orthogonal to one another.

The history $\psi_0 \odot \bar{c} \odot e$ has zero weight. To see this, construct the chain ket starting with

$$\bar{c}|\psi_0\rangle = (|c\bar{c}\rangle + |d\bar{c}\rangle))/\sqrt{3} = (|c\rangle + |d\rangle) \otimes |\bar{c}\rangle/\sqrt{3}. \tag{25.9}$$

When $T(t_2, t_1)$ is applied to this, the result — see (25.2) — will be $|f\rangle$ times a ket for the b particle, and applying the projector e to it yields zero. As a consequence, since $\psi_0 \odot \bar{d} \odot e$ has finite weight, one has

$$\Pr(\bar{d}, t_1 \mid e, t_2) = 1, \tag{25.10}$$

where the times t_1 and t_2 associated with the events $\bar{d}$ and e are indicated explicitly, rather than by subscripts as in earlier chapters. Thus if particle a emerges in e at time t_2, one can be sure that particle b was in the $\bar{d}$ arm at time t_1.

A similar result is obtained if instead of (25.8) one uses the family

$$\mathcal{F}_1' : \ \Psi_0 \odot \{\bar{c}, \bar{d}\} \odot \{E, F\}, \tag{25.11}$$

with events at times t_0, t_1, and t_3, where

$$|\Psi_0\rangle = |\psi_0\rangle \otimes |M\bar{M}\rangle \tag{25.12}$$

includes the initial states of the measuring devices. The fact that $\Psi_0 \odot \bar{c} \odot E$ has zero weight implies that

$$\Pr(\bar{d}, t_1 \mid E, t_3) = 1. \tag{25.13}$$

Of course, (25.13) is what one would expect, given (25.10), and vice versa: the measuring device shows that particle a emerged in the e channel if and only if this was actually the case. In the discussion which follows we will, because it is somewhat simpler, use families of the type $\mathcal{F}_1$ which do not include any measuring devices. But the same sort of argument will work if instead of e, $\bar{e}$, etc. one uses measurement outcomes E, $\bar{E}$, etc.

By symmetry it is clear that the family

$$\mathcal{F}_2 : \ \psi_0 \odot \{c, d\} \odot \{\bar{e}, \bar{f}\}, \tag{25.14}$$

obtained by interchanging the role of particles a and b in (25.8), is consistent. Since the history $\psi_0 \odot c \odot \bar{e}$ has zero weight, it follows that

$$\Pr(d, t_1 \mid \bar{e}, t_2) = 1, \tag{25.15}$$

or, if measurements are included,

$$\Pr(d, t_1 \mid \bar{E}, t_3) = 1. \tag{25.16}$$

That is, if particle b emerges in channel $\bar{e}$ (the measurement result is $\bar{E}$), then particle a was earlier in the d and not the c arm of its interferometer.

To complete the paradox, we need two additional families. Using

$$\mathcal{F}_3 : \ \psi_0 \odot I \odot \{e\bar{e}, e\bar{f}, f\bar{e}, f\bar{f}\}, \tag{25.17}$$

one can show, see (25.4), that

$$\Pr(e\bar{e}, t_2) = 1/12. \tag{25.18}$$

Finally, the family

$$\mathcal{F}_4 : \ \psi_0 \odot \{c\bar{c}, c\bar{d}, d\bar{c}, d\bar{d}\} \odot I \tag{25.19}$$

yields the result

$$\Pr(d\bar{d}, t_1) = 0, \tag{25.20}$$

because $|d\bar{d}\rangle$ occurs with zero amplitude in $|\psi_0\rangle$, (25.1).

Hardy's paradox can be stated in the following way. Whenever a emerges in the e channel we can be sure, (25.10), that b was earlier in the $\bar{d}$ arm, and whenever b emerges in the $\bar{e}$ channel we can be sure, (25.15), that a was earlier in the d arm. The probability that a will emerge in e at the same time that b emerges in $\bar{e}$ is $1/12$, (25.18), and when this happens it must be true that a was earlier in d and b was earlier in $\bar{d}$. But given the initial state $|\psi_0\rangle$, it is impossible for a to be in d at the same time that b is in $\bar{d}$, (25.20), so we have reached a contradiction.

Here is a formal argument using probability theory. First, (25.10) implies that

$$\Pr(\bar{d}, t_1 \mid e\bar{e}, t_2) = 1, \tag{25.21}$$

because if a conditional probability is equal to 1, it will also be equal to 1 if the condition is made more restrictive, assuming the new condition has positive probability. In the case at hand the condition e is replaced with $e\bar{e}$, and the latter has a probability of $1/12$, (25.18). In the same way,

$$\Pr(d, t_1 \mid e\bar{e}, t_2) = 1 \tag{25.22}$$

is a consequence of (25.15). Combining (25.21) and (25.22) leads to

$$\Pr(d\bar{d}, t_1 \mid e\bar{e}, t_2) = 1, \qquad (25.23)$$

and therefore, in light of (25.18),

$$\Pr(d\bar{d}, t_1) \geq 1/12. \qquad (25.24)$$

In Hardy's original paper this version of the paradox was constructed in a somewhat different way. Rather than using conditional probabilities to infer properties at earlier times, Hardy reasoned as follows, employing the version of the gedanken experiment in which there is a final measurement. Suppose that both interferometers are extremely large, so that the difference $t_3 - t_1$ is small compared to the time required for light to travel from the source to one of the beam splitters, or from one beam splitter to the other. (The choice of t_1 in our analysis is somewhat arbitrary, but there is nothing wrong with choosing it to be just before the particles arrive at their respective beam splitters.) In this case there is a moving coordinate system or Lorentz frame in which relativistic effects mean that the detection of particle a in the e channel occurs (in this Lorentz frame) at a time when the b particle is still inside its interferometer. In this case, the inference from E to $\bar{d}$ can be made using wave function collapse, Sec. 18.2. By using a different Lorentz frame in which the b particle is the first to pass through its beam splitter, one can carry out the corresponding inference from $\bar{E}$ to d. Next, Hardy made the assumption that inferences of this sort which are valid in one Lorentz frame are valid in another Lorentz frame, and this justifies the analogs of (25.13) and (25.16). With these results in hand, the rest of the paradox is constructed in the manner indicated earlier, with a few obvious changes, such as replacing $e\bar{e}$ in (25.18) with $E\bar{E}$.

25.3 Analysis of the first paradox

In order to arrive at the contradiction between (25.24) and (25.20), it is necessary to combine probabilities obtained using four different frameworks, $\mathcal{F}_1$–$\mathcal{F}_4$ (or their counterparts with the measuring apparatus included). While there is no difficulty doing so in classical physics, in the quantum case one must check that the corresponding frameworks are compatible, that is, there is a single consistent family which contains all of the histories in $\mathcal{F}_1$–$\mathcal{F}_4$. However, it turns out that *no two of these frameworks are mutually compatible.*

One way to see this is to note that the family

$$\mathcal{J}_1 : \ \psi_0 \odot \{c, d\} \odot \{e, f\} \qquad (25.25)$$

is *inconsistent*, as one can show by working out the chain kets and showing that they are not orthogonal. This inconsistency should come as no surprise in view of the discussion of interference in Ch. 13, since the histories in $\mathcal{J}_1$ contain projectors indicating both which arm of the interferometer particle a is in at time t_1 *and* the channel in which it emerges at t_2. To be sure, the initial condition $|\psi_0\rangle$ is more complicated than its counterpart in Ch. 13, but it would have to be of a fairly special form in order not to give rise to inconsistencies. (It can be shown that each of the four histories in (25.25) is *intrinsically* inconsistent in the sense that it can never occur in a consistent family, Sec. 11.8.) Similarly, the family

$$\mathcal{J}_2 : \ \psi_0 \odot \{\bar{c}, \bar{d}\} \odot \{\bar{e}, \bar{f}\} \tag{25.26}$$

is inconsistent.

A comparison of $\mathcal{F}_1$ and $\mathcal{F}_2$, (25.8) and (25.14), shows that a common refinement will necessarily include all of the histories in $\mathcal{J}_1$, since c and d occur in $\mathcal{F}_2$ at t_1, and e and f in $\mathcal{F}_1$ at t_2. Therefore no common refinement can be a consistent family, and $\mathcal{F}_1$ and $\mathcal{F}_2$ are incompatible. In the same way, with the help of $\mathcal{J}_1$ and $\mathcal{J}_2$ one can show that both $\mathcal{F}_1$ and $\mathcal{F}_2$ are incompatible with $\mathcal{F}_3$ and $\mathcal{F}_4$, and that $\mathcal{F}_3$ is incompatible with $\mathcal{F}_4$. As a consequence of these incompatibilities, the derivation of (25.21) from (25.10) is invalid, as is the corresponding derivation of (25.22) from (25.15).

Although $\mathcal{F}_3$ and $\mathcal{F}_4$ are incompatible, there is a consistent family

$$\mathcal{F}_5 : \ \psi_0 \odot \{d\bar{d}, I - d\bar{d}\} \odot \{e\bar{e}, e\bar{f}, f\bar{e}, f\bar{f}\} \tag{25.27}$$

from which one can deduce both (25.18) and (25.20). Consequently, the argument which results in a paradox can be constructed by combining results from only three incompatible families, $\mathcal{F}_1$, $\mathcal{F}_2$, and $\mathcal{F}_5$, rather than four. But three is still two too many.

It is worth pointing out that the defect we have uncovered in the argument in Sec. 25.2, the violation of consistency conditions, has nothing to do with any sort of mysterious long-range influence by which particle b or a measurement carried out on particle b somehow influences particle a, even when they are far apart. Instead, the basic incompatibility is to be found in the fact that $\mathcal{J}_1$, a family which involves *only* properties of particle a after the initial time t_0, is inconsistent. Thus the paradox arises from ignoring the quantum principles which govern what one can consistently say about the behavior of a *single* particle.

A similar comment applies to Hardy's original version of the paradox, for which he employed different Lorentz frames. Although relativistic quantum theory is outside the scope of this book, it is worth remarking that there is nothing wrong with Hardy's *conclusion* that the measurement outcome E for particle a implies that particle b was in the $\bar{d}$ arm of the interferometer before reaching the beam splitter $\bar{B}$, even if some of the *assumptions* used in his argument, such as wave function collapse and the Lorentz invariance of quantum theory, might be open to dispute.

For his conclusion is the same as (25.13), a result obtained by straightforward application of quantum principles, without appealing to wave function collapse or Lorentz invariance. Thus the paradox does not, in and of itself, provide any indication that quantum theory is incompatible with special relativity, or that Lorentz invariance fails to hold in the quantum domain.

25.4 The second paradox

In this formulation we assume that both beam splitters are in place, and then make a counterfactual comparison with situations in which one or both of them are absent in order to produce a paradox. In order to model the counterfactuals, we suppose that two quantum coins are connected to servomechanisms in the manner indicated in Sec. 19.4, one for each beam splitter. Depending on the outcome of the coin toss, each servomechanism either leaves the beam splitter in place or removes it at the very last instant before the particle arrives.

Consider a family of histories with support

$$\Phi_0 \odot I \odot \{B\bar{B}, B\bar{O}, O\bar{B}, O\bar{O}\} \odot \{E\bar{E}, E\bar{F}, F\bar{E}, F\bar{F}\} \qquad (25.28)$$

at times $t_0 < t_1 < t_2 < t_3$, where t_1 is a time before the quantum coin is tossed, t_2 a time after the toss and after the servomechanisms have done their work, but before the particles reach the beam splitters (if still present), and t_3 a time after the detection of each particle in one of the output channels. Note that the definition of t_2 differs from that used in Sec. 25.3. The initial state $|\Phi_0\rangle$ includes the quantum coins, servomechanisms, and beam splitters, along with $|\psi_0\rangle$, (25.1), for particles a and b.

Various probabilities can be computed with the help of the unitary transformations given in (25.4), (25.6), and (25.7). For our purposes we need only the following results:

$$\Pr(E\bar{E}, t_3 \mid B\bar{B}, t_2) = 1/12, \qquad (25.29)$$
$$\Pr(E\bar{F}, t_3 \mid B\bar{O}, t_2) = 0, \qquad (25.30)$$
$$\Pr(F\bar{E}, t_3 \mid O\bar{B}, t_2) = 0, \qquad (25.31)$$
$$\Pr(E\bar{E}, t_3 \mid O\bar{O}, t_2) = 0. \qquad (25.32)$$

These probabilities can be used to construct a counterfactual paradox in the following manner.

H1. Consider a case in which $B\bar{B}$ occurs as a result of the quantum coin tosses, and the outcome of the final measurement on particle a is E.

H2. Suppose that instead of being present, the beam splitter $\bar{B}$ had been absent, $B\bar{O}$. The removal of a distant beam splitter at the last moment could not possibly have affected the outcome of the measurement on particle a, so E would have occurred in case $B\bar{O}$, just as it did in case $B\bar{B}$.

H3. Since by (25.30) $E\bar{F}$ is impossible in this situation, $E\bar{E}$ would have occurred in the case $B\bar{O}$.

H4. Given that $\bar{E}$ would have occurred in the case $B\bar{O}$, it would also have occurred with both beam splitters absent, $O\bar{O}$, since, once again, the removal of a distant beam splitter B at the last instant could not possibly have affected the outcome of a measurement on particle b.

H5. It follows from H1–H4 that if E occurs in the case $B\bar{B}$, then $\bar{E}$ would have occurred, in this particular experiment, if the quantum coin tosses had resulted in both beam splitters being absent, $O\bar{O}$, rather than present.

H6. Upon interchanging the roles of particles a and b in H1–H4, we conclude that if $\bar{E}$ occurs in the case $B\bar{B}$, then E would have been the case had both beam splitters been absent, $O\bar{O}$.

H7. Consider a situation in which both E and $\bar{E}$ occur in the case $B\bar{B}$; note that the probability for this is greater than 0, (25.29). Then in the counterfactual situation in which $O\bar{O}$ was the case rather than $B\bar{B}$, we can conclude using H5 that $\bar{E}$, and using H6 that E, would have occurred. That is, the outcome of the measurements would have been $E\bar{E}$ had the quantum coin tosses resulted in $O\bar{O}$.

H8. But according to (25.32), $E\bar{E}$ cannot occur in the case $O\bar{O}$, so we have reached a contradiction.

25.5 Analysis of the second paradox

A detailed analysis of H1–H4 is a bit complicated, since both H2 and H4 involve counterfactuals, and the conclusion, stated in H5, comes from chaining together two counterfactual arguments. In order not to become lost in intricate details of how one counterfactual may be combined with another, it is best to focus on the end result in H5, which can be restated in the following way: If in the actual world the quantum coin tosses result in $B\bar{B}$ and the measurement outcome is E, then in a counterfactual world in which the coin tosses had resulted in $O\bar{O}$, particle b would have triggered detector $\bar{E}$.

To support this argument using the scheme of counterfactual reasoning discussed in Sec. 19.4, we need to specify a *single* consistent family which contains the events we are interested in, which are the outcomes of the coin tosses and at least some of the outcomes of the final measurements, together with some event (or perhaps events) at a time earlier than when the quantum coins were tossed, which can serve as a suitable pivot. The framework might contain more than this, but it must contain at least this much. (Note that the pivot event or events can make reference to both particles, and could be more complicated than simply the product of a projector for a times a projector for b.) From this point of view, the intermediate steps in the argument — for example, H2, in which only one of the beam splitters is removed — can be thought of as a method for finding the final framework and pivot through a series of intermediate steps. That is, we may be able to find a framework and pivot which will justify H2, and then modify the framework and choose another pivot, if necessary, in order to incorporate H3 and H4, so as to arrive at the desired result in H5.

We shall actually follow a somewhat different procedure: make a guess for a framework which will support the result in H5, and then check that it works. An intelligent guess is not difficult, for E in case $B\bar{B}$ implies that the b particle was earlier in the $\bar{d}$ arm of its interferometer, (25.13), and when the beam splitter $\bar{B}$ is out of the way, a particle in $\bar{d}$ emerges in the $\bar{e}$ channel, which will result in $\bar{E}$. This suggests taking a look at the consistent family containing the following histories, in which the alternatives $\bar{c}$ and $\bar{d}$ occur at t_1:

$$\Phi_0 \odot \begin{cases} \bar{c} \odot \begin{cases} B\bar{B} \odot F, \\ O\bar{O} \odot \bar{F}. \end{cases} \\ \bar{d} \odot \begin{cases} B\bar{B} \odot \{E, F\}, \\ O\bar{O} \odot \bar{E}. \end{cases} \end{cases} \tag{25.33}$$

The $B\bar{O}$ and $O\bar{B}$ branches have been omitted from (25.33) in order to save space and allow us to concentrate on the essential task of finding a counterfactual argument which leads from $B\bar{B}$ to $O\bar{O}$. Including these other branches terminated by a noncommittal I at t_3 will turn (25.33) into the support of a consistent family without having any effect on the following argument.

The consistency of (25.33) can be seen in the following way. The events $B\bar{B}$ and $O\bar{O}$ are macroscopically distinct, hence orthogonal, and since they remain unchanged from t_2 to t_3, we only need to check that the chain operators for the two histories involving $O\bar{O}$ are orthogonal to each other — as is obviously the case, since the final $\bar{E}$ and $\bar{F}$ are orthogonal — and the chain operators for the three histories involving $B\bar{B}$ are mutually orthogonal. The only conceivable problem arises because two of the $B\bar{B}$ histories terminate with the same projector F. However, because at earlier times these histories involve orthogonal states $\bar{c}$ and $\bar{d}$ of particle b, and F has to do with particle a (that is, a measurement on particle a), rather than b, the chain operators are, indeed, orthogonal. The reader can check this by working out the chain kets.

One can use (25.33) to support the conclusion of H5 in the following way. The outcome E in the case $B\bar{B}$ occurs in only one history, on the third line in (25.33). Upon tracing this outcome back to $\bar{d}$ as a pivot, and then moving forward in time on the $O\bar{O}$ branch we come to $\bar{E}$ as the counterfactual conclusion. Having obtained the result in H5, we do not need to discuss H2, H3, and H4. However, it is possible to justify these statements as well by adding a $B\bar{O}$ branch to (25.33) with suitable measurement outcomes at t_3 in place of the noncommittal I, and then adding some more events involving properties of particle a at time t_1 in order to construct a suitable pivot for the argument in H2. As the details are not essential for the present discussion, we leave them as a (nontrivial) exercise for anyone who wishes to explore the argument in more depth.

By symmetry, H6 can be justified by the use of a consistent family (with, once again, the $B\bar{O}$ and $O\bar{B}$ branches omitted)

$$\Phi_0 \odot \begin{cases} c \odot \begin{cases} B\bar{B} \odot \bar{F}, \\ O\bar{O} \odot F, \end{cases} \\ d \odot \begin{cases} B\bar{B} \odot \{\bar{E}, \bar{F}\}, \\ O\bar{O} \odot E, \end{cases} \end{cases} \tag{25.34}$$

which is (25.33) with the roles of a and b interchanged. However, H7, which combines the results of H5 and H6, is not valid, because the family (25.33) on which H5 is based is incompatible with the family (25.34) on which H6 is based. The problem with combining these two families is that when one introduces the events E and F at t_3 in the $B\bar{B}$ branch of a family which contains c and d at an earlier time, it is essentially the same thing as introducing e and f to make the inconsistent family $\mathcal{J}_1$, (25.25). In the same way, introducing $\bar{E}$ and $\bar{F}$ in the $B\bar{B}$ branch following an earlier $\bar{c}$ and $\bar{d}$ leads to trouble. Even the very first statement in H7, that $E\bar{E}$ occurs in case $B\bar{B}$ with a positive probability, requires the use of a family which is incompatible with both (25.33) and (25.34)! Thus the road to a contradiction is blocked by the single-framework rule.

This procedure for blocking the second form of Hardy's paradox is very similar to the one used in Sec. 25.3 for blocking the first form of the paradox. Indeed, for the case $B\bar{B}$ we have used essentially the same families; the only difference comes from the (somewhat arbitrary) decision to word the second form of the paradox in terms of measurement outcomes, and the first in a way which only makes reference to particle properties.

The second form of Hardy's paradox, like the first, cannot be used to justify some form of quantum nonlocality in the sense of some mysterious long-range influence of the presence or absence of a beam splitter in the path of one particle on the behavior of the other particle. Locality was invoked in H2 and H4 (and at the corresponding points in H6). But H2 and H4, as well as the overall conclusion in H5, can be supported by using a suitable framework and pivots. (We have only given the explicit argument for H5.) Thus, while our analysis does not prove that the locality assumptions entering H2 and H4 are correct, it shows that there is no reason to suspect that there is anything wrong with them. The overall argument, H1–H8, results in a contradiction. However, the problem lies not in the locality assumptions in the earlier statements, but rather in the quantum incompatibility overlooked when writing down the otherwise plausible H7. This incompatibility, as noted earlier, has to do with the way a single particle is being described, so it cannot be blamed on anything nonlocal.

Our analysis of H1–H6 was based upon particular frameworks. As there are a large number of different possible frameworks, one might suppose that an alternative choice might be able to support the counterfactual arguments and lead to a contradiction. There is, however, a relatively straightforward argument to demonstrate that no single framework, and thus no set of compatible frameworks, could possibly support the argument in H1–H7. Consider any framework which contains $E\bar{E}$ at t_3 both in the case $B\bar{B}$ and also in the case $O\bar{O}$. In this framework both (25.29) and (25.32) are valid: $E\bar{E}$ occurs with finite probability in case $B\bar{B}$, and with zero probability in case $O\bar{O}$. The reason is that even though (25.29) and (25.32) were obtained using the framework (25.28), it is a general principle of quantum reasoning, see Sec. 16.3, that the probability assigned to a collection of events in one framework will be precisely the same in *all* frameworks which contain these events and the same initial data (Φ_0 in the case at hand). But in any single framework in which $E\bar{E}$ occurs with probability 0 in the case $O\bar{O}$ it is clearly impossible to reach the conclusion at the end of a series of counterfactual arguments that $E\bar{E}$ would have occurred with both beam splitters absent had the outcomes of the quantum coin tosses been different from what actually occurred.

To be more specific, suppose one could find a framework containing a pivot P at t_1 with the following properties: (i) P must have occurred if $B\bar{B}$ was followed by $E\bar{E}$; (ii) if P occurred and was then followed by $O\bar{O}$, the measurement outcome would have been $E\bar{E}$. These are the properties which would permit this framework to support the counterfactual argument in H1–H7. But since $B\bar{B}$ followed by $E\bar{E}$ has a positive probability, the same must be true of P, and therefore $O\bar{O}$ followed by $E\bar{E}$ would also have to occur with a finite probability. (A more detailed analysis shows that $\Pr(E\bar{E}, t_3 \mid O\bar{O}, t_2)$ would have to be at least as large as $\Pr(E\bar{E}, t_3 \mid B\bar{B}, t_2)$.) However, since $O\bar{O}$ is, in fact, never followed by $E\bar{E}$, a framework and pivot of this kind does not exist.

The conclusion is that it is impossible to use quantum reasoning in a consistent way to arrive at the conclusion H7 starting from the assumption H1. In some respects the analysis just presented seems too simple: it says, in effect, that if a counterfactual argument of the form H1–H7 arrives at a contradiction, then this very fact means there is some way in which this argument violates the rules of quantum reasoning. Can one dispose of a (purported) paradox in such a summary fashion? Yes, one can. The rule requiring that quantum reasoning of this type employ a *single* framework means that the usual rules of ordinary (classical) reasoning and probability theory can be applied as long as one sticks to this particular framework, and there can be no contradiction. To put the matter in a different way, if there is some very clever way to produce this paradox *using only one framework*, then there will also be a corresponding "classical" paradox, and whatever it is that is paradoxical will not be unique to quantum theory.

Nonetheless, there is some value in our working out specific aspects of the paradox using the explicit families (25.33) and (25.34), for they indicate that the basic difficulty with the argument in H1–H8 lies in an implicit assumption that the different frameworks are compatible, an assumption which is easy to make because it is always valid in classical mechanics. Incompatibility rather than some mysterious nonlocality is the crucial feature which distinguishes quantum from classical physics, and ignoring it is what has led to a paradox.

26

Decoherence and the classical limit

26.1 Introduction

Classical mechanics deals with objects which have a precise location and move in a deterministic way as a function of time. By contrast, quantum mechanics uses wave functions which always have some finite spatial extent, and the time development of a quantum system is (usually) random or stochastic. Nonetheless, most physicists regard classical mechanics as an approximation to quantum mechanics, an approximation which works well when the object of interest contains a large number of atoms. How can it be that classical mechanics emerges as a good approximation to quantum mechanics in the case of large objects?

Part of the answer to the question lies in the process of *decoherence* in which a quantum object or system interacting with a suitable environment (which is also quantum mechanical) loses certain types of quantum coherence which would be present in a completely isolated system. Even in classical physics the interaction of a system with its environment can have significant effects. It can lead to irreversible processes in which mechanical energy is turned into heat, with a resulting increase of the total entropy. Think of a ball rolling along a smooth, flat surface. Eventually it comes to rest as its kinetic energy is changed into heat in the surrounding air due to viscous effects, or dissipated as vibrational energy inside the ball or in the material which makes up the surface. (From this perspective the vibrational modes of the ball form part of its "environment".) While decoherence is (by definition) quantum mechanical, and so lacks any exact analog in classical physics, it is closely related to irreversible effects.

In this chapter we explore a very simple case, one might even think of it as a toy model, of a quantum particle interacting with its environment as it passes through an interferometer, in order to illustrate some of the principles which govern decoherence. In the final section there are some remarks on how classical mechanics emerges as a limiting case of quantum mechanics, and the role which decoherence

349

plays in relating classical and quantum physics. The discussion of decoherence
and of the classical limit of quantum mechanics presented here is only intended as
an introduction to a complex subject. The bibliography indicates some sources of
additional material.

26.2 Particle in an interferometer

Consider a particle passing through an interferometer, shown schematically in
Fig. 26.1, in which an input beam in channel a is separated by a beam splitter
into two arms c and d, and then passes through a second beam splitter into two
output channels e and f. While this has been drawn as a Mach–Zehnder interferometer similar to the interferometers considered in earlier chapters, it is best to
think of it as a neutron interferometer or an interferometer for atoms. The principles of interference for photons and material particles are the same, but photons
tend to interact with their environment in a different way.

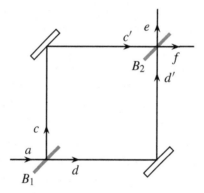

Fig. 26.1. Particle passing through an interferometer.

Let us suppose that the interferometer is set up so that a particle entering through
channel a always emerges in the f channel due to interference between the waves
in the two arms c and d. As discussed in Ch. 13, this interference disappears if there
is a measurement device in one or both of the arms which determines which arm
the particle passes through. Even in the absence of a measuring device, the particle
may interact with something, say a gas molecule, while traveling through one arm
but not through the other arm. In this way the interference effect will be reduced if
not entirely removed. One refers to this process as *decoherence* since it removes,
or at least reduces the interference effects resulting from a *coherent superposition*
of the two wave packets in the two arms. Sometimes one speaks metaphorically of
the environment "measuring" which arm the particle passes through.

Assume that at the first beam splitter the particle state undergoes a unitary time development

$$|a\rangle \mapsto (|c\rangle + |d\rangle)/\sqrt{2}, \tag{26.1}$$

while passage through the arms of the interferometer results in

$$|c\rangle \mapsto |c'\rangle, \quad |d\rangle \mapsto |d'\rangle. \tag{26.2}$$

Here $|a\rangle$ is a wave packet in the input channel at time t_0, $|c\rangle$ and $|d\rangle$ are wave packets emerging from the first beam splitter in the c and d arms of the interferometer at time t_1, and $|c'\rangle$ and $|d'\rangle$ are the corresponding wave packets at time t_2 just before they reach the second beam splitter. The effect of passing through the second beam splitter is represented by

$$|c'\rangle \mapsto \big(|e\rangle + |f\rangle\big)/\sqrt{2}, \quad |d'\rangle \mapsto \big(-|e\rangle + |f\rangle\big)/\sqrt{2}, \tag{26.3}$$

where $|e\rangle$ and $|f\rangle$ are wave packets in the output channels of the second beam splitter at time t_3. The notation is chosen to resemble that used for the toy models in Sec. 12.1 and Ch. 13.

Next assume that while inside the interferometer the particle interacts with something in the environment in a way which results in a unitary transformation of the form

$$|c\rangle|\epsilon\rangle \mapsto |c'\rangle|\epsilon'\rangle, \quad |d\rangle|\epsilon\rangle \mapsto |d'\rangle|\epsilon''\rangle, \tag{26.4}$$

on the Hilbert space $\mathcal{A} \otimes \mathcal{E}$ of the particle $\mathcal{A}$ and environment $\mathcal{E}$, where $|\epsilon\rangle$ is the normalized state of $\mathcal{E}$ at time t_1, and $|\epsilon'\rangle$ and $|\epsilon''\rangle$ are normalized states at t_2. For example, it might be the case that if the particle passes through the c arm some molecule is scattered from it resulting in the change from $|\epsilon\rangle$ to $|\epsilon'\rangle$, whereas if the particle passes through the d arm there is no scattering, and the change in the environment from $|\epsilon\rangle$ to $|\epsilon''\rangle$ is the same as it would have been in the absence of the particle. The complex number

$$\alpha = \langle\epsilon''|\epsilon'\rangle = \alpha' + i\alpha'', \tag{26.5}$$

with real and imaginary parts α' and α'', plays an important role in the following discussion. The final particle wave packets $|c'\rangle$ and $|d'\rangle$ in (26.4) are the same as in the absence of any interaction with the environment, (26.2). That is, we are assuming that the scattering process has an insignificant influence upon the center of mass of the particle itself as it travels through either arm of the interferometer. This approximation is made in order to simplify the following discussion; one could, of course, explore a more complicated situation.

The complete unitary time evolution of the particle and its environment as the particle passes through the interferometer is given by

$$|\psi_0\rangle = |a\rangle|\epsilon\rangle \mapsto (|c\rangle + |d\rangle)|\epsilon\rangle/\sqrt{2} \mapsto$$
$$|\psi_2\rangle = (|c'\rangle|\epsilon'\rangle + |d'\rangle|\epsilon''\rangle)/\sqrt{2} \mapsto \qquad (26.6)$$
$$|\psi_3\rangle = [|e\rangle(|\epsilon'\rangle - |\epsilon''\rangle) + |f\rangle(|\epsilon'\rangle + |\epsilon''\rangle)]/2,$$

where we assume that the environment state $|\epsilon\rangle$ does not change between t_0 and t_1. (This is not essential, and one could assume a different state, say $|\bar{\epsilon}\rangle$, at t_0, which develops unitarily into $|\epsilon\rangle$ at t_1.) Therefore, in the family with support $[\psi_0] \odot I \odot I \odot \{[e], [f]\}$ the probabilities for the particle emerging in each of the output channels are given by

$$\Pr(e) = \tfrac{1}{4}(\langle\epsilon'| - \langle\epsilon''|) \cdot (|\epsilon'\rangle - |\epsilon''\rangle)$$
$$= \tfrac{1}{4}(\langle\epsilon'|\epsilon'\rangle + \langle\epsilon''|\epsilon''\rangle - \langle\epsilon'|\epsilon''\rangle - \langle\epsilon''|\epsilon'\rangle) = \tfrac{1}{2}(1 - \alpha'), \qquad (26.7)$$
$$\Pr(f) = \tfrac{1}{4}(\langle\epsilon'| + \langle\epsilon''|) \cdot (|\epsilon'\rangle + |\epsilon''\rangle) = \tfrac{1}{2}(1 + \alpha').$$

Because the states entering the inner product in (26.5) are normalized, $|\alpha|$ cannot be greater than 1. If $|\epsilon''\rangle = |\epsilon'\rangle$, then $\alpha' = \alpha = 1$ and there is no decoherence: the interference pattern is the same as in the absence of any interaction with the environment, and the particle always emerges in f. The interference effect disappears when $\alpha' = 0$, and the particle emerges with equal probability in e or f. This could happen even with $|\alpha|$ rather large, for example, $\alpha = i$. But in such a case there would still be a substantial coherence between the wave packets in the c and d arms, and the corresponding interference effect could be detected by shifting the second beam splitter by a small amount so as to change the difference in path length between the c and d arms by a quarter wavelength. Hence it seems sensible to use $|\alpha|$ rather than α' as a measure of coherence between the two arms of the interferometer, and $1 - |\alpha|$ as a measure of the amount of decoherence.

26.3 Density matrix

In a situation in which one is interested in what happens to the particle after it passes through the second beam splitter without reference to the final state of the environment, it is convenient to use a density matrix ρ_2 for the particle at the intermediate time t_2 in (26.6), just before the particle passes through the second beam splitter. By taking a partial trace over the environment $\mathcal{E}$ in the manner indicated in Sec. 15.3, one obtains

$$\rho_2 = \text{Tr}_\mathcal{E}(|\psi_2\rangle\langle\psi_2|) = \tfrac{1}{2}(|c'\rangle\langle c'| + |d'\rangle\langle d'| + \alpha|c'\rangle\langle d'| + \alpha^*|d'\rangle\langle c'|). \qquad (26.8)$$

This has the form

$$\rho_2 = \begin{pmatrix} 1/2 & \alpha/2 \\ \alpha^*/2 & 1/2 \end{pmatrix} \tag{26.9}$$

when written as a matrix in the basis $\{|c'\rangle, |d'\rangle\}$, with $\langle c'|\rho_2|c'\rangle$ in the upper left corner. If we think of ρ_2 as a pre-probability, see Sec. 15.2, its diagonal elements represent the probability that the particle will be in the c or the d arm. Twice the magnitude of the off-diagonal elements serves as a convenient measure of coherence between the two arms of the interferometer, and thus $1 - |\alpha|$ is a measure of the decoherence.

After the particle passes through the second beam splitter, the density matrix is given by (see Sec. 15.4)

$$\rho_3 = T_A(t_3, t_2)\rho_2 T_A(t_2, t_3), \tag{26.10}$$

where $T_A(t_3, t_2)$ is the unitary transformation produced by the second beam splitter, (26.3), and we assume that during this process there is no further interaction of the particle with the environment. The result is

$$\rho_3 = \tfrac{1}{2}\left[(1 - \alpha')|e\rangle\langle e| + (1 + \alpha')|f\rangle\langle f| + i\alpha''\big(|e\rangle\langle f| - |f\rangle\langle e|\big)\right]. \tag{26.11}$$

The diagonal parts of ρ_3, the coefficients of $|e\rangle\langle e|$ and $|f\rangle\langle f|$, are the probabilities that the particle will emerge in the e or the f channel, and are, of course, identical with the expressions in (26.7).

Using a density matrix is particularly convenient for discussing a situation in which the particle interacts with the environment more than once as it passes through the c or the d arm of the interferometer. The simplest situation to analyze is one in which each of these interactions is independent of the others, and they do not alter the wave packet of the particle. In particular, let the environment consist of a number of separate pieces (e.g., separate molecules) with a Hilbert space

$$\mathcal{E} = \mathcal{E}_1 \otimes \mathcal{E}_2 \otimes \mathcal{E}_3 \otimes \cdots \mathcal{E}_n \tag{26.12}$$

and an initial state

$$|\epsilon\rangle = |\epsilon_1\rangle \otimes |\epsilon_2\rangle \otimes |\epsilon_3\rangle \otimes \cdots |\epsilon_n\rangle \tag{26.13}$$

at time t_1. The jth interaction results in $|\epsilon_j\rangle$ changing to $|\epsilon_j'\rangle$ if the particle is in the c arm, or to $|\epsilon_j''\rangle$ if the particle is in the d arm. Thus the net effect of all of these interactions as the particle passes through the interferometer is

$$\begin{aligned} |c\rangle|\epsilon\rangle &= |c\rangle|\epsilon_1\rangle|\epsilon_2\rangle \cdots |\epsilon_n\rangle \mapsto |c'\rangle|\epsilon'\rangle = |c'\rangle|\epsilon_1'\rangle|\epsilon_2'\rangle \cdots |\epsilon_n'\rangle, \\ |d\rangle|\epsilon\rangle &= |d\rangle|\epsilon_1\rangle|\epsilon_2\rangle \cdots |\epsilon_n\rangle \mapsto |d'\rangle|\epsilon''\rangle = |d'\rangle|\epsilon_1''\rangle|\epsilon_2''\rangle \cdots |\epsilon_n''\rangle. \end{aligned} \tag{26.14}$$

The reduced density matrix ρ_2 for the particle just before it passes through the second beam splitter is again of the form (26.8) or (26.9), with

$$\alpha = \langle \epsilon'' | \epsilon' \rangle = \alpha_1 \alpha_2 \cdots \alpha_n, \qquad (26.15)$$

where

$$\alpha_j = \langle \epsilon_j'' | \epsilon_j' \rangle, \qquad (26.16)$$

and ρ_3, when the particle has passed through the second beam splitter, is again given by (26.11).

In a typical situation one would expect the α_j to be less than 1, though not necessarily small. Note that if there are a large number of collisions, α in (26.15) can be very small, even if the individual α_j are not themselves small quantities. Thus repeated interactions with the environment will in general lead to greater decoherence than that produced by a single interaction, and if these interactions are of roughly the same kind, one expects the coherence $|\alpha|$ to decrease exponentially with the number of interactions.

Even if the different interactions with the environment are not independent of one another, the net effect may well be much the same, although it might take more interactions to produce a given reduction of $|\alpha|$. In any case, what happens at the second beam splitter, in particular the probability that the particle will emerge in each of the output channels, depends only on the density matrix ρ for the particle when it arrives at this beam splitter, and not on the details of all the scattering processes which have occurred earlier. For this reason, a density matrix is very convenient for analyzing the nature and extent of decoherence in this situation.

26.4 Random environment

Suppose that the environment which interacts with the particle is itself random, and that at time t_1, when the particle emerges from the first beam splitter, it is described by a density matrix R_1 which can be written in the form

$$R_1 = \sum_j p_j |\epsilon^j\rangle\langle\epsilon^j|, \qquad (26.17)$$

where $\{|\epsilon^j\rangle\}$ is an orthonormal basis of $\mathcal{E}$, and $\sum p_j = 1$. Although it is natural, and for many purposes not misleading to think of the environment as being in the state $|\epsilon^j\rangle$ with probability p_j, we shall think of R_1 as simply a pre-probability (Sec. 15.2). Assume that while the particle is inside the interferometer, during the time interval from t_1 to t_2, the interaction with the environment gives rise to unitary transformations

$$|c\rangle|\epsilon^j\rangle \mapsto |c'\rangle|\zeta^j\rangle, \quad |d\rangle|\epsilon^j\rangle \mapsto |d'\rangle|\eta^j\rangle, \qquad (26.18)$$

where we are again assuming, as in (26.4) and (26.14), that the environment has a negligible influence on the particle wave packets $|c'\rangle$ and $|d'\rangle$. Because the time evolution is unitary, $\{|\zeta^j\rangle\}$ and $\{|\eta^j\rangle\}$ are orthonormal bases of $\mathcal{E}$.

Let the state of the particle and the environment at t_1 be given by a density matrix

$$\Psi_1 = [\bar{a}] \otimes R_1, \tag{26.19}$$

where $|\bar{a}\rangle = (|c\rangle + |d\rangle)/\sqrt{2}$ is the state of the particle when it emerges from the first beam splitter. At t_2, just before the particle leaves the interferometer, the density matrix resulting from unitary time evolution of the total system will be

$$\Psi_2 = \tfrac{1}{2} \sum_j p_j \big[|c'\rangle\langle c'| \otimes |\zeta^j\rangle\langle \zeta^j| + |d'\rangle\langle d'| \otimes |\eta^j\rangle\langle \eta^j|$$
$$+ |c'\rangle\langle d'| \otimes |\zeta^j\rangle\langle \eta^j| + |d'\rangle\langle c'| \otimes |\eta^j\rangle\langle \zeta^j| \big]. \tag{26.20}$$

Taking a partial trace gives the expression

$$\rho_2 = \mathrm{Tr}_{\mathcal{E}}(\Psi_2) = \tfrac{1}{2}\big[|c'\rangle\langle c'| + |d'\rangle\langle d'| + \alpha|c'\rangle\langle d'| + \alpha^*|d'\rangle\langle c'| \big], \tag{26.21}$$

for the reduced density matrix of the particle at time t_2, where

$$\alpha = \sum_j p_j \langle \eta^j | \zeta^j \rangle. \tag{26.22}$$

The expression (26.21) is formally identical to (26.8), but the complex parameter α is now a weighted average of a collection of complex numbers, the inner products $\{\langle \eta^j | \zeta^j \rangle\}$, each with magnitude less than or equal to 1. Consider the case in which

$$\langle \eta^j | \zeta^j \rangle = e^{i\phi_j}, \tag{26.23}$$

that is, the interaction with the environment results in nothing but a phase difference between the wave packets of the particle in the c and d arms. Even though $|\langle \eta^j | \zeta^j \rangle| = 1$ for every j, the sum (26.22) will in general result in $|\alpha| < 1$, and if the sum includes a large number of random phases, $|\alpha|$ can be quite small. Hence a random environment can produce decoherence even in circumstances in which a nonrandom environment (as discussed in Secs. 26.2 and 26.3) does not.

The basis $\{|\epsilon^j\rangle\}$ in which R_1 is diagonal is useful for calculations, but does not actually enter into the final result for ρ_2. To see this, rewrite (26.18) in the form

$$|c\rangle \otimes |\epsilon\rangle \mapsto |c'\rangle \otimes U_c|\epsilon\rangle, \quad |d\rangle \otimes |\epsilon\rangle \mapsto |d'\rangle \otimes U_d|\epsilon\rangle, \tag{26.24}$$

where $|\epsilon\rangle$ is any state of the environment, and U_c and U_d are unitary transformations on $\mathcal{E}$. Then (26.22) can be written in the form

$$\alpha = \mathrm{Tr}_{\mathcal{E}}\big(R_1 U_d^\dagger U_c\big), \tag{26.25}$$

which makes no reference to the basis $\{|\epsilon^j\rangle\}$.

26.5 Consistency of histories

Consider a family of histories at times $t_0 < t_1 < t_2 < t_3$ with support

$$Y^{ce} = [\psi_0] \odot [c] \odot [c'] \odot [e],$$
$$Y^{de} = [\psi_0] \odot [d] \odot [d'] \odot [e],$$
$$Y^{cf} = [\psi_0] \odot [c] \odot [c'] \odot [f],$$
$$Y^{df} = [\psi_0] \odot [d] \odot [d'] \odot [f],$$

(26.26)

where $[\psi_0]$ is the initial state $|a\rangle|\epsilon\rangle$ in (26.6), and the unitary dynamics is that of Sec. 26.2. The chain operators for histories which end in $[e]$ are automatically orthogonal to those of histories which end in $[f]$. However, when the final states are the same, the inner products are

$$\langle K(Y^{df}), K(Y^{cf}) \rangle = \alpha/4 = -\langle K(Y^{de}), K(Y^{ce}) \rangle,$$

(26.27)

where α is the parameter defined in (26.5), which appears in the density matrix (26.8) or (26.9). Equation (26.27) is also valid in the case of multiple interactions with the environment, where α is given by (26.15). And it holds for the random environment discussed in Sec. 26.4, with α defined in (26.22), provided one re-defines the histories in (26.26) by eliminating the initial state $[\psi_0]$, so that each history begins with $[c]$ or $[d]$ at t_1, and uses the density matrix Ψ_1, (26.19), as an initial state at time t_1 in the consistency condition (15.48). In this case the operator inner product used in (26.27) is $\langle , \rangle_{\Psi_1}$, as it involves the density matrix Ψ_1, see (15.48).

If there is no interaction with the environment, then $\alpha = 1$ and (26.27) implies that the family (26.26) is not consistent. However, if α is very small, even though it is not exactly zero, one can say that the family (26.26) is *approximately consistent*, or consistent for all practical purposes, for the reasons indicated at the end of Sec. 10.2: one expects that by altering the projectors by small amounts one can produce a nearby family which is exactly consistent, and which has essentially the same physical significance as the original family.

This shows that the presence of decoherence may make it possible to discuss the time dependence of a quantum system using a family of histories which in the absence of decoherence would violate the consistency condtions and thus not make sense. This is an important consideration when one wants to understand how the classical behavior of macroscopic objects is consistent with quantum mechanics, which is the topic of the next section.

26.6 Decoherence and classical physics

The simple example discussed in the preceding sections of this chapter illustrates two important consequences of decoherence: it can destroy interference effects, and it can render certain families of histories of a subsystem consistent, or at least approximately consistent, when in the absence of decoherence such a family is inconsistent. There is an additional important effect which is not part of decoherence as such, for it can arise in either a classical or a quantum system interacting with its environment: the environment perturbs the motion of the system one is interested in, typically in a random way. (A classical example is Brownian motion, Sec. 8.1.)

The laws of classical mechanics are simple, have an elegant mathematical form, and are quite unlike the laws of quantum mechanics. Nonetheless, physicists believe that classical laws are only an approximation to the more fundamental quantum laws, and that quantum mechanics determines the motion of macroscopic objects made up of many atoms in the same way as it determines the motion of the atoms themselves, and that of the elementary particles of which the atoms are composed. However, showing that classical physics is a limiting case of quantum physics is a nontrivial task which, despite considerable progress, is not yet complete, and a detailed discussion lies outside the scope of this book. The following remarks are intended to give a very rough and qualitative picture of how the correspondence between classical and quantum physics comes about. More detailed treatments will be found in the references listed in the bibliography.

A macroscopic object such as a baseball, or even a grain of dust, is made up of an enormous number of atoms. The description of its motion provided by classical physics ignores most of the mechanical degrees of freedom, and focuses on a rather small number of *collective coordinates*. These are, for example, the center of mass and the Euler angles for a rigid body, to which may be added the vibrational modes for a flexible object. For a fluid, the collective coordinates are the hydrodynamic variables of mass and momentum density, thought of as obtained by "coarse graining" an atomic description by averaging over small volumes which still contain a very large number of atoms. It is important to note that the classical description employs a very special set of quantities, rather than using all the mechanical degrees of freedom.

It is plausible that properties represented by classical collective coordinates, such as "the mass density in region X has the value Y", correspond to projectors onto subspaces of a suitable Hilbert space. These subspaces will have a very large dimension, because the classical description is relatively coarse, and there will not be a unique projector corresponding to a classical property, but instead a collection of projectors (or subspaces), all of which correspond within some approximation to the same classical property.

In the same way, a classical property which changes as a function of time will be associated with different projectors as time progresses, and thus with a quantum history. The continuous time variable of a classical description can be related to the discrete times of a quantum history in much the same way as a continuous classical mass distribution is related to the discrete atoms of a quantum description. Just as a given classical property will not correspond to a unique quantum projector, there will be many quantum histories, and families of histories, which correspond to a given classical description of the motion, and represent it to a fairly good approximation. The term "quasi-classical" is used for such a quantum family and the histories which it contains.

In order for a quasi-classical family to qualify as a genuine quantum description, it must satisfy the consistency conditions. Can one be sure that this is the case? Gell-Mann, Hartle, Brun, and Omnès (see references in the bibliography) have studied this problem, and concluded that there are some fairly general conditions under which one can expect consistency conditions to be at least approximately satisfied for quasi-classical families of the sort one encounters in hydrodynamics or in the motion of rigid objects. That such quasi-classical families will turn out to be consistent is made plausible by the following consideration. Any system of macroscopic size is constantly in contact with an environment. Even a dust particle deep in interstellar space is bombarded by the cosmic background radiation, and will occasionally collide with atoms or molecules. In addition to an external environment of this sort, macroscopic systems have an internal environment constituted by the degrees of freedom left over when the collective coordinates have been specified. Both the external and the internal environment can contribute to processes of decoherence, and these can make it very hard to observe quantum interference effects. While the absence of interference, which is signaled by the fact that the density matrix of the subsystem is (almost) diagonal in a suitable representation, is not the same thing as the consistency of a suitable family of histories, nonetheless the two are related, as suggested by the example considered earlier in this chapter, where the same parameter α characterizes both the degree of coherence of the particle when it leaves the interferometer, and also the extent to which certain consistency conditions are not fulfilled. The effectiveness of this kind of decoherence is what makes it very difficult to design experiments in which macroscopic objects, even those no bigger than large molecules, exhibit quantum interference.

If a quasi-classical family can be shown to be consistent, will the histories in it obey, at least approximately, classical equations of motion? Again, this is a nontrivial question, and we refer the reader to the references in the bibliography for various studies. For example, Omnès has published a fairly general argument that classical and quantum mechanics give similar results if the quantum projectors correspond (approximately) to a cell in the classical phase space which is not too small and has a fairly regular shape, provided that during the time interval of interest the classical equations of motion do not result in too great a distortion of this cell. This last condition can break down rather quickly in the presence of chaos, a situation in which the motion predicted by the classical equations depends in a very sensitive way upon initial conditions.

Classical equations of motion are deterministic, whereas a quantum description employing histories is stochastic. How can these be reconciled? The answer is that the classical equations are idealizations which in appropriate circumstances work rather well. However, one must expect the motion of any real macroscopic system to show some effects of a random environment. The deterministic equations one usually writes down for classical collective coordinates ignore these environmental effects. The equations can be modified to allow for the effects of the environment by including stochastic noise, but then they are no longer deterministic, and this narrows the gap between classical and quantum descriptions. It is also worth keeping in mind that under appropriate conditions the quantum probability associated with a suitable quasi-classical history of macroscopic events can be very close to 1. These considerations would seem to remove any conflict between classical and quantum physics with respect to determinism, especially when one realizes that the classical description must in any case be an approximation to some more accurate quantum description.

In conclusion, even though many details have not been worked out and much remains to be done, there is no reason at present to doubt that the equations of classical mechanics represent an appropriate limit of a more fundamental quantum description based upon a suitable set of consistent histories. Only certain aspects of the motion of macroscopic physical bodies, namely those described by appropriate collective coordinates, are governed by classical laws. These laws provide an approximate description which, while quite adequate for many purposes, will need to be supplemented in some circumstances by adding a certain amount of environmental or quantum noise.

27

Quantum theory and reality

27.1 Introduction

The connection between human knowledge and the real world to which it is (hopefully) related is a difficult problem in philosophy. The purpose of this chapter is not to discuss the general problem, but only some aspects of it to which quantum theory might make a significant contribution. In particular, we want to discuss the question as to how quantum mechanics requires us to revise pre-quantum ideas about the nature of physical reality. This is still a very large topic, and space will permit no more than a brief discussion of some of the significant issues.

Physical theories should not be confused with physical reality. The former are, at best, some sort of abstract or symbolic representation of the latter, and this is as true of classical physics as of quantum physics. The phase space used to represent a classical system and the Hilbert space used for a quantum system are both mathematical constructs, not physical objects. Neither planets nor electrons integrate differential equations in order to decide where to go next. Wave functions exist in the theorist's notebook and not, unless in some metaphorical sense, in the experimentalist's laboratory. One might think of a physical theory as analogous to a photograph, in that it contains a representation of some object, but is not the object itself. Or one can liken it to a map of a city, which symbolizes the locations of streets and buildings, even though it is only made of paper and ink.

We can comprehend (to some extent) with our minds the mathematical and logical structure of a physical theory. If the theory is well developed, there will be clear relationships among the mathematical and logical elements, and one can discuss whether the theory is coherent, logical, beautiful, etc. The question of whether a theory is true, its relationship to the real world "out there", is more subtle. Even if a theory has been well confirmed by experimental tests, as in the case of quantum mechanics, believing that it is (in some sense) a true description of the real world requires a certain amount of faith. A decision to accept a theory as an adequate, or even as an approximate representation of the world is a matter of judgment which must inevitably move beyond issues of mathematical proof, logical rigor, and agreement with experiment.

If a theory makes a certain amount of sense and gives predictions which agree reasonably well with experimental or observational results, scientists are inclined to believe that its logical and mathematical structure reflects the structure of the real world in some way, even if philosophers will remain permanently sceptical. Granted that all theories are eventually shown to have limitations, we nonetheless think that Newton's mechanics is a great improvement over that of Aristotle, because it is a much better reflection of what the real world is like, and that relativity theory improves upon the science of Newton because space-time actually does have a structure in which light moves at the same speed in any inertial coordinate system. Theories such as classical mechanics and classical electromagnetism do a remarkably good job within their domains of applicability. How can this be understood if not by supposing that they reflect something of the real world in which we live?

The same remarks apply to quantum mechanics. Since it has a consistent mathematical and logical structure, and is in good agreement with a vast amount of observational and experimental data, it is plausible that quantum theory is a better reflection of what the real world is like than the classical theories which preceded it, and which could not explain many of the microscopic phenomena that are now understood using quantum methods. The faith of the physicist is that the real world is something like our best theories, and at the present time it is universally agreed that quantum mechanics is a very good theory of the physical world, better than any other currently available to us.

27.2 Quantum vs. classical reality

What are the main respects in which quantum mechanics differs from classical mechanics? To begin with, quantum theory employs wave functions belonging to a Hilbert space, rather than points in a classical phase space, in order to describe a physical system. Thus a quantum particle, in contrast to a classical particle, Secs. 2.3 and 2.4, does not possess a precise position or a precise momentum. In addition, the precision with which either of these quantities can be defined is limited by the Heisenberg uncertainty principle, (2.22). This does not mean that quantum entities are "fuzzy" and ill-defined, for a ray in the Hilbert space is as precise a specification as a point in phase space. What it does mean is that the classical concepts of position and momentum can only be used in an approximate way when applied to the quantum domain. As pointed out in Sec. 2.4, the uncertainty principle refers primarily to the fact that quantum entities are described by a very different mathematical structure than are classical particles, and only secondarily to issues associated with measurements. The limitations on measurements come about because of the nature of quantum reality, and the fact that what does not exist cannot be measured.

A second respect in which quantum mechanics is fundamentally different from classical mechanics is that the basic classical dynamical laws are deterministic, whereas quantum dynamical laws are, in general, stochastic or probabilistic, so that the future behavior of a quantum system cannot be predicted with certainty, even when given a precise initial state. It is important to note that in quantum theory this unpredictability in a system's time development is an intrinsic feature of the world, in contrast to examples of stochastic time development in classical physics, such as the diffusion of a Brownian particle (Sec. 8.1). Classical unpredictability arises because one is using a coarse-grained description where some information about the underlying deterministic system has been thrown away, and there is always the possibility, in principle, of a more precise description in which the probabilistic element is absent, or at least the uncertainties reduced to any extent one desires. By contrast, the Born rule or its extension to more complicated situations, Chs. 9 and 10, enters quantum theory as an axiom, and does not result from coarse graining a more precise description. To be sure, there have been efforts to replace the stochastic structure of quantum theory with something more akin to the determinism of classical physics, by supplementing the Hilbert space with hidden variables. But these have not turned out to be very fruitful, and, as discussed in Ch. 24, the Bell inequalities indicate that such theories can only restore determinism at the price of introducing nonlocal influences violating the principles of special relativity.

Of course, there is no reason to suppose that quantum mechanics as understood at the present time is the ultimate theory of how the world works. It could be that at some future date its probabilistic laws will be derived from a superior theory which returns to some form of determinism, but it is equally possible that future theories will continue to incorporate probabilistic time development as a fundamental feature. The fact that it was only with great reluctance that physicists abandoned classical determinism in the course of developing a theory capable of explaining experimental results in atomic physics strongly suggests, though it does not prove, that stochastic time development is part of physical reality.

27.3 Multiple incompatible descriptions

The feature of quantum theory which differs most from classical physics is that it allows one to describe a physical system in many different ways which are *incompatible* with one another. Under appropriate circumstances two (or more) incompatible descriptions can be said to be true in the sense that they can be derived in different incompatible frameworks starting from the same information about the system (the same initial data), but they cannot be combined in a single description, see Sec. 16.4. There is no really good classical analog of this sort of incompatibility, which is very different from what we find in the world of everyday experience, and it suggests that reality is in this respect very different from anything dreamed of prior to the advent of quantum mechanics.

As a specific example, consider the situation discussed in Sec. 18.4 using Fig. 18.4, where a nondestructive measurement of S_z is carried out on a spin-half particle by one measuring device, and this is followed by a later measurement of S_x using a second device. There is a framework $\mathcal{F}$, (18.31), in which it is possible to infer that at the time t_1 when the particle was between the two measuring devices it had the property $S_z = +1/2$, and another, incompatible framework $\mathcal{G}$, (18.33), in which one can infer the property $S_x = +1/2$ at t_1. But there is no way in which these inferences, even though each is valid in its own framework, can be combined, for in the Hilbert space of a spin-half particle there is no subspace which corresponds to $S_z = +1/2$ AND $S_x = +1/2$, see Sec. 4.6. Thus we have two descriptions of the same quantum system which because of the mathematical structure of quantum theory cannot be combined into a single description.

It is not the multiplicity of descriptions which distinguishes quantum from classical mechanics, for multiple descriptions of the same object occur all the time in classical physics and in everyday life. A teacup has a different appearance when viewed from the top or from the side, and the side view depends on where the handle is located, but there is never any problem in supposing that these different descriptions refer to the same object. Or consider a macroscopic body which is spinning. One description might specify the z-component L_z of its angular momentum, and another the x-component L_x. In classical physics, two correct descriptions of a single object can always be combined to produce a single, more precise description, and if this process is continued using all possible descriptions, the result will be a *unique exhaustive description* which contains each and every detail of every true description. In the case of a mechanical system at a single time, the unique exhaustive description corresponds to a single point in the classical phase space. Any true description can be obtained from the unique exhaustive description by coarsening it, that is, by omitting some of the details. Thus specifying a region in the phase space rather than a single point produces a coarser description of a mechanical system.

For the purposes of the following discussion it is convenient to refer to the idea that there exists a unique exhaustive description as the *principle of unicity*, or simply *unicity*. This principle implies that every conceivable property of a particular physical system will be either true or false, since it either is or is not contained in, or implied by, the unique exhaustive description. Thus unicity implies the existence of a universal truth functional as defined in Sec. 22.4. But as was pointed out in that section, there cannot be a universal truth functional for a quantum Hilbert space of dimension greater than 2. This is one of several ways of seeing that quantum theory is inconsistent with the principle of unicity, so that unicity is not part of quantum reality. It is the incompatibility of quantum descriptions which prevents them from being combined into a more precise description, and thus makes it impossible to create a unique exhaustive description.

The difference between classical and quantum mechanics in this respect can be seen by considering a nondestructive measurement of L_z for a macroscopic spinning body, followed by a later measurement of L_x. Combining a description based upon the first measurement with one based on the second takes one two thirds of the way towards a unique exhaustive description of the angular momentum vector. But trying to combine S_z and S_x values for a spin-half particle is, as already noted, an impossibility, and this means that these two descriptions cannot be obtained by coarsening a unique exhaustive quantum description, and therefore no such description exists.

In order to describe a quantum system, a physicist must, of necessity, adopt some framework and this means choosing among many incompatible frameworks, no one of which is, from a fundamental point of view, more appropriate or more "real" than any other. This freedom of choice on the part of the physicist has occasionally been misunderstood, so it is worth pointing out some things which it does *not* mean.

First, the freedom to use different incompatible frameworks in order to construct different incompatible descriptions does not make quantum mechanics a subjective science. Two physicists who employ the same framework will reach identical conclusions when starting from the same initial data. More generally, they will reach the same answers to the same physical questions, even when some question can be addressed using more than one framework; see the consistency argument in Sec. 16.3. To use an analogy, if one physicist discusses L_z for a macroscopic spinning object and another physicist L_x, their descriptions cannot be compared with each other, but if both of them describe the same component of angular momentum and infer its value from the same initial data, they will agree. The same is true of S_z and S_x for a spin-half particle.

Second, what a physicist happens to be thinking about when choosing a framework in order to construct a quantum description does not somehow influence reality in a manner akin to psychokinesis. No one would suppose that a physicist's choosing to describe L_z rather than L_x for a macroscopic spinning body was somehow influencing the body, and the same holds for quantum descriptions of microscopic objects. Choosing an S_z, rather than, say, an S_x framework makes it possible to discuss S_z, but does not determine its value. Once the framework has been adopted it may be possible by logical reasoning, given suitable data, to infer that $S_z = +1/2$ rather than $-1/2$, but this is no more a case of mind influencing matter than would be a similar inference of a value of L_z for a macroscopic body.

Third, choosing a framework $\mathcal{F}$ for constructing a description does not mean that some other description constructed using an incompatible framework $\mathcal{G}$ is false. Quantum incompatibility is very different from the notion of mutually exclusive descriptions, where the truth of one implies the falsity of the other. Once again the analogy of classical angular momentum is helpful: a description which assigns a value to L_z does not in any way render false a description which assigns a value to L_x, even though it does exclude a description that assigns a *different* value to L_z. The same comments apply to S_z and S_x in the quantum case.

In order to avoid the mistake of supposing that incompatible descriptions are mutually exclusive, it is helpful to think of them as referring to *different aspects* of a quantum system. Thus using the S_z framework allows the physicist to describe the "S_z aspect" of a spin-half particle, which is quite distinct from the "S_x aspect". To be sure, one still has to remember that, unlike the situation in classical physics, two incompatible aspects cannot both enter a single description of a quantum system. While using an appropriate terminology and employing classical analogies are helpful for understanding the concept of quantum incompatibility, it remains true that this is one feature of quantum reality which is far easier to represent in mathematical terms than by means of a physical picture.

27.4 The macroscopic world

Our most immediate contact with physical reality comes from our sensory experience of the macroscopic world: what we see, hear, touch, etc. A fundamental physical theory should, at least in principle, be able to explain the macroscopic phenomena we encounter in everyday life. But there is no reason why it must be built up entirely out of concepts from everyday experience, or restricted to everyday language. Modern physical theories posit all sorts of strange things, from quarks to black holes, that are totally alien to everyday experience, and whose description often requires some rather abstract mathematics. There is no reason to deny that such objects are part of physical reality, as long as they form part of a coherent theoretical structure which can relate them, even somewhat indirectly, to things which are accessible to our senses.

Two considerations suggest that quantum mechanics can (in principle) explain the world of our everyday experience in a satisfactory way. First, the macroscopic world can be described very well using classical physics. Second, as discussed in Sec. 26.6, classical mechanics is a good approximation to a fully quantum mechanical description of the world in precisely those circumstances in which classical physics is known to work very well. This quantum description employs a quasi-classical framework in which appropriate macro projectors represent properties of macroscopic objects, and the relevant histories, which are well-approximated by solutions of classical equations of motion, are rendered consistent by a process of decoherence, that is, by interaction with the (internal or external) environment of the system whose motion is being discussed.

It is important to note that all of the phenomena of macroscopic classical physics can be described using a *single* quasi-classical quantum framework. Within a single framework the usual rules of classical reasoning and probability theory apply, and quantum incompatibility, which has to do with the relationship between *different* frameworks, never arises. In this way one can understand why quantum incompatibility is completely foreign to classical physics and invisible in the everyday world. (As pointed out in Sec. 26.6, there are actually many different quasi-classical frameworks, each of which gives approximately the same results for the macroscopic variables of classical physics. This multiplicity does not alter the validity of the preceding remarks, since a description can employ any one of these frameworks and still lead to the same classical physics.)

Stochastic quantum dynamics can be reconciled with deterministic classical dynamics by noting that the latter is in many circumstances a rather good approximation to a quasi-classical history that the quantum system follows with high probability. Classical chaotic motion is an exception, but in this case classical dynamics, while in principle deterministic, is as a practical matter stochastic, since small errors in initial conditions are rapidly amplified into large and observable differences in the motion of the system. Thus even in this instance the situation is not much different from quantum dynamics, which is intrinsically stochastic.

The relationship of quantum theory to pre-quantum physics is in some ways analogous to the relationship between special relativity and Newtonian mechanics. Space and time in relativity theory are related to each other in a very different way than in nonrelativistic mechanics, in which time is absolute. Nonetheless, as long as velocities are much less than the speed of light, nonrelativistic mechanics is an excellent approximation to a fully relativistic mechanics. One never even bothers to think about relativistic corrections when designing the moving parts of an automobile engine. The same theory of relativity that shows that the older ideas of physical reality are very wrong when applied to bodies moving at close to the speed of light also shows that they work extremely well when applied to objects which move slowly. In the same way, quantum theory shows us that our notions of pre-quantum reality are entirely inappropriate when applied to electrons moving inside atoms, but work extremely well when applied to pistons moving inside cylinders.

However, quantum mechanics also allows the use of non-quasi-classical frameworks for describing macroscopic systems. For example, the macroscopic detectors which determine the channel in which a spin-half particle emerges from a Stern–Gerlach magnet, as discussed in Secs. 17.3 and 17.4, can be described by a quasi-classical framework $\mathcal{F}$, such as (17.25), in which one or the other detector detects the particle, or by a non-quasi-classical framework $\mathcal{G}$ in which the initial state develops unitarily into a macroscopic quantum superposition (MQS) state of the detector system. Is it a defect of quantum mechanics as a fundamental theory that it allows the physicist to use either of the incompatible frameworks $\mathcal{F}$ and $\mathcal{G}$ to construct a description of this situation, given that MQS states of this sort are never observed in the laboratory?

One must keep in mind the fact mentioned in the previous section that two incompatible quantum frameworks $\mathcal{F}$ and $\mathcal{G}$ do not represent mutually-exclusive possibilities in the sense that if the world is correctly described by $\mathcal{F}$ it cannot be correctly described by $\mathcal{G}$, and vice versa. Instead it is best to think of $\mathcal{F}$ and $\mathcal{G}$ as means by which one can describe different aspects of the quantum system, as suggested at the end of Sec. 27.3. To discuss which detector has detected the particle one must employ $\mathcal{F}$, since the concept makes no sense in $\mathcal{G}$, whereas the "MQS aspect" or "unitary time development aspect" for which $\mathcal{G}$ is appropriate makes no sense in $\mathcal{F}$. Either framework can be employed to answer those questions for which it is appropriate, but the answers given by the two frameworks cannot be combined or compared. (Also see the discussion of Schrödinger's cat in Sec. 9.6.)

If one were trying to set up an experiment to detect the MQS state, then one would want to employ the framework $\mathcal{G}$, or, rather, its extension to a framework which included the additional measuring apparatus which would be needed to determine whether the detector system was in the MQS state or in some state orthogonal to it. In fact, by using the principles of quantum theory one can argue that actual observations of MQS states are extremely difficult, even if "macroscopic" is employed somewhat loosely to include even an invisible grain of material containing a few million atoms. The process of decoherence in such situations is extremely fast, and in any case constructing some apparatus sensitive to the relative phases in a macroscopic superposition is a practical impossibility. It may be helpful to draw an analogy with the second law of thermodynamics. Whereas there is nothing in the laws of classical (or quantum) mechanics which prevents the entropy of a system from decreasing as a function of time, in practice this is never observed, and the principles of statistical mechanics provide a plausible explanation through assigning an extremely small probability to violations of the second law. In a similar way, quantum mechanics can explain why MQS states are never observed in the laboratory, even though they are very much a part of the fundamental theory, and hence also part of physical reality to the extent that quantum theory reflects that reality.

The difficulty of observing MQS states also explains why violations of the principle of unicity (see the previous section) are not seen in macroscopic systems, even though readily apparent in atoms. The breakdown of unicity is only apparent when one constructs descriptions using different incompatible frameworks, so it is never apparent if one restricts attention to a single framework. As noted earlier, classical physics works very well for a macroscopic system precisely because it is a good approximation to a quantum description based on a single quasi-classical framework. Hence even though quantum mechanics violates the principle of unicity, quantum mechanics itself provides a good explanation as to why that principle is always obeyed in classical physics, and its violation was neither observed nor even suspected before the advent of the scientific developments which led to quantum theory.

27.5 Conclusion

Quantum mechanics is clearly superior to classical mechanics for the description of microscopic phenomena, and in principle works equally well for macroscopic phenomena. Hence it is at least plausible that the mathematical and logical structure of quantum mechanics better reflect physical reality than do their classical counterparts. If this reasoning is accepted, quantum theory requires various changes in our view of physical reality relative to what was widely accepted before the quantum era, among them the following:

1. Physical objects never possess a completely precise position or momentum.
2. The fundamental dynamical laws of physics are stochastic and not deterministic, so from the present state of the world one cannot infer a unique future (or past) course of events.
3. The principle of unicity does not hold: there is not a unique exhaustive description of a physical system or a physical process. Instead, reality is such that it can be described in various alternative, incompatible ways, using descriptions which cannot be combined or compared.

All of these, and especially the third, represent radical revisions of the pre-quantum view of physical reality based upon, or at least closely allied to classical mechanics. At the same time it is worth emphasizing that there are other respects in which the development of quantum theory leaves previous ideas about physical reality unchanged, or at least very little altered. The following is not an exhaustive list, but indicates a few of the ways in which the classical and quantum viewpoints are quite similar:

1. Measurements play no fundamental role in quantum mechanics, just as they play no fundamental role in classical mechanics. In both cases, measurement apparatus and the process of measurement are described using the same basic mechanical principles which apply to all other physical objects and physical processes. Quantum measurements, when interpreted using a suitable framework, can be understood as revealing properties of a measured system before the measurement took place, in a manner which was taken for granted in classical physics. See the discussion in Chs. 17 and 18. (It may be worth adding that there is no special role for human consciousness in the quantum measurement process, again in agreement with classical physics.)

2. Quantum mechanics, like classical mechanics, is a local theory in the sense that the world can be understood without supposing that there are mysterious influences which propagate over long distances more rapidly than the speed of light. See the discussion in Chs. 23–25 of the EPR paradox, Bell's inequalities, and Hardy's paradox. The idea that the quantum world is permeated by superluminal influences has come about because of an inadequate understanding of quantum measurements — in particular, the assumption that wave function collapse is a physical process — or through assuming the existence of hidden variables instead of (or in addition to) the quantum Hilbert space, or by employing counterfactual arguments which do not satisfy the single-framework rule. By contrast, a consistent application of quantum principles provides a positive demonstration of the *absence* of nonlocal influences, as in the example discussed in Sec. 23.4.

3. Both quantum mechanics and classical mechanics are consistent with the notion of an *independent reality*, a real world whose properties and fundamental laws do not depend upon what human beings happen to believe, desire, or think. While this real world contains human beings, among other things, it existed long before the human race appeared on the surface of the earth, and our presence is not essential for it to continue.

The idea of an independent reality had been challenged by philosophers long before the advent of quantum mechanics. However, the difficulty of interpreting quantum theory has sometimes been interpreted as providing additional reasons for doubting that such a reality exists. In particular, the idea that measurements collapse wave functions can suggest the notion that they thereby bring reality into existence, and if a conscious observer is needed to collapse the wave function (MQS state) of a measuring apparatus, this could mean that consciousness somehow plays a fundamental role in reality. However, once measurements are understood as no more than particular examples of physical processes, and wave function collapse as nothing more than a computational tool, there is no reason to suppose that quantum theory is incompatible with an independent reality, and one is back to the situation which preceded the quantum era. To be sure, neither quantum nor classical mechanics provides watertight arguments in favor of an independent reality. In the final analysis, believing that there is a real world "out there", independent of ourselves, is a matter of faith. The point is that quantum mechanics is just as consistent with this faith as was classical mechanics. On the other hand, quantum theory indicates that the *nature* of this independent reality is in some respects quite different from what was earlier thought to be the case.

Bibliography

References are listed here by chapter. An alphabetical list will be found at the end. The abbreviation WZ stands for the source book by Wheeler and Zurek (1983).

Ch. 1. Introduction

Someone unfamiliar with quantum mechanics may wish to look at one of the many textbooks which provide an introduction to the subject. Here are some possibilities: Feynman *et al.* (1965), Cohen-Tannoudji *et al.* (1977), Bransden and Joachain (1989), Schwabl (1992), Shankar (1994), Merzbacher (1998). A useful history of the development of quantum mechanics will be found in Jammer (1974), and the collection edited by Wheeler and Zurek (1983) contains reprints of a number of important papers. For Bohr's account of his discussions with Einstein, Sec. 1.7, see Bohr (1951), reprinted in WZ.

Ch. 2. Wave functions

A discussion of the Hilbert space of square integrable functions will be found in Reed and Simon (1972).

Ch. 3. Linear algebra in Dirac notation

Introductory texts on linear algebra in finite-dimensional spaces at an appropriate level include Halmos (1958) and Strang (1988). The more advanced books by Horn and Johnson (1985, 1991) are useful references. The bra and ket notation is due to Dirac, see Dirac (1958). Properties of infinite-dimensional Hilbert spaces are discussed by Halmos (1957), Reed and Simon (1972), and Blank *et al.* (1994). Gieres (2000) discusses some of the mistakes which can arise when the rules for finite-dimensionsal spaces are applied to an infinite-dimensional Hilbert space.

The proof that the summands are orthogonal when a projector is a sum of other projectors, (3.46), will be found in Halmos (1957), §28 theorem 2.

Ch. 4. Physical properties

The connection between physical properties and subspaces of the Hilbert space is discussed in Ch. III, Sec. 5 of von Neumann (1932). Birkhoff and von Neumann (1936) contains their proposal for the conjunction and disjunction of quantum properties, and the corresponding quantum logic.

The concept of incompatibility used here goes back to Griffiths (1984) and Omnès (1992); for a more recent discussion see Griffiths (1998a). There is a connection with Bohr's (not very precise) notion of "complementary", for which see the relevant chapters in Jammer (1974).

Ch. 5. Probabilities and physical variables

There are many books on probability theory which present the subject at a level more than adequate for our purposes, among them: Feller (1968), DeGroot (1986), and Ross (2000) (or any of numerous editions).

Ch. 6. Composite systems and tensor products

Tensor products are also known as *direct products* or *Kronecker products*. The discussion in Halmos (1958) is a bit abstract, and that in Horn and Johnson (1991) is somewhat technical. Nielsen and Chuang (2000) take an approach similar to that adopted here, and on p. 109 give a straightforward derivation of the Schmidt decomposition (6.18).

Identical particles are discussed in the quantum textbooks listed above under Ch. 1. For a more advanced approach, see Ch. 1 of Fetter and Walecka (1971).

Ch. 7. Unitary dynamics

For properties of unitary matrices, see Horn and Johnson (1985) or Lütkepohl (1996). Unitary time development operators are discussed in many quantum textbooks: Cohen-Tannoudji *et al.* (1977) p. 308, Shankar (1994) p. 51, Feynman *et al.* (1965), Ch. 8.

The essence of a toy model is a quantum system which is discrete in space and time. The idea has probably occurred independently to a large number of people. Some earlier work of which I am aware is found in Gudder (1989) and Meyer (1996).

Ch. 8. Stochastic histories

Brownian motion is discussed in Ch. 22 of Wannier (1966), and from a more mathematical point of view in Sec. 10.1 of Ross (2000).

The concept of a quantum history in the sense used in this chapter goes back to Griffiths (1984), and was extended by Omnès (1992, 1994, 1999) and by Gell-Mann and Hartle (1990, 1993). The treatment here follows Griffiths (1996). That histories can be associated in a natural way with projectors on a tensor product space was first pointed out by Isham (1994).

Ch. 9. The Born rule

Classical random walks are discussed in many places, including Feller (1968) and Ross (2000).

Born (1926) is his original proposal for introducing probabilities in quantum mechanics; an English translation is in WZ, p. 52. Also note the discussion on pp. 38ff of Jammer (1974). Schrödinger (1935) discusses the infamous cat; an English translation is in WZ.

Ch. 10. Consistent histories

The operator inner product $\langle A, B \rangle$ is mentioned in Lütkepohl (1996) p. 104 and will be found as an exercise on p. 332 of Horn and Johnson (1985). It is sometimes called a Hilbert–Schmidt inner product, but several other names seem equally appropriate; see Horn and Johnson (1985), p. 291.

Consistency conditions for families of histories were first proposed by Griffiths (1984). Simpler conditions were introduced by Gell-Mann and Hartle (1990, 1993), and have been adopted here as an orthogonality condition, for which see McElwaine (1996). Omnès (1999) uses the formulation of Gell-Mann and Hartle.

Section 4 of Dowker and Kent (1996) contains their analysis of approximate consistency referred to at the end of Sec. 4.3.

Ch. 12. Examples of consistent families

Chapter VI of von Neumann (1932) contains his theory of quantum measurements.

Ch. 13. Quantum interference

Feynman's masterful discussion of interference from two slits (or two holes) is found in Ch. 1 of Feynman *et al.* (1965). The Mach–Zehnder interferometer is described in Sec. 7.5.7 of Born and Wolf (1980). For a discussion of neutron interference effects, see the review article by Greenberger (1983).

Ch. 14. Dependent (contextual) events

Branch-dependent histories, Sec. 14.2, were introduced by Gell-Mann and Hartle (1993).

Ch. 16. Quantum reasoning

The importance of having appropriate rules for quantum reasoning has been understood (at least in some quarters) ever since the work of Birkhoff and von Neumann (1936), and has been stressed by Omnès (1992, 1994, 1999). The formulation given here is based on Griffiths (1996, 1998a).

Chs. 17 and 18. Measurements

A very readable account of the problems encountered in trying to base the interpretation of quantum theory upon measurements will be found in Wigner (1963). Despite a great deal of effort in the quantum foundations community there has been essentially no progress in solving these problems; see the protest by Bell (1990) and the pessimistic assessment by Mittelstaedt (1998).

Chapter VI of von Neumann (1932) contains his theory of quantum measurements. That the *absence* of some occurrence can constitute a "measurement" was pointed out by Renninger (1960), although according to Jammer (1974), p. 495, this difficulty was known much earlier. The Lüders rule is found in Lüders (1951).

The nondestructive measurement apparatus in Fig. 18.2 was inspired by the one in Fig. 5-3 of Feynman *et al.* (1965).

Ch. 19. Coins and counterfactuals

The importance of counterfactuals for understanding quantum theory has been stressed by d'Espagnat (1984, 1989). Philosophers have analyzed counterfactual reasoning with somewhat inconclusive results; the work of Lewis (1973, 1986) is often cited. The approach used in this chapter extends Griffiths (1999a).

Ch. 20. Delayed choice paradox

The paradox in the form considered here is due to Wheeler (1978, 1983). The analysis uses material from Griffiths (1998b). For experimental realizations see Alley *et al.* (1983, 1987) and Hellmuth *et al.* (1987).

Ch. 21. Indirect measurement paradox

Aside from dramatization, the paradox stated here is that of Elitzur and Vaidman (1993). However, the basic idea seems to go back much earlier; see p. 495 of Jammer (1974). A somewhat similar paradox was put forward by Hardy (1992b). The term "interaction-free" appears to have originated with Dicke (1981).

Ch. 22. Incompatibility paradoxes

The impossibility of simultaneously assigning values to all quantum variables was pointed out by Bell (1966) and by Kochen and Specker (1967). See the helpful discussion of these and other results in Mermin (1993). The two-spin paradox of Mermin (1990) was inspired by earlier work by Peres (1990). See Mermin (1993) for a very clear presentation of this and related paradoxes.

The discussion of truth functionals in this chapter draws on Griffiths (2000a,b). The paradox of three boxes in Sec. 22.5 was introduced by Aharonov and Vaidman (1991); the discussion given in this chapter extends that in Griffiths (1996).

Ch. 23. Singlet state correlations

The original paper by Einstein, Podolsky, and Rosen (1935) has given rise to an enormous literature. A few of the items that I myself have found helpful are: Fine (1986), Greenberger *et al.* (1990), and Hajek and Bub (1992).

Bohm's study of the EPR problem using a singlet state of two spin-half particles appeared in Ch. 22 of Bohm (1951). The analysis of spin correlations given here extends an earlier discussion in Griffiths (1987).

Ch. 24. EPR paradox and Bell inequalities

See the references above for Ch. 23.

Bohm's hidden variable theory was introduced in Bohm (1952). For recent formulations, see Bohm and Hiley (1993) and Berndl *et al.* (1995). A serious problem with the Bohm theory was pointed out by Englert *et al.* (1992); see Griffiths (1999b) for references to the ensuing discussion.

Mermin's model for hidden variables appeared in Mermin (1981, 1985). A number of Bell's papers have been reprinted in Bell (1987), which is an excellent source for his work. These include the original inequality paper, Bell (1964). The CHSH inequality was published by Clauser *et al.* (1969), and a derivation will be found in Ch. 4 of Bell (1987). There is a vast literature devoted to Bell inequalities and their significance. In addition to the sources already mentioned, the reader may find it helpful to consult Ch. 4 of Redhead (1987), Shimony (1990), and Greenberger *et al.* (1990).

Ch. 25. Hardy's paradox

Hardy (1992a) is the original publication of his paradox. Mermin (1994) gives a very clear exposition of the basic idea. The GHZ paradox is explained in Greenberger *et al.* (1990). The counterfactual analysis in Sec. 25.5 extends Griffiths (1999a).

Ch. 26. Decoherence and the classical limit

A great deal has been written on the topic of decoherence and its relationship to the emergence of classical physics from quantum theory. For an introduction to the subject, see Zurek (1991). The book by Giulini *et al.* (1996) has contributions from diverse points of view and extensive references to earlier work. For work from a perspective close to the point of view found here, see Gell-Mann and Hartle (1993), Omnès (1999) (which gives references to earlier work), and Brun (1993, 1994).

For the argument of Omnès on the relationship of classical and quantum mechanics referred to in Sec. 26.6, see Chs. 10 and 11 of Omnès (1999), and his references to earlier work.

Ch. 27. Quantum theory and reality

On the difficulty of observing MQS states, Sec. 27.4, see Omnès (1994), Ch. 7, Sec. 8.

References

Note: WZ refers to the source book by Wheeler and Zurek (1983).

Aharonov, Y. and L. Vaidman (1991). Complete description of a quantum system at a given time. *J. Phys. A* **24**, 2315.

Alley, C. O., O. Jakubowicz, C. A. Steggerda, and W. C. Wickes (1983). A delayed random choice quantum mechanics experiment with light quanta, in *Proc. Int. Symp. Foundations of Quantum Mechanics*, ed. S. Kamefuchi *et al.* (Physical Society of Japan, Tokyo), p. 158.

Alley, C. O., O. Jakubowicz, and W. C. Wickes (1987). Results of the delayed-random-choice quantum mechanics experiment with light quanta, in *Proc. 2nd Int. Symp. Foundations of Quantum Mechanics*, ed. M. Namiki (Physical Society of Japan, Tokyo), p. 36.

Bell, J. S. (1964). On the Einstein Podolsky Rosen paradox. *Physics* **1**, 195. Reprinted in Ch. 2 of Bell (1987).

Bell, J. S. (1966). On the problem of hidden variables in quantum mechanics. *Rev. Mod. Phys.* **38**, 447.

Bell, J. S. (1987). *Speakable and Unspeakable in Quantum Mechanics* (Cambridge University Press, Cambridge, UK).

Bell, J. S. (1990). Against measurement, in *Sixty-Two Years of Uncertainty*, edited by A. I. Miller (Plenum Press, New York), p. 17.

Berndl, K., M. Daumer, D. Dürr, S. Goldstein, and N. Zanghí (1995). A survey of Bohmian mechanics. *Nuovo Cimento B* **110**, 737.

Birkhoff, G. and J. von Neumann (1936). The logic of quantum mechanics. *Annals Math.* **37**, 823. Reprinted in *John von Neumann Collected Works*, ed. A. H. Taub (Macmillan, New York, 1962), Vol. IV, p. 105.

Blank, B., P. Exner, and M. Havlíček (1994). *Hilbert Space Operators in Quantum Physics* (American Institute of Physics, New York).

Bohm, D. (1951). *Quantum Theory* (Prentice Hall, New York).

Bohm, D. (1952). A suggested interpretation of the quantum theory in terms of "hidden" variables, I and II. *Phys. Rev.* **85**, 166. Reprinted in WZ p. 369.

Bohm, D. and B. J. Hiley (1993). *The Undivided Universe* (Routledge, London).

Bohr, N. (1951). Discussion with Einstein on epistemological problems in atomic physics, in *Albert Einstein: Philosopher Scientist*, 2d edition, ed. P. A. Schilpp (Tudor Publishing, New York), p. 199. Reprinted in WZ p. 9.

Born, M. (1926). Zur Quantenmechanik der Stoßvorgänge. *Z. Phys.* **37**, 863. An English translation is in WZ, p. 52.

Born, M. and E. Wolf (1980). *Principles of Optics*, 6th edition (Pergamon Press, Oxford).

Bransden, B. H. and C. J. Joachain (1989). *Introduction to Quantum Mechanics* (Longman, London).

Brun, T. A. (1993). Quasiclassical equations of motion for nonlinear Brownian systems. *Phys. Rev. D* **47**, 3383.

Brun, T. A. (1994). Applications of the Decoherence Formalism, PhD dissertation, California Institute of Technology.

Clauser, J. F., M. A. Horne, A. Shimony, and R. A. Holt (1969). Proposed experiment to test local hidden-variable theories. *Phys. Rev. Lett.* **23**, 880.

Cohen-Tannoudji, C., B. Diu, and F. Laloë (1977). *Quantum Mechanics* (Hermann, Paris).

DeGroot, M. H. (1986). *Probability and Statistics*, 2d edition (Addison-Wesley, Reading, Massachusetts).

d'Espagnat, B. (1984). Nonseparability and the tentative descriptions of reality. *Phys. Repts.* **110**, 201.

d'Espagnat, B. (1989). *Reality and the Physicist* (Cambridge University Press, Cambridge, UK).

Dicke, R. H. (1981). Interaction-free quantum measurements: a paradox? *Am. J. Phys.* **49**, 925.

Dirac, P. A. M. (1958). *The Principles of Quantum Mechanics*, 4th edition (Oxford University Press, Oxford).

Dowker, F. and A. Kent (1996). On the consistent histories approach to quantum mechanics. *J. Stat. Phys.* **82**, 1575.

Einstein, A., B. Podolsky, and N. Rosen (1935). Can quantum-mechanical description of physical reality be considered complete? *Phys. Rev.* **47**, 777.

Elitzur, A. C. and L. Vaidman (1993). Quantum mechanical interaction-free measurements. *Found. Phys.* **23**, 987.

Englert, B. G., M. O. Scully, G. Sussmann, and H. Walther (1992). Surrealistic Bohm trajectories. *Z. Naturforsch.* **47a**, 1175.

Feller, W. (1968). *An Introduction to Probability Theory and Its Applications*, Vol. 1, 3rd edition (John Wiley & Sons, New York).

Fetter, A. L. and J. D. Walecka (1971). *Quantum Theory of Many-Particle Systems* (McGraw-Hill, New York).

Feynman, R. P., R. B. Leighton, and M. Sands (1965). *The Feynman Lectures on Physics*, Vol. III (Addison-Wesley, Reading, Massachusetts).

Fine, A. (1986). *The Shaky Game: Einstein Realism and the Quantum Theory* (University of Chicago Press, Chicago). A similar perspective will be found in the 2nd edition (1996).

Gell-Mann, M. and J. B. Hartle (1990). Quantum mechanics in the light of quantum cosmology, in *Complexity, Entropy, and the Physics of Information*, ed. W. Zurek (Addison Wesley, Reading, Massachusetts), p. 425; also in *Proceedings of the 25th International Conference on High Energy Physics, Singapore, 1990*, eds. K. K. Phua and Y. Yamaguchi (World Scientific, Singapore).

Gell-Mann, M. and J. B. Hartle (1993). Classical equations for quantum systems. *Phys. Rev. D* **47**, 3345.

Gieres, F. (2000). Mathematical surprises and Dirac's formalism in quantum mechanics. *Rep. Prog. Phys.* **63**, 1893.

Giulini, D., E. Joos, C. Kiefer, J. Kupsch, I.-O. Stamatescu, and H. D. Zeh (1996). *Decoherence and the Appearance of a Classical World in Quantum Theory* (Springer,

Berlin).

Greenberger, D. M. (1983). The neutron interferometer as a device for illustrating the strange behavior of quantum systems. *Rev. Mod. Phys.* **55**, 875.

Greenberger, D. M., M. A. Horne, A. Shimony, and A. Zeilinger (1990). Bell's theorem without inequalities. *Amer. J. Phys.* **58**, 1131.

Griffiths, R. B. (1984). Consistent histories and the interpretation of quantum mechanics. *J. Stat. Phys.* **36**, 219.

Griffiths, R. B. (1987). Correlations in separated quantum systems: a consistent history analysis of the EPR problem. *Am. J. Phys.* **55**, 11.

Griffiths, R. B. (1996). Consistent histories and quantum reasoning. *Phys. Rev. A* **54**, 2759.

Griffiths, R. B. (1998a). Choice of consistent family, and quantum incompatibility. *Phys. Rev. A* **57**, 1604.

Griffiths, R. B. (1998b). Consistent histories and quantum delayed choice. *Fortschr. Phys.* **46**, 741.

Griffiths, R. B. (1999a). Consistent quantum counterfactuals. *Phys. Rev. A* **60**, 5.

Griffiths, R. B. (1999b). Bohmian mechanics and consistent histories. *Phys. Lett. A* **261**, 227.

Griffiths, R. B. (2000a). Consistent histories, quantum truth functionals, and hidden variables. *Phys. Lett. A* **265**, 12.

Griffiths, R. B. (2000b). Consistent quantum realism: A reply to Bassi and Ghirardi. *J. Stat. Phys.* **99**, 1409.

Gudder, S. (1989). Realism in quantum mechanics. *Found. Phys.* **19**, 949.

Hajek, A. and J. Bub (1992). EPR. *Found. Phys.* **22**, 313.

Halmos, P. R. (1957). *Introduction to Hilbert Space and the Theory of Spectral Multiplicity*, 2nd edition (Chelsea Publishing, New York).

Halmos, P. R. (1958). *Finite-Dimensional Vector Spaces*, 2nd edition (Van Nostrand, Princeton).

Hardy, L. (1992a). Quantum mechanics, local realistic theories and Lorentz-invariant realistic theories. *Phys. Rev. Lett.* **68**, 2981.

Hardy, L. (1992b). On the existence of empty waves in quantum theory. *Phys. Lett. A* **167**, 11.

Hellmuth, T., H. Walther, A. Zajonc, and W. Schleich (1987). Delayed-choice experiments in quantum interference. *Phys. Rev. A* **35**, 2532.

Horn, R. A. and C. R. Johnson (1985). *Matrix Analysis* (Cambridge University Press, Cambridge, UK).

Horn, R. A. and C. R. Johnson (1991). *Topics in Matrix Analysis* (Cambridge University Press, Cambridge, UK).

Isham, C. J. (1994). Quantum logic and the histories approach to quantum theory. *J. Math. Phys.* **35**, 2157.

Jammer, M. (1974). *The Philosophy of Quantum Mechanics* (Wiley, New York).

Kochen, S. and E. P. Specker (1967). The problem of hidden variables in quantum mechanics. *J. Math. Mech.* **17**, 59.

Lewis, D. (1973). *Counterfactuals* (Harvard University Press, Cambridge, Massachusetts).

Lewis, D. (1986). *Philosophical Papers*, Vol. 2 (Oxford University Press, New York).

Lüders, G. (1951). Über die Zustandsänderung durch den Meßprozeß. *Ann. Phys. (Leipzig)* **8**, 322.

Lütkepohl, H. (1996). *Handbook of Matrices* (Wiley, New York).

McElwaine, J. N. (1996). Approximate and exact consistency of histories. *Phys. Rev. A* **53**, 2021.

Mermin, N. D. (1981). Bringing home the atomic world: quantum mysteries for anybody. *Am. J. Phys.* **49**, 940.

Mermin, N. D. (1985). Is the moon there when nobody looks? Reality and the quantum theory. *Physics Today* **38**, April, p. 38.

Mermin, N. D. (1990). Simple unified form for the major no-hidden-variables theorems. *Phys. Rev. Lett.* **65**, 3373.

Mermin, N. D. (1993). Hidden variables and the two theorems of John Bell. *Rev. Mod. Phys.* **65**, 803.

Mermin, N. D. (1994). Quantum mysteries refined. *Am. J. Phys.* **62**, 880.

Meyer, D. A. (1996). On the absence of homogeneous scalar unitary cellular automata. *Phys. Lett. A* **223**, 337.

Merzbacher, E. (1998). *Quantum Mechanics*, 3nd edition (Wiley, New York).

Mittelstaedt, P. (1998). *The Interpretation of Quantum Mechanics and the Measurement Process* (Cambridge University Press, Cambridge, UK).

Nielsen, M. A. and I. L. Chuang (2000). *Quantum Computation and Quantum Information* (Cambridge University Press, Cambridge, UK).

Omnès, R. (1992). Consistent interpretations of quantum mechanics. *Rev. Mod. Phys.* **64**, 339.

Omnès, R. (1994). *The Interpretation of Quantum Mechanics* (Princeton University Press, Princeton).

Omnès, R. (1999). *Understanding Quantum Mechanics* (Princeton University Press, Princeton).

Peres, A. (1990). Incompatible results of quantum measurements. *Phys. Lett. A* **151**, 107.

Redhead, M. (1987). *Incompleteness, Nonlocality, and Realism* (Oxford University Press, Oxford).

Reed, M. and B. Simon (1972). *Methods of Modern Mathematical Physics. I: Functional Analysis* (Academic Press, New York).

Renninger, M. (1960). Messungen ohne Störung des Meßobjekts. *Z. Phys.* **158**, 417.

Ross, S. M. (2000). *Introduction to Probability Models*, 7th edition (Harcourt Academic Press, San Diego).

Schrödinger, E. (1935). Die gegenwärtige Situation in der Quantenmechanik. *Naturwissenschaften* **23** 807, 823, 844. An English translation is in WZ, p. 152.

Schwabl, F. (1992). *Quantum Mechanics* (Springer-Verlag, Berlin).

Shankar, R. (1994). *Principles of Quantum Mechanics*, 2nd edition (Plenum Press, New York).

Shimony, A. (1990). An exposition of Bell's theorem, in *Sixty-Two Years of Uncertainty*, ed. A. I. Miller, (Plenum Press, New York) p. 33. Reprinted in A. Shimony, *Search for a Naturalistic World View*, Vol. II (Cambridge University Press, Cambridge, UK, 1993), p. 90.

Strang, G. (1988). *Linear Algebra and Its Applications*, 3rd edition (Harcourt, Brace and Jovanovich, San Diego).

von Neumann, J. (1932). *Mathematische Grundlagen der Quantenmechanik* (Springer-Verlag, Berlin). English translation: *Mathematical Foundations of Quantum Mechanics* (Princeton University Press, Princeton, 1955).

Wannier, G. H. (1966). *Statistical Physics* (Wiley, New York).

Wheeler, J. A. (1978). The "past" and the "delayed-choice" double-slit experiment, in *Mathematical Foundations of Quantum Theory*, ed. A. R. Marlow (Academic Press, New York), p. 9.

Wheeler, J. A. (1983). Law without law, in WZ, p. 182.

Wheeler, J. A. and W. H. Zurek (editors) (1983). *Quantum Theory and Measurement*

(Princeton University Press).

Wigner, E. P. (1963). The problem of measurement. *Am. J. Phys.* **31**, 6.

Zurek, W. H. (1991). Decoherence and the transition from quantum to classical. *Phys. Today* **44**, Oct., p. 36.

Index